With best wishes papa!

Excitatory-Inhibitory Balance

Excitatory-Inhibitory Balance
Synapses, Circuits, Systems

Edited by

Takao K. Hensch and Michela Fagiolini
RIKEN Brain Science Institute
Wako-shi, Japan

Kluwer Academic / Plenum Publishers
New York, Boston, Dordrecht, London, Moscow

Library of Congress Cataloging-in-Publication Data

Excitatory-inhibitory balance: synapses, circuits, systems/edited by Takao K. Hensch ... [et al.].
    p.   cm.
Includes bibliographical references and index.
ISBN 0-306-47962-1
    1. Neural transmission—Regulation. 2. Neural circuitry—Adaptation. I. Hensch, Takao K., 1966–

QP364.5.E96 2004
612.8—dc22

2003060288

ISBN 0-306-47962-1

©2004 Kluwer Academic/Plenum Publishers, New York
233 Spring Street, New York, New York 10013

http://www.wkap.com

10 9 8 7 6 5 4 3 2 1

A C.I.P. record for this book is available from the Library of Congress

All rights reserved

No part of this book may be reproduced, stored in a retrieval system, or transmitted in any form or by any means, electronic, mechanical, photocopying, microfilming, recording, or otherwise, without written permission from the Publisher, with the exception of any material supplied specifically for the purpose of being entered and executed on a computer system, for exclusive use by the purchaser of the work

Permissions for books published in Europe: *permissions@wkap.nl*
Permissions for books published in the United States of America: *permissions@wkap.com*

Printed in the United States of America

PREFACE

An optimal excitatory and inhibitory balance is the basis for proper brain function and plasticity. Yet, the study of neuron function has largely neglected the fact that cells are embedded in networks of both elements. While remarkable progress has been made in our understanding of excitatory synaptic transmission and its modification, the study of inhibitory interneurons and their plasticity has lagged behind. This book provides an integrated view of neuron function, operating in a balanced regime of excitation and inhibition. It emphasizes how this equilibrium is established, maintained and modified at various levels across several brain regions.

Two recent developments now make a more balanced view possible: a surge in research activity focused on inhibitory interneurons, and numerous technological advances. Improved imaging methods, visualized patch-clamp recording, multiplex single-cell PCR, proteomic analysis and gene-targeted deletion or knock-in mice are just some of the novel tools featured here. A broad spectrum of topics from molecular to cellular and systems neuroscience yields a new perspective and novel principles at each level through a deeper understanding of the interaction and integration of excitation and inhibition.

The book takes the reader beyond simple models of plasticity at single excitatory synapses and considers the full impact of neuronal behavior as a consequence of excitatory-inhibitory balance *in vivo*. It will be a valuable reference source for those who must deal with complex brain systems as a whole. We thank the many active young investigators from around the world for contributing their latest research findings in this emerging area. We hope this book will aid them and others in educating the brain scientists of tomorrow, by appealing to a wide audience of advanced graduate students, post-docs, faculty and clinicians.

Wako-shi
May, 2003

T.K.H.
M.F.

Invited Contributors

Yang Dan
ydan@uclink4.berkeley.edu

Michela Fagiolini
fagiolin@postman.riken.go.jp

Daniel E. Feldman
dfeldman@biomail.ucsd.edu

Jean-Marc Fritschy
fritschy@pharma.unizh.ch

Seth G.N. Grant
sgrant@srv0.bio.ed.ac.uk

Yasunori Hayashi
yhayashi-tky@umin.ac.jp

Takao K Hensch
hensch@postman.riken.go.jp

Shaul Hestrin
shaul.hestrin@stanford.edu

Masao Ito
masao@brain.riken.go.jp

Masanobu Kano
mkano@med.m.kanazawa-u.ac.jp

Yasuo Kawaguchi
yasuo@nips.ac.jp

Yukio Komatsu
komatsu@riem.nagoya-u.ac.jp

Pierre-Marie Lledo
pmlledo@pasteur.fr

Henry Markram
henry.markram@epfl.ch

David A. McCormick
david.mccormick@yale.edu

Hanns Möhler
mohler@pharma.unizh.ch

Mu-ming Poo
mpoo@uclink.berkeley.edu

Uwe Rudolph
rudolph@pharma.unizh.ch

Antoine Triller
triller@biologie.ens.fr

Gina G. Turrigiano
turrigiano@brandeis.edu

Weimin Zheng
zheng@nsi.edu

CONTENTS

Introduction
    Historical Overview: Search for inhibitory neurons and
    their function   1
    Masao Ito

I. Synapses

1. The organization and integrative function of the
   post-synaptic proteome   13
   Seth Grant

2. Dynamism of postsynaptic proteins as the mechanism
   of synaptic plasticity   45
   Kenny Futai and Yasunori Hayashi

3. Construction, stability and dynamics of the inhibitory
   postsynaptic membrane   59
   Christian Vannier and Antoine Triller

4. Long-term modification at visual cortical inhibitory synapses   75
   Yukio Komatsu and Yumiko Yoshimura

5. Activity-dependent modification of cation-chloride cotransporters underlying plasticity of GABAergic transmission   89
   Melanie Woodin and Mu-ming Poo

6. Endocannabinoid-mediated modulation of excitatory and
   inhibitory synaptic transmission   99
   *Masanobu Kano, Takako Ohno-Shosaku, Takashi Maejima
   and Takayuki Yoshida*

II. Circuits

7. Balanced recurrent excitation and inhibition in local
   cortical networks   113
   David A. McCormick, You-Sheng Shu and Andrea Hasenstaub

8. Local circuit neurons in the frontal cortico-striatal system 125
   Yasuo Kawaguchi

9. Interneuron heterogeneity in neocortex 149
   Anirudh Gupta, Maria Toledo-Rodriguez, Gilad Silberberg and Henry Markram

10. Fast spiking cells and the balance of excitation and inhibition in the neocortex 173
    Mario Galarreta and Shaul Hestrin

11. Homeostatic regulation of excitatory-inhibitory balance 187
    Gina Turrigiano

12. Adult neurogenesis controls excitatory-inhibitory balance in the olfactory bulb 197
    Pierre-Marie Lledo, Armen Saghatelyan and Gilles Gheusi

III. Systems

13. $GABA_A$ receptor subtypes: memory function and neurological disorders 215
    Jean-Marc Fritschy, Uwe Rudolph, Florence Crestani and Hanns Mohler

14. LTD, spike timing and somatosensory barrel cortex plasticity 229
    Daniel E. Feldman, Cara B. Allen and Tansu Celikel

15. Maintaining stability and promoting plasticity: context-dependent functions of inhibition 241
    Weimin Zheng

16. Spike timing and visual cortical plasticity 255
    Yu-Xi Fu and Yang Dan

17. Excitatory-inhibitory balance controls critical period plasticity 269
    Michela Fagiolini and Takao K. Hensch

Index 283

Introduction

HISTORICAL OVERVIEW: THE SEARCH FOR INHIBITORY NEURONS AND THEIR FUNCTION

Masao Ito[*]

Even though inhibition had long been recognized as a distinct process in the nervous system (see Sherrington[1]), the discovery of inhibitory synapses had to wait until the middle of the 20th century. In mammalian motoneurons, Eccles and his associates recorded inhibitory postsynaptic potentials (IPSPs) that form a mirror image of excitatory postsynaptic potentials (EPSPs)[2]. Furthermore, they identified Renshaw cells and Ia inhibitory neurons specifically mediating the generation of IPSPs in motoneurons (Fig. 1). Renshaw cells are activated by impulses of recurrent axon

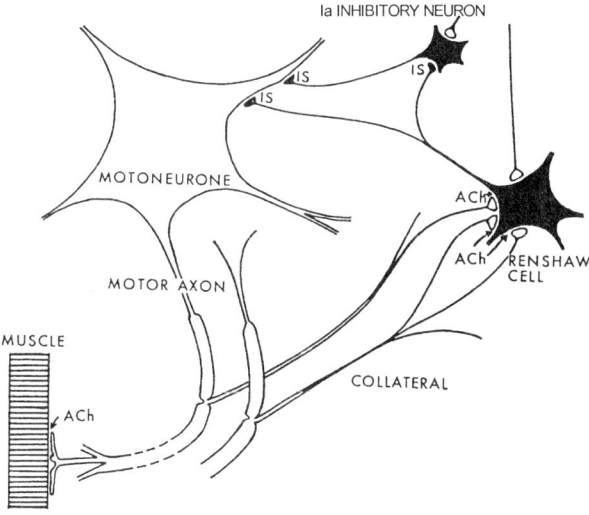

Fig. 1 Inhibitory connections to spinal motoneurons. In this as well as in the succeeding figures, excitatory neurons and synapses are shown as hollow structures, while inhibitory ones are filled in black[3]. IS, inhibitory synapse.

---

[*]M. Ito, RIKEN Brain Science Institute, 2-1 Hirosawa, Wako-shi, Saitama 351-0198, Japan

collaterals of motoneurons, thereby generating IPSPs in motoneurons in a feedback manner[4]. Ia inhibitory neurons are activated by signals in group Ia muscle afferents, thereby generating IPSPs in motoneurons in a feedforward manner[5]. Both Renshaw cells and Ia inhibitory neurons are small neurons locally extending short axons. However, cerebellar Purkinje cells that extend long axons were also found to induce generation of IPSPs in their target neurons[6]. When a neuron generates IPSPs in a target, it also generates IPSPs in other targets; hence, short- or long-axon inhibitory neurons can be defined distinctly from excitatory neurons that produce EPSPs in their targets. While the chemical transmitter in Renshaw cells and Ia inhibitory neurons was identified to be glycine, as suggested first by Aprison and Werman[7], that in Purkinje cells was identified to be GABA[8] as in the peripheral inhibitory synapses of Crustacea[9].

Since then, many neurons of the central nervous system have been identified as inhibitory neurons operating with either glycine or GABA as a transmitter and supplying inhibitory synapses comparable in number with excitatory synapses to tissues of the central nervous system. Three types of GABA receptors have been identified, ionotropic $GABA_A$ and $GABA_C$, which are ligand-gated chloride channels, and metabotropic $GABA_B$ (see Watanabe et al.,[10]). Inhibitory neurons that were initially regarded as minor components of central neurons have thus been recognized as major components, which play diverse roles in neuronal circuits of the central nervous system.

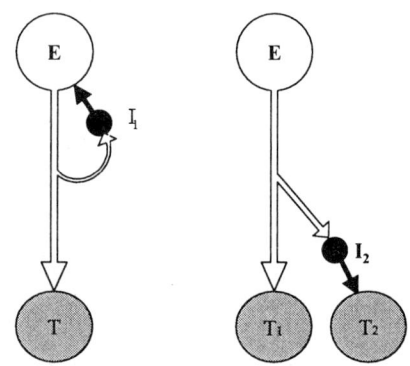

Fig. 2 Inhibitory neurons as a commutator in neuronal circuit. E, excitatory neurons. I_1, feedback inhibitory neuron. I_2, feedforward inhibitory neuron. T, T1, T2, target neurons, which can be either excitatory or inhibitory.

Commutator   Renshaw cells commuting excitatory signals of motoneurons to signals that inhibit the motoneurons should provide negative feedback regulation to stabilize the activities of motoneurons, particularly tonic-type motoneurons[11] (Fig. 2A). Connnections of Renshaw cells to other neurons and their firing behavior have extensively been studied (see Windhorst[12]). Various roles of Renshaw cells have been proposed, but a recent simulation study suggests that a major role of Renshaw cells is to decorrelate firing of motoneurons in a pool, thus diminishing physiological tremor[13]. Impulses of group Ia afferents from a muscle excite its motoneurons as well as its agonist muscles, and at the same time inhibit the motoneurons of its antagonistic muscles via Ia inhibitory neurons; Ia inhibitory neurons function as a commutator of synaptic effects of muscle afferents and produce a pair of excited and inhibited motoneurons (Fig. 2B). Renshaw cells and Ia inhibitory neurons represent commutation in the feedback and feedforward manners, respectively (Figs. 2A and B).

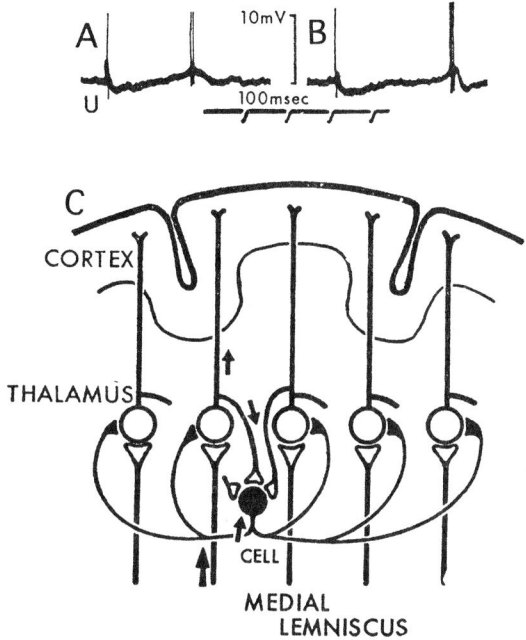

Fig. 3 Oscillatory discharges caused by recurrent inhibition. A, discharges and intervening recurrent inhibition of a thalamocortical neuron induced by peripheral nerve stimulation (A) and cortical stimulation (B). C, neuronal circuit underlying the oscillation (reproduced by permission of Nature)[14].

Oscillator Two types of neuronal network containing inhibitory neurons have been known to generate a rhythm. One is the inhibitory phasing by a feedback inhibition[14]. When a neuron is in a tonically depolarized state, its recovery from inhibition induces its excitation. Hence, in a neuron equipped with feedback inhibition, rhythmic oscillation occurs as the initial excitation leads to recurrent inhibition that, at its recovery, causes another excitation, and so on (Fig. 3). That is, the recovery phase of the inhibitory feedback provides a delayed positive feedback that generates oscillation. The recurrent inhibition of thalamocortical neurons via reticular nucleus neurons in fact induces oscillation (Fig. 3) and has been assumed as the mechanism that generates EEG alpha waves[14].

The other is the reciprocal inhibition between two neurons. When these neurons are tonically driven from common sources, the excitation of one neuron causes the inhibition of its partner neuron (Fig. 4). When the excitation that prevails in one neuron weakens due to fatigue, the inhibition that prevails in the other neuron also weakens, so that the excitation of the latter neuron will prevail and exert inhibition on the former neuron. This alternating excitation and inhibition will continue, that is, when a cell inhibits its partner and is inhibited by it, the net effect is a positive feedback that induces oscillation[15]. In locomotion and stepping, a pair of extensor and flexor muscles generates reciprocally alternating contraction and relaxation. This rhythm was postulated to be generated by the mutual inhibition between segmental neurons in the spinal cord mediating descending excitatory commands to the paired flexor and extensor motoneurons[16]. The right and left coordinations in aquatic vertebrates and mammals are also mediated by mutual inhibition through commissural fibers crossing the midline of the spinal cord[17-18]. Mutual inhibition between GABA neurons is also assumed to generate a theta rhythm in the hippocampus[19]. An interesting theoretical possibility of a mutually inhibitory neuron pair is that, when the strength of the mutual inhibitory connection is subjected to activity-dependent plasticity, an oscillator will emerge from an assembly of initially randomly interconnected inhibitory neurons[20].

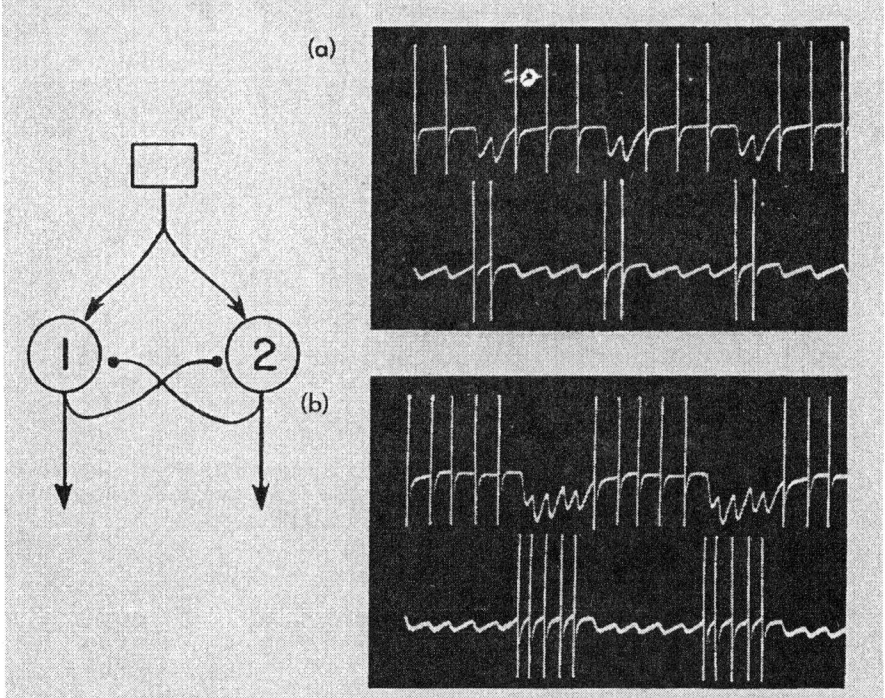

Fig. 4 Alternating bursts in two reciprocally inhibiting neurons. A two-membered (1 and 2) reciprocal inhibition network produces alternating bursts of impulses. Increasing inputs generate longer bursts, but at a lower rate of alternation. Simulation study (a and b)[15]

Pacemaker neurons are also an important element of rhythm generators of the nervous system (see Selverston and Moulin[21], Arshavsky[22]). The oscillation arising from the endogenous pacemaker cell mechanism synergistically interacts with that arising from a network of a group of neurons. In most central pattern generators controlling automatic movements, a rhythm originates from individual pacemaker neurons, and network properties shape the final motor output.

Integrator Integral calculus is an important element of neuronal circuit operations. It can be performed by neurons equipped with a positive feedback through excitatory recurrent connections[23], but a mutually inhibiting neuron pair can also constitute an integrator circuit because the inhibition cascade (see above) acts as a positive feedback. The parameters of the circuit could be set such that the paired neurons gradually build up excitation in an integral calculus manner, in contrast to the quick switching to full excitation and the subsequent decay assumed above for oscillators. The neuronal circuit for vestibuloocular reflex includes a neural integrator, which converts eye velocity signals to eye position signals. This integrator has been modeled using a pool of mutually inhibiting neurons[24-25]. The presence of inhibitory commissural connections between vestibular nuclei that mediate vestibuloocular reflex has been confirmed[26].

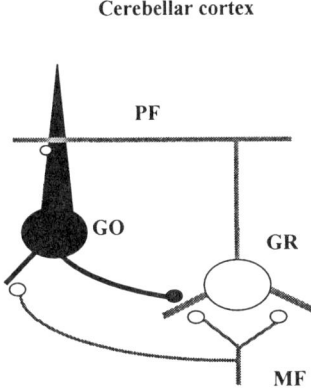

Fig. 5 Connections to and from Golgi cell in the cerebellar cortex. Go, Golgi cell. PF, parallel fiber. MF, mossy fiber. GR, granule cell.

Golgi cells are inhibitory neurons attached to the mossy fiber-granule cell pathway in the cerebellar cortex in both feedforward and feedback manners (Fig. 5). In his adaptive filter model of the cerebellum, Fujita[27] uniquely assumed that Golgi cells act as a leaky integrator with a time constant of several seconds, and consequently as a phase lag element. Under this assumption, a mossy fiber - granule cell - Golgi cell network converts a mossy fiber signal to signals of a set of granule cells with various phase leads or lags, from which Purkinje cells learn to choose. This adaptive filter model can explain learning in a temporal domain, while the models proposed by Marr[28] and Albus[29] explain learning in a spatial domain.

A delay line has been assumed to provide a mechanism for determining the timing of an event in a neuronal circuit. For example, if an impulse travels along a long axon that formes excitatory synapses with a number of neurons one after another at constant intervals, these neurons will sequentially fire with a constant delay. If an impulse travels a long chain composed of many neurons serially connected via excitatory synapses, it also constitutes a delay line. Braitenberg and Onesto[30] proposed that parallel fiber axons of granule cells in the cerebellar cortex act as a delay line and induce Purkinje cells to fire sequentially. However, since the total time required for an impulse to travel along a parallel fiber is expected to be as brief as a few milliseconds, this delay line does not have a sufficiently long time range to be utilized for cerebellar control of movements. By contrast, the above-mentioned integrator action of Golgi cells would provide a delay of several seconds. Fujita[27] predicts that the delay time for each granule cell varies according to the strength of its synaptic connection efficacy to and from a Golgi cell, and hence considers that a set of a number of granule cells produce a range of delay time. The delay line composed of mossy fiber-parallel fiber- Golgi cell could be a basis for fine tuning of timing in movements by the cerebellum.

Motion detector In the mechanisms of direction-selective motion detection in the visual system, inhibitory neurons have been assumed to account for division-like and subsequent subtraction-like interactions between neighboring visual elements (see Clifford and Ibbotson[31]). Division-like operation may be exerted by the shunting of excitatory synaptic currents through the increased membrane conductance of inhibitory synapses[32]. Interaction between excitatory and inhibitory synapses in dendrites has long been considered as an important mechanism of computation in neurons[33].

Lateral inhibition Inhibitory neurons attached to a pool of relay neurons in a sensory pathway such as thalamocortical neurons play a role in lateral inhibition. These inhibitory neurons are activated by the convergence of recurrent axons of relay neurons and divergently inhibit relay neurons in a feedback manner (Fig. 3). The excitation of relay neurons will be sharply focused on those relay neurons receiving the strongest drive, while the others are suppressed. Lateral inhibition may be exerted also by the convergence of presynaptic impulses to the relay neurons and then divergently distribute inhibitory signals to the relay neurons in a feedforward manner.

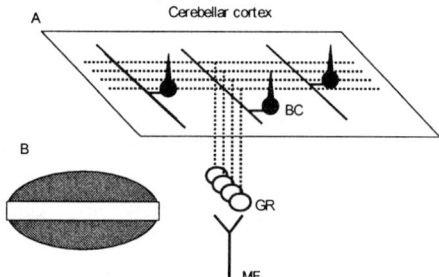

Fig. 6 Feedforward inhibition in the cerebellar cortex. A, connections. BC, basket cell. B, lateral inhibition pattern suggested by the excitatory and inhibitory connections.

The feedforward inhibition via basket and stellate cells in the cerebellar cortex has been suggested to represent lateral inhibition in Purkinje cells[34]. Conducting impulses in a bundle of parallel fibers would form a narrow band of excited Purkinje cells flanked with side bands of inhibited Purkinje cells (Fig. 6). However, such a pattern has never been observed under natural stimulation conditions in an *in vivo* cerebellum, and hence the lateral inhibition pattern in the cerebellar cortex could be an artifact to be induced by electrical stimulation of a parallel fiber bundle.

Inhibitory modulation by the cerebellum   Purkinje cells are inhibitory neurons that provide the sole output from the cerebellar cortex to cerebellar nuclei and certain vestibular nuclei (Fig. 7). Afferent fibers arising from numerous precerebellar nuclei in the brainstem and spinal cord supply excitatory synapses to nuclear neurons and pass to the cerebellar cortex as mossy fibers. Mossy fiber signals are relayed by granule cells that in turn activate Purkinje cells. Hence, the fundamental operation of the cerebellar circuit is inhibitory modulation of various command signals (for example, for a movement) by Purkinje cells mediated by mossy fiber-nuclear neuron pathways.

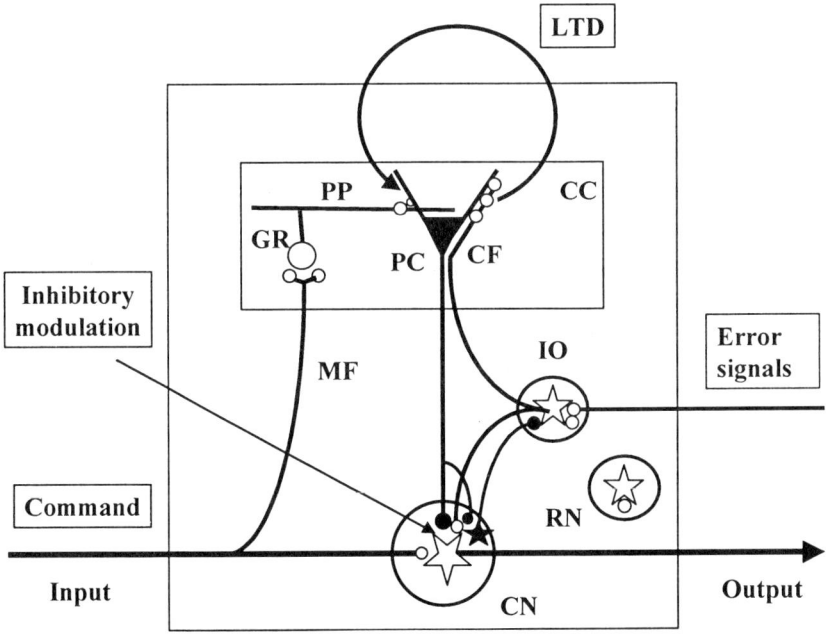

Fig. 7 Neuronal circuit in the cerebellum. Slightly modified from Ito[35].

Purkinje cells also receive climbing fiber afferents arising from the inferior olive. Climbing fibers convey error signals, which induce long-term depression (LTD) in granule cell - Purkinje cell synapses receiving command signals. LTD thus removes those synapses involved in the unsuccessful execution of command signals. During repeated trials, mossy fiber - granule cell - Purkinje cell pathways are shaped by elimination of erroneous connections, so that the Purkinje cell modulation of command signals becomes optimal. This appears to be the mechanism of motor learning (see ref. 35).

Disinhibitory modulation by the basal ganglia The major circuit in basal ganglia (Fig. 8) is a cascade of two types of inhibitory neuron, one in the striatum (putamen and caudate nucleus) and the other in the substantia nigra pars reticulata (SNr)[36-37]. SNr neurons are activated by sidepaths involving the subthalamic nucleus and the external segment of the globus pallidus and tonically inhibit their target neurons in the thalamus and brainstem. When inhibitory neurons in the striatum are activated, SNr inhibition is removed so that its target neurons are released to action. With various inputs from the cerebral cortex, striatum neurons may compete with each other. The dominant group of striatum neurons would suppress the surrounding striatum neurons in a center (excitation)-surround (inhibition) manner[38] so that only a limited number of SNr neurons are inhibited and release their target system from inhibition. The interaction of dopaminergic projections from the substantia nigra pars compacta and cortical input in the striatum is considered a learning mechanism. This selection-based stabilization has been investigated in studies of saccadic eye movements (see ref. 39), and appears to be the principal mechanism of the basal ganglion cascade.

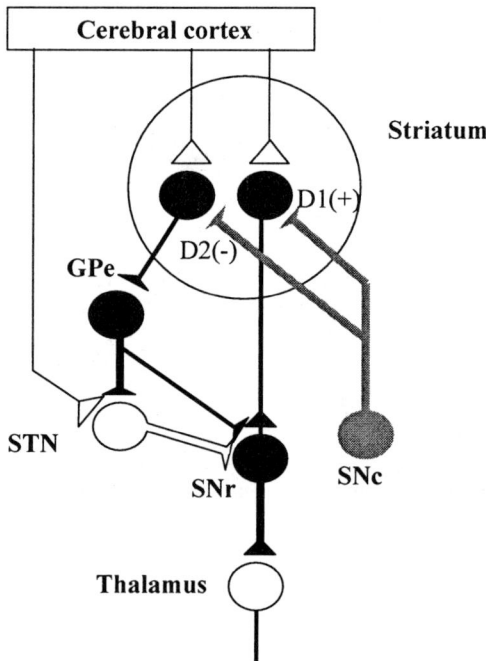

Fig. 8 Neuronal circuit in the basal ganglia. Slightly modified from Hikosaka et al.[40].

State controller If an inhibitory neuron supplies synapses diffusely to neurons in a pool, it may control the general activity state of this neuron pool. Neurons in the reticular formation extend long axons that extensively branch in the medulla and spinal cord[41]. The inhibitory neurons located in the medullary reticular formation and extending long descending axons to the spinal cord[42] are a candidate for such a state-controlling neuron. However, the general motor inhibition induced in the spinal cord by stimulation of the medulla oblongata may be mediated by segmental inhibitory neurons activated by long descending excitatory axons[43]. The concept of state-control inhibitory neurons still remains hypothetical.

I mentioned above a list of the major roles of inhibitory neurons in neuronal circuits of the brain conceived during the past five decades. More recent knowledge of inhibitory synapses and circuitry function involving inhibitory neurons is compiled in this monograph, which represents a big step toward a major goal of neuroscience, that is, the reconstruction of neuronal circuit structures for the entire central nervous system with identified excitatory and inhibitory neurons.

REFERENCES

1. Sherrington, C.S. (1906) *Integrative Action of the Nervous System.* (Yale Univ. Press, New Haven).
2. Brock, L.G., Coombs, J.S. and Eccles, J.C. (1952) The recording of potentials from motoneurones with an intracellular electrode. *J. Physiol. (Lond.)* **117**, 431-460.
3. Eccles, J.C. (1962) Spinal neuron: synaptic connexions in relation to chemical transmitters and pharmacological responses. *Proc. First Intern. Pharmacol. Meeting* **8**, 157-182.
4. Eccles, J.C., Fatt, P. and Koketsu, K. (1954) Cholinergic and inhibitory synapses in a pathway from motor axon collaterals to motoneurons. *J. Physiol. (Lond.)* **126**, 524-564.
5. Eccles, J.C., Fatt, P. and Landgren, S. (1956) Central pathway for direct inhibitory action on impulses in large afferent nerve fibres to muscle. *J. Neurophysiol.* **19**, 75-98,
6. Ito, M. and Yoshida, M. (1964) The cerebellar-evoked monosynaptic inhibition of Deiters neurones. *Experientia.* **40**, 762-764.
7. Aprison, M.H. and Werman, R. (1965) The distribution of glycine in cat spinal cord and roots. *Life Sci.* **4**, 2075-2083.
8. Obata, K. Ito, M,, Ochi, R. and Sato, N. (1967) Pharmacological properties of the postsynaptic inhibition by Purkinje cell axons and the action of g-aminobutryic acid on Deiters neurons. *Exp. Brain Res.* **4**, 43-57
9. Kuffler, S.W. and Edwards, C. (1958) Mechanism of gamma-aminobutyric acid (GABA) action and its relation to synaptic inhibition. *J. Neurophysiol.* **21**, 589-610.
10. Watanabe. M,, Maemura. K,, Kanbara, K., Tamayama, T. and Hayasaki, H. (2002) GABA and GABA receptors in the central nervous system and other organs. *Int. Rev. Cytol.* **213**, 1-47.
11. Eccles, J.C., Eccles, R.M., Iggo, A. and Ito, M. (1961) Distribution of recurrent inhibition among motoneurones. *J. Physiol. (Lond.)* **159**, 479-499.
12. Windhorst, U. (1990) Activation of Renshaw cells. *Prog. Neurobiol.* **35**, 135-179.
13. Maltenfort, M.G., Heckman, C.J. and Rymer, W.Z. (1998) Decorrelating actions of renshaw interneurons on the firing of spinal motoneurons within a motor nucleus: a simulation study. *J. Neurophysiol.* **80**, 309-323.
14. Andersen, P. and Eccles, J.C. (1962) Inhibitory phasing of neuronal discharge. *Nature* **196**, 645-647.
15. Wilson, D.M. and Wardron, I. (1968) Models for the generation of the motor ouutput pattern in flying locusts. *Proc. IEEE* **56**, 1058-1064.
16. Lundberg, A. (1981) Half-center revisited. In: *Regulatory Pathway of the CNS Principles of Motion and Organization*, eds. J. Szentagothai, M. Parkovits, and J. Hamori. *Adv. Physiol. Sci.* **1**, 155-167.
17. Parker, D. and Grillner, S. (2000) Neuronal mechanisms of synaptic and network plasticity in the lamprey spinal cord. *Prog. Brain Res.* **125**, 381-398.
18. Butt, S.J.B., Lebret, J.M. and Liehn, O. (2002) Organization of left-right coordination in the mammaian locomotor network. *Brain Res. Rev.* **40**, 107-117.

19. Traub, R.D., Whittington, M.A., Colling, S.B., Buzsaki, G. and Jefferys, J.G. (1996) Analysis of gamma rhythms in the rat hippocampus in vitro and in vivo. *J. Physiol. (Lond.)* **493**, 471-484.
20. Soto-Trevino, C., Thoroughman, K.A., Marder, E. and Abbott, L.F. (2001) Activity-dependent modification of inhibitory synapses in models of rhythmic neural networks. *Nat. Neurosci.* **4**, 297-303.
21. Selverston AI, Moulins M. (1985) Oscillatory neural networks. *Annu Rev Physiol.* **47**, 29-48.
22. Arshavsky, Y.I., (2003) Cellular and network properties in the functioning of the nervous system : from central patter generators to cognition. *Brain.Res. Rec* **41**, 229-267.
23. Koulakov, A.A., Raghavachari, S., Kepecs, A. and Lisman, J.E. (2002) Model for a robust neural integrator. *Nat. Neurosci.* **5**, 775-782.
24. Cannon, S.C., Robinson, D.A. and Shamma, S. (1983) A proposed neural network for the integrator of the oculomotor system. *Biol. Cybern.* **49**, 127-136.
25. Cannon, S.C. and Robinson, D.A. (1985.25) An improved neural-network model for the neural integrator of the oculomotor system: more realistic neuron behavior. *Biol. Cybern.* **53**, 93-108.
26. Holstein, G.R., Martinelli, G.P. and Cohen, B. (1999) The ultrastructure of GABA-immunoreactive vestibular commissural neurons related to velocity storage in the monkey. *Neuroscience* **93**, 171-181.
27. Fujita M. (1982) Adaptive filter model of the cerebellum. *Biol. Cybern.* **45**, 195-206.
28. Marr, D.A. (1969) A theory of cerebellar cortex. *J. Physiol (Lond.)* **202** , 437-470.
29. Albus, J.S. (1971) A theory of cerebellar function. *Math. Biosci.* **10**, 25-26.
30. Braitenberg, V. and Onesto, N. (1962) The cerebellar cortex as a timing organ. Discussion of a hypothesis. *Proc. 1st Intern. Conf. Med. Cybern.* (Giannini: Naples, Italy), pp 1-19.
31. Clifford, C.W.G. and Ibbotson, M.R. (2003) Fundamental mechanisms of visual motion detection: models, cells and functions. *Progr. Neurobiol.* **68**, 409-437.
32. Torre, V. and Poggio, T. (1978) A synaptic mechanism possibly underlying directional selectivity to motion. *Proc. R. Soc. Lond. B* **202**, 409-416.
33. Koch, C., Poggio, T. and Torre, V. (1983) Non-linear interactions in a dendritic tree: localization, timing and role in information processing. *Proc. Natl. Acad. Sci. USA* **80**, 2799-2802.
34. Eccles, M., Ito, M. and Szentagothai J. (1967) *The Cerebellum as a Neuronal Machine.* (Springer-Verlag: New York), p 209.
35. Ito, M. (2002) Controller-regulator model of the central nervous system. *J. Integrative Neurosci.* 1, 129-143.
36. Yoshida, M. and Precht, W. (1971) Monosynaptic inhibition of neurons in the substantia nigra by caudate-nigral fibers. *Brain Res.* **32**, 225-228.
37. Ueki, A., Uno, M., Anderson, M. and Yoshida, M. (1977) Monosynaptic inhibition of thalamic neurons produced by stimulation of the substantia nigra. *Experientia* **33**, 1480-1481.
38. Kaji R. (2001) Basal ganglia as a sensory gating devise for motor control, *J. Med. Invest.* **48**, 142-146.
39. Hikosaka O., Takikawa Y., Kawagoe R. (2000) Role of the basal ganglia in the control of purposive saccadic eye movements. *Physiol. Rev.* **80**, 953-978.
40. Hikosaka, O., Takikawa, Y., and Kawagoe, R. (2000) Role of the basal ganglia in the control of purposive saccadic eye movements. *Physiol. Rev.* **80**, 953-978.
41. Scheibel, M.F. and Scheibel, A.B. (1958) Structural substrates for integrating patterns in the brain stem reticular core. In: *Reticular Formation of the Brain*, eds. H.H. Jasper et al. Little, Brown & Co. Boston
42. Ito, M., Udo, M. and Mano, N. (1970) Long inhibitory and excitatory pathways converging onto cat reticular and Deiters' neurons and their relevance to reticulofugal axons. *J. Neurophysiol.* **33**, 210-226.
43. Takakusaki, K., Kohyama, J., Matsuyama, K. and Mori, S. (2001) Medullary reticulospinal tract mediating the generalized motor inhibition in cats: parallel inhibitory mechanisms acting on motoneurons and on interneuronal transmission in reflex pathways. *Neuroscience* **103**, 511-527.

Excitatory-Inhibitory Balance:

Synapses

THE ORGANIZATION AND INTEGRATIVE FUNCTION OF THE POST-SYNAPTIC PROTEOME

S.G.N. Grant, H. Husi, J. Choudhary, M. Cumiskey, W. Blackstock and J.D. Armstrong[*]

1. INTRODUCTION

The postsynaptic terminal is an example of signal transduction specialisation *par excellence*. Signaling is essentially found at two levels; at the level of transmitting electrical activity between nerve cells and converting electrical activity into molecular signals via intracellular signal transduction. A wealth of information on the molecular composition and electophysiological properties of the post synaptic terminal has raised new and crucial questions for the neurobiology of nerve cells and behaviour.

This paper will attempt to define these new questions and present emerging strategies aimed at addressing their solutions. In doing so, the molecular composition and electrophysiological properties of synapses will be discussed. However, this will not be a review aimed at simply updating on the latest set of molecules or physiological phenomena. Moreover, the discussion will be restricted in two ways: First, synapses will be discussed in their generic form, although the description is of the excitatory synapses of the mammalian brain, which have been the most intensively studied. Second, the presynaptic (axon) terminal of the synapse and its role in regulating vesicular release of neurotransmitter, which is a formidable subject itself, can be found elsewhere. Here, the postsynaptic terminal, which is found on dendrites and is the site at which axons terminate and transmit signals to postsynaptic neurotransmitter receptors will be discussed.

---

[*] SGN Grant, H Husi, Division of Neuroscience, Edinburgh University, 1 George Square, Edinburgh, EH8 9JZ; J Choudhary, W Blackstock, Cellzome, UK; M Cumiskey, JD Armstrong, School of Informatics, Edinburgh University, UK.

Excitatory-Inhibitory Balance: Synapses, Circuits, Systems
Edited by Hensch and Fagiolini, Kluwer Academic/Plenum Publishers, 2003

An overview of the molecular challenges of the postsynaptic terminal.

How the postsynaptic terminal works as a molecular machine composed of myriads of interacting molecules in a highly localised environment is the central problem of synaptic biology. Superimposed on this is the need to explain physiology, behaviour and disease. As a result of proteomic studies, it is now evident that the postsynapatic terminal is composed of ~1000 proteins, many of which are known to be involved with specific physiological functions and diseases. Recent advances in bioinformatic tools and mathematical theories allow this apparently bewildering complexity to be simplified into a molecular architecture. This architecture provides a new strategy for experimental design as well as provides insights into the evolution of signalling in multicellular organisms.

Functions of the postsynaptic terminal.

Sherrington's description of synaptic transmission[1] and Adrian's discovery that sensory information is encoded in the pattern of action potentials[2] underpins the importance of the synapse in reliably transmitting electrical activity from the presynaptic to the postsynaptic neuron. In this way, information is transferred through circuits in the nervous system. A second function of synapses is to detect and monitor these patterns of activity as they pass by, and then activate intrasynaptic signal transduction pathways that lead to changes in the structure and function of the synapse and neuron. This property of synapses is loosely known as 'synaptic plasticity'.

The concept that electrical activity could be converted into biochemical and structural changes was famously exposited by Hebb[3] in his neurophysiological postulate: 'Let us assume then that the persistence or repetition of a reverberatory activity (or 'trace') tends to induce lasting cellular changes that add to its stability. The assumption can be precisely stated as follows: When an axon of cell A is near enough to excite a cell B and repeatedly or persistently takes part in firing it, *some growth process or metabolic change* takes place in one of both cells such that A's efficiency, as one of the cell firing B, is increased.' In the ensuing 50 years, the intellectual framework of Hebb's model has furnished an extraordinary amount of experimental investigation, much of which was aimed at finding evidence for activity-dependent cellular changes and suitable experimental models. The terms synaptic transmission and synaptic plasticity will be used to distinguish these core functions of the synapse.

From function to molecule

Since studies of synaptic biology have been historically dominated by electrophysiological studies, most molecular biological investigations have focused on identifying essential molecules for these higher phenomena. This reductionist strategy aims at finding mechanistic simplification – the discovery of some core process, and has been very successful in many areas of physiology. This hold true for synaptic transmission, where the fundamentals of neurotransmitter synthesis, packaging release and receptor mediated transmission are now understood. In contrast, the study of synaptic plasticity, which is dominated by studies of postsynaptic function, has progressed from early molecular models of striking simplicity through a period of molecular discovery to problematic complexity. This complexity has arisen by the parallel progress in identifying postsynaptic proteins and the testing of their function.

Early (1980s) models of synaptic plasticity involved signaling pathways of three protein steps comprising a receptor to kinase to receptor[3]. With the advent of mouse gene targeting strategies, it was possible to test the function of any synaptic protein and this resulted in new 'plasticity molecules' being identified[4]. The ensuing decade has lead to the discovery of many dozens of proteins required for plasticity, which fundamentally challenges the idea that synaptic plasticity is a property of the simplest pathways[5].

In addition to an increase in the molecular complexity of synaptic plasticity, it has also become clear that plasticity molecules are not restricted to any one category of molecular function. For example, there are plasticity proteins that are membrane channels and receptors, transcription factors, translation factors, ubiquitination proteins and many other distinct intracellular cell biological processes. This contrasts with the mechanistically more simplified view of neurotransmitter vesicle release.

Another clue to this molecular puzzle is found in the details of experiments on plasticity where a single gene or protein has been knocked out with either mutations or drugs. Although there are many molecules involved, essentially all of them are dispensable. For example, when synaptic plasticity (long term potentiation of synaptic transmission) is induced using different trains of electrical activity, it was found that some of these trains produce changes in synaptic strength that are interfered with by molecular disruption, but other trains are not affected. In other words, synaptic plasticity is a remarkably robust and the loss of any molecule only produces are partial effect (see examples [4,6-10]).

2. MOLECULAR COMPOSITION OF THE POSTSYNAPTIC TERMINAL

A major challenge for biology is to devise strategies that extract emergent physiological properties and simple biological principles from highly complex genomic and proteomic datasets, and shed light on both mechanisms and disease processes. The synapse proteome is a suitable prototype to explore these general issues because; i) it contains a highly localized set of proteins found in dendritic spines; ii) an important role for signaling complexes and pathways has been established, iii) signaling can be exquisitely studied using patterns of action potentials, iv) genetic and pharmacological perturbations result in behavioural changes. However, the global composition and organization is poorly characterized.

The proteome of the postsynaptic terminal of excitatory synapses in the mouse central nervous system was analysed using a combination of proteomic and bioinformatics strategies. The neurotransmitter glutamate is the major excitatory neurotransmitter at vertebrate excitatory synapses and activates two major subclasses of glutamate receptors, the ionotropic N-methyl-D-aspartate (NMDA), α-amino-3-hydroxy-5-methyklisoxazole-4-proprionic acid (AMPA) and Kainate receptors and metabotropic (mGluR) receptors found in the postsynaptic membrane. The cytoplasmic C-terminus of NMDA receptor subunits (NR2A/ε1, NR2B/ε2) binds PDZ domains of Post Synaptic Density 95 (PSD-95), a MAGUK (Membrane Associated Guanylate Kinase) protein, which acts as an adaptor/scaffold to organize signal transduction and plasticity mediated by the NMDA receptor[7]. Previous proteomic studies showed that NMDAR-PSD95 complexes were 2-3MDa particles comprising 77 proteins including mGluR receptors and that AMPA receptors were in different complexes[11,12]. The NMDAR-PSD95 and AMPA complexes are embedded in the postsynaptic terminal of excitatory synapses with other postsynaptic proteins in a structure known as the Post Synaptic Density (PSD)(Fig. 1a). The PSD, which was originally observed as an electron dense structure beneath the postsynaptic membrane can be isolated by biochemical fractionation and was shown to comprise at least 27 major proteins[13,14].

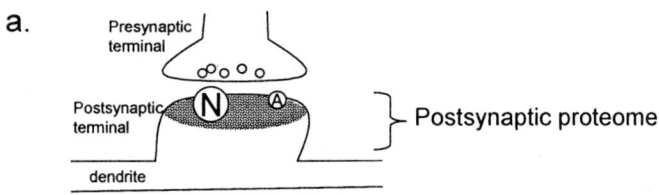

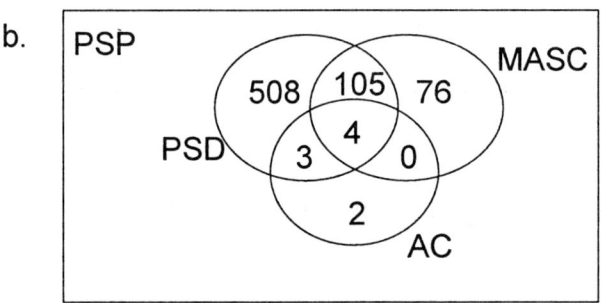

	PSD #	MASC #	AMPA #	PSP #
Channels and Receptors	60	12	5	62
MAGUKs / Adaptors / Scaffolders	31	21		37
Ser/Thr Kinases	23	8		23
Tyr Kinase	3	2		4
Protein Phosphatases	12	7		15
G-proteins and Modulators	39	19		50
Signalling molecules and Enzymes	166	51		202
Transcription and Translation	36	5		38
Cytoskeletal and Cell Adhesion Molecules	100	35	3	112
Synaptic Vesicles and Protein Transport	119	22	1	122
unknown	31	3		34
Summary	620	185	9	699

Fig. 1. Organisation of the postsynaptic proteome
Schematic diagram of synapse and postsynaptic proteome. A single synapse is illustrated with presynaptic and postsynaptic terminal. The NMDA receptor-PSD-95 complex (N) or MASC and AMPA complex (A) are physically separate and embedded in the PostSynaptic Density (speckled). Each of these complexes/fractions were isolated and proteins identified using mass spectrometry.
Venn diagram of proteomic analysis of protein identities in MASC, AC and PSD sets and overall PostSynaptic Proteome (PSP). Summary table of protein classifications. Details for each protein are found in Table 1 (at end).

We isolated MAGUK associated signaling complexes (MASCs) using a previously described peptide affinity method[12]. These complexes are similar in composition to NMDA receptor complexes and MAGUK complexes isolated with antibodies[11,12]. AMPA complexes (ACs) and PSDs were isolated from mouse brain extracts using established methods[11,12] and proteins identified using mass spectrometry. Together these proteins provide a general composition of the postsynaptic proteome and totaled 698 distinct proteins (Fig. 1b,c & Table 1 at end of chapter). Of the 698 proteins, 184 were found in MASCs, 10 in ACs and 620 in PSDs with the main overlap between MASC and PSD of 108 proteins (Fig. 1b).

To facilitate analysis of the organization of the PSP at the level of individual proteins, their domains, complexes and networks, we constructed a Proteome Analysis DataBase (PADB) and several bioinformatics tools. PADB was designed to represent and compare proteomes within and between eukaryotic species (yeast, worm, fly, mouse, rat, human) and is described in detail elsewhere (in preparation). A major feature of PADB is a mammalian (mouse, rat, human) Protein-Protein Interaction Database (PPID) describing approximately 4000 protein interactions for > 1700 proteins (www.PPID.org).

The simplest approach to characterizing the organization of the postsynaptic proteome was to classify the proteins into categories that describe their known functional properties (Tables 1 & Fig.1c). Within the overall postsynaptic proteome, representatives of many classes of proteins representing a broad range of cell biological functions were observed: membrane bound receptors, adhesion proteins and channels; a plethora of signaling proteins and adaptors; and proteins involved with regulation of transport, RNA metabolism, translation and transcription. In addition to the MASC and AMPA complexes, components of other complexes such as cell adhesion, growth factor, cytoskeletal, transport and ribosomal complexes were detected. This is consistent with a postsynaptic organization where the MASC signaling complex is connected to multiple cell biological effector mechanisms organized into their respective complexes.

3. FUNCTIONAL SUBSETS IN SYNAPSE PROTEOMES

We next asked if synapse proteomes comprise particular functional subsets of proteins by examining their protein domain and motif profile using the Interpro resource (www.ebi.ac.uk/interpro) (Table 2). All domains were identified in the MASC proteins and ranked by frequency (% of proteins in MASC with a specific domain) and compared with the mouse proteome[15]. Kinase domains were found in 12% of MASC proteins compared with 1.6% in the proteome[16], a 7-fold enrichment. This enrichment in kinase domains was correlated with enrichment in other domains commonly found in protein kinases, which allow them to interact in networks (e.g. SH2, SH3, PDZ interaction domains; Table 2)[16]. Moreover, domains involved with Ca^{2+}-dependent signaling, a major feature of postsynaptic signal transduction, were also highly abundant (C2, calcium-binding EF hand, IQ motif). Members of the Ras GTPase superfamily were enriched by over 10-fold. Conversely, some domains that are highly abundant in the mouse proteome were absent from the MASC, such as Rhodopsin-like GPCR superfamily Zn-finger and homeobox domains (2nd, 4th and 6th most abundant domains respectively). The enrichment in the synapse of sets of domains involved with signaling also parallels the phylogenetic expansion of these domains in evolution of metazoans[16].

Table 2. Domain profile in MASC.
The 10 most frequent domains in MASC are listed with their identifiers (Interpro), MASC rank and rank in the mouse proteome (www.ebi.ac.uk/interpro). The % of proteins in MASC (MASC %) and the mouse proteome (Proteome %) and the ratio of MASC to proteome, which is a measure of enrichment in MASC is indicated. The note column indicates: * kinase or phosphatase; ** domain commonly found in kinases[16]; *** calcium signaling domain.

Interpro	MASC rank	Proteome rank	Description	MASC %	Proteome %	Ratio	Note
IPR000719	1	3	Eukaryotic protein kinase	11.9	1.6	7.4	*
IPR002290	2	5	Serine/Threonine protein kinase	10.8	1.1	9.8	*
IPR001452	3	16	SH3 domain	6.5	0.6	10.8	**
IPR001245	4	9	Tyrosine protein kinase	5.9	0.7	8.5	*
IPR001806	5	26	Ras GTPase superfamily	5.4	0.4	13.5	
IPR001849	6	17	Pleckstrin-like	4.9	0.5	9.7	**
IPR001478	7	24	PDZ/DHR/GLGF domain	4.9	0.5	9.7	**
IPR002048	8	14	Calcium-binding EF-hand	3.8	0.7	5.4	***
IPR000008	9	38	C2 domain	3.8	0.3	12.6	***
IPR000048	10	48	IQ calmodulin-binding region	3.2	0.2	16.2	***

Large scale mapping of protein interactions in the yeast proteome reveals highly complex networks of protein interactions[17-21]. These networks have not only been useful for clustering functionally related proteins, but were recently subject to analysis with mathematical tools developed for theoretical and non-molecular models such as social networks and the world wide web[22,23]. These studies reveal that the apparent complexity in diagrams of network connectivity can be defined with simple algorithms and can be used to predict biological properties. Networks are represented in graph theory by nodes connected by links, and the simplest networks comprise randomly linked nodes, where each node has a similar number of links and is known as a random network[24]. A Network that includes occasional long-range connections and a small number of highly connected nodes (hubs) produces a very different network, which is heterogeneous and known as a scale-free network[25]. These two classes of network can be readily identified by graphing the distribution of the probability, P(k), that a node has k links; a random network follows a Poisson distribution whereas a scale-free network has no peak, but follows a power-law ($P(k) \approx k^{-\gamma}$) where it appears as a negatively sloped straight line on a log-log plot[25,26].

Large datasets of binary protein interactions have been obtained and represented in complex networks for *Saccharomyces cerevisiae* (1870 nodes, 2240 links)[27]. For these unicellular organisms the probability that a given protein interacts with k other proteins was shown to follow a power-law typical of scale-free networks. We therefore asked if a similar topology was found in mammalian protein interaction networks using PPID. A network of 1729 proteins and 3860 links was graphed (Figure 2a) and found to follow scale-free network architecture (P<0.00001). These data suggests that the design principles underlying protein interaction networks in yeast are evolutionary conserved to humans.

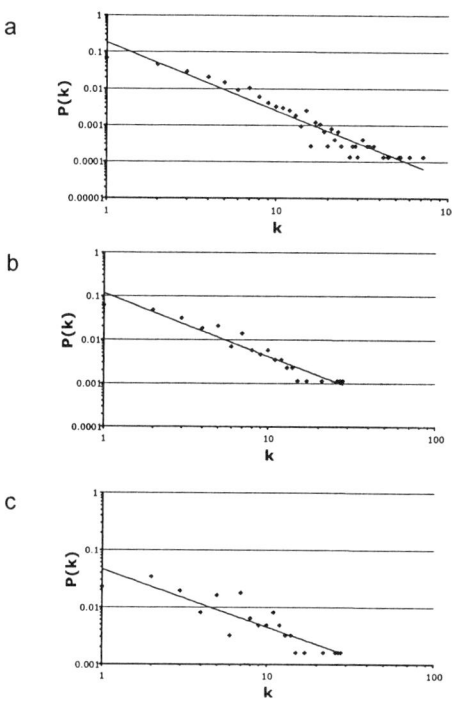

Fig. 2. Scale-free architecture of mammalian proteomes.
Connectivity distribution P(k) for (a) mammalian proteome, (b) Post-synaptic Proteome (PSP), (c) MASC.

We next analysed the networks of the postsynaptic proteome. Graphing the distribution of links shows a scale-free architecture in the overall postsynaptic proteome ($P(k)\sim k^{-1.46}$, 207 proteins, 889 links ($p<0.00001$) (Fig. 2b) and MASC ($P(k)\sim k^{-1.21}$, 103 proteins, 455 links)($P<0.00001$)(Fig. 2c), whereas the AMPA complex has too few proteins to assign this topology (data not shown). A striking feature of scale-free networks is their 'small world' property, which has been used to explain the short distances between individuals in social networks and pages on the world wide web amongst others[22,23]. This property can be quantified using the network diameter, which is the average of the shortest path (number of links) between all pairs of nodes in a network. Using our datasets we found that the diameter of the MASC was 3.41 and the postsynaptic proteome 3.69. While this figure may be an underestimate limited by the available amount of interaction data, it is worth noting that diameter increases as a logarithmic function, and the diameter for the available mammalian proteome dataset is 4.69. This suggests that the dendritic spine, which is the subcellular compartment that harbors the postsynaptic proteome, is a small world where its proteins are on average less than 4 connections from any other and organized into a scale-free network. The binary interactions of the postsynaptic proteome network were illustrated and show a dense pattern of interconnectivity with MASC and AMPA subsets (Fig. 3).

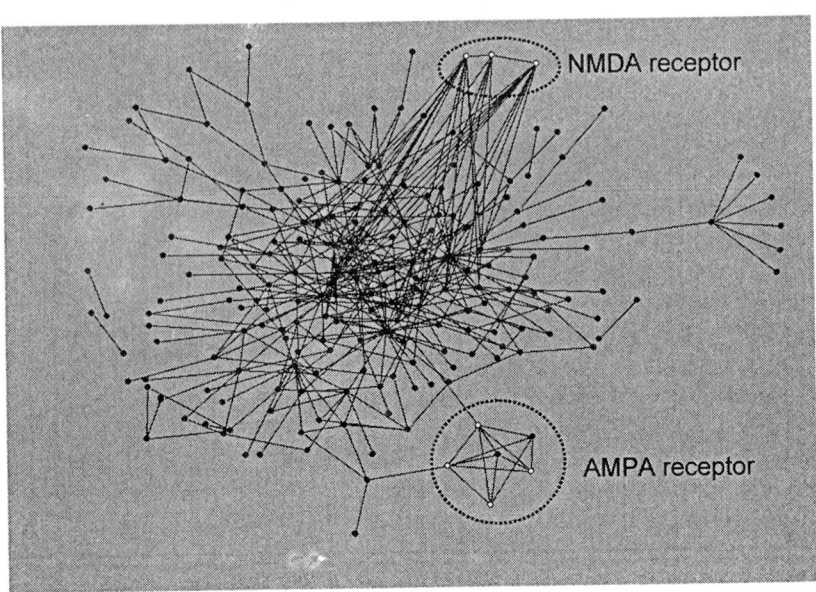

Fig. 3. Connection diagram of PSP.
210 PSP proteins with known binary interactions were mapped and key subunits of the NMDA receptor (NR1, NR2A, NR2B) and AMPA receptor (GluR1,2,3,4) highlighted.

There are no reports on the properties of scale-free networks and their implications for signal transduction complexes or pathways. Moreover, there are several features of highly connected networks that may be relevant to the postsynaptic proteome and its signaling roles. The first is that stability and robustness to perturbation is maintained by the pattern of connectivity. Importantly, scale-free networks, unlike random networks, are intrinsically more robust to loss of nodes[25,27]. However, the disruption of highly connected nodes (hubs) has a greater effect on network diameter in a scale-free network. Thus, the scale-free network is more tolerant to random errors, a feature used to explain the intrinsic robustness of *S. cerevisiae* to mutation affecting cell growth[27]. A second and related property is that highly interconnected networks provide multiple alternative routes between 2 nodes, which allows redundancy in pathways. We therefore focused on the MASC network and used the NMDA receptor and its known signaling role as a model to explore the network model.

To test the robustness to random disruption we simulated the effect of knocking out each node from the MASC network and measured the change in network diameter (Fig. 4). We found a highly significant relationship between the number of links (k) and the increase in diameter ($P<0.0001$). Of the most highly connected (hub) proteins in the MASC (actin, 21 links; calmodulin, 17; PSD-95, 16; Tubulin, 14; NR2B, 12; Src, 12; CamKIIα, 11) several are neuronal (PSD-95, NR2B, CamKIIα) whereas the others are common to most cells. These 'neural hubs' may be important in distinguishing the scale-free molecular networks of synapse proteomes from other cell-type specific signaling proteomes.

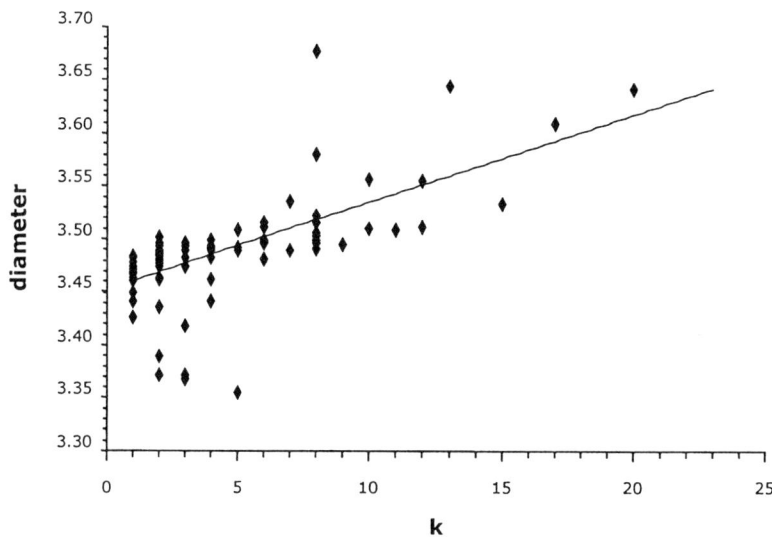

Fig. 4. Simulated disruption of MASC
The network diameter (y-axis) is the average of the shortest distance (number of links) between all pairs of proteins. The effect on diameter was measured for disruption of each proteins in figure 3. There was a strong correlation between the number of connections and its impact on the network diameter upon removal (P<0001).

As a first step toward assessing if the MASC network node disruption was relevant to physiological signaling, we searched literature for genetic and pharmacological evidence that their disruption interferes with synaptic plasticity (long-term potentiation, LTP and long-term depression, LTD) (Table 3). We found 32 proteins were essential – without these proteins synaptic plasticity is impaired. This number can be increased to 45 MASC proteins if additional proteins that were implicated are included. These numbers are likely to be an underestimate since the majority of the genes have not been tested.

The simulation of node disruption and the schematic representation of the network indicates that individual knockouts do not abolish all pathways between any two proteins. For example, the NMDA receptor (NR1 subunit) is connected by 2 paths of 4 interactions, 19 of 5 interactions and 161 of 6 interaction pathways to the AMPA receptor (GluR1 subunit) within PSP.

These observations may serve as a useful model for two important puzzles in synaptic plasticity: namely that there are many dozens of implicated proteins[5,12] and that knockout (mutation or drug) does not abolish plasticity, but only partially impairs it (see for examples[4,6,8-10]. Another feature of the network is that it connects (in short paths) many of the key regulators of plasticity effector mechanisms gene transcription, local protein translation and cytoskeleton that together result in the expression of synaptic plasticity.

One mechanistic implications of the MASC signaling network is that disruption of individual proteins will result in cognitive deficits. We therefore searched for papers reporting the involvement of specific proteins/genes in learning or conditioning paradigms (Table 3). The brain systems involved in learning and fear conditioning include the amygdala and hippocampus and we annotated papers into those affecting spatial learning (hippocampus dependent) and cue/context conditioning (amygdala) and other behavioural paradigms[28]. Overall there were 37 proteins involved with behaviour of which 29 with learning (21 spatial, 18 cue/context with 10 in both).

Rodent behavioural abnormalities resulting from interference with MASC networks raises the possibility that human orthologues may be involved with psychiatric disorders. We searched Medline abstracts and found 44 MASC proteins were in mental illness (Table 3). Although all mental disorders were searched we found 23 in schizophrenia, 17 in mental retardation, 5 in Bipolar disorder and 4 in depressive illness. It is notable that schizophrenia and mental disorders have a major cognitive component to the symptoms. These data include a wide variety of results such as evidence of mendelian inheritance and change in protein level in brain tissue. Whilst it is considered that these disorders are complex or multigenic, and that any one of these reported changes is not sufficient to account for the disorder, it is conceivable that the disruption of the network architecture at multiple nodes may be relevant to the behavioural disorder. Indeed, simulations show combinations of node disruptions produce more severe effects on the network function (data not shown). The network topology will allow testing of the importance of specific combinations of mutant alleles in those genes already implicated in mental illness. We recognize that this analysis was with the proteomic data reported here, and can be extended to include additional data from other studies. For example, PSD-95 directly binds neuroligins[29], ERB4[30] (neuregulin receptor) and huntingtin[31] which are genes involved with autism[32], schizophrenia[33] and Huntingdon's disease[34] respectively.

We recognize that this model of synapse proteome organization into scale-free networks is a draft limited by available data and will be enhanced by further systematic studies of protein interactions, which should increase in number, and thus maintain the essential design of the small-world network. In addition, more detailed biological annotation of the relative importance of nodes, such as quantitative comparison of phenotypes will also refine this model.

4. CONCLUDING REMARKS

The proteomic studies described here identify ~700 proteins comprising the proteome of the postsynaptic terminal of central nervous system synapses. This composition represents a 'draft' of a map of the overall composition of the synapse. Beyond identifying specific proteins, which should be very useful for many types of studies, we present a molecular network model for the organization and function of the proteome within the postsynaptic terminal. The Synaptic Proteome Network (SynProNet) has a scale-free architecture that is evolutionary conserved. Activation of channels and receptors (such as NMDA receptors or Voltage Dependent Calcium channels) could initiate signaling in the MASC, which then orchestrates the multiple pathways and cellular mechanisms for the expression of plasticity. The scale-free network architecture provides robustness and complexity as observed in molecular studies of synaptic plasticity. Moreover, the network of interactions may allow distinct patterns of activity (e.g. frequencies of action potentials) to preferentially activate some areas of the network. These properties provide a new model for understanding the emergence of cognitive function from simple proteomes and the molecular basis of mental illness, and could be used to design diagnostic and therapeutic strategies. It may be interesting to explore hierarchies of interacting 'small-world' networks in brain and behaviour, where the SynProNet is the basic molecular network, underpinning neural networks of local circuits, which are organized into brain networks that influence behaviour relevant to social interactions.

5. ACKNOWLEDGMENTS

Dr. B. Webber, Dr. D. Barber, Mr. T. Theodosiou, Ms. Ada Delaney, Dr.P.Visscher for bioinformatics and statistical assistance., and J. Turner for editorial assistance. SGNG, HH, MC, DA were supported by the Wellcome Trust Genes to Cognition programme.

6. REFERENCES

1. Sherrington, C. S. (1906) *The integrative action of the nervous system* (C. Scribner's sons, New York).
2. Adrian, E.D. (1928) *The basis of sensation, the action of the sense organs* (W. W. Norton & company inc., New York).
3. Hebb, D. O. (1949) *The organization of behavior; a neuropsychological theory* (Wiley, New York).
4. Grant, S. G. et al. (1992) Impaired long-term potentiation, spatial learning, and hippocampal development in fyn mutant mice [see comments]. *Science* 258, 1903-10.
5. Sanes, J. R. & Lichtman, J. W. (1999) Can molecules explain long-term potentiation? *Nat Neurosci* 2, 597-604.
6. Opazo, P., Watabe, A. M., Grant, S. G. & O'Dell, T. J. (2003) Phosphatidylinositol 3-Kinase Regulates the Induction of Long-Term Potentiation through Extracellular Signal-Related Kinase-Independent Mechanisms. *J Neurosci* 23, 3679-88.
7. Migaud, M. et al. (1998) Enhanced long-term potentiation and impaired learning in mice with mutant postsynaptic density-95 protein [see comments]. *Nature* 396, 433-9.
8. Komiyama, N. H. et al. (2002) SynGAP regulates ERK/MAPK signaling, synaptic plasticity, and learning in the complex with postsynaptic density 95 and NMDA receptor. *J Neurosci* 22, 9721-32.
9. Watabe, A. M., Zaki, P. A. & O'Dell, T. J. (2000) Coactivation of beta-adrenergic and cholinergic receptors enhances the induction of long-term potentiation and synergistically activates mitogen-activated protein kinase in the hippocampal CA1 region. *J Neurosci* 20, 5924-31.
10. Yasuda, H., Barth, A. L., Stellwagen, D. & Malenka, R. C. (2003) A developmental switch in the signaling cascades for LTP induction. *Nat Neurosci* 6, 15-6.
11. Husi, H., Ward, M. A., Choudhary, J. S., Blackstock, W. P. & Grant, S. G. (2000) Proteomic analysis of NMDA receptor-adhesion protein signaling complexes [see comments]. *Nat Neurosci* 3, 661-9.
12. Husi, H. & Grant, S. G. (2001) Isolation of 2000-kDa complexes of N-methyl-D-aspartate receptor and postsynaptic density 95 from mouse brain. *J Neurochem* 77, 281-91.
13. Yamauchi, T. (2002) Molecular constituents and phosphorylation-dependent regulation of the post-synaptic density. *Mass Spectrom Rev* 21, 266-86.
14. Walikonis, R. S. et al. (2000) Identification of proteins in the postsynaptic density fraction by mass spectrometry. *J Neurosci* 20, 4069-80.
15. Waterston, R. H. et al. (2002) Initial sequencing and comparative analysis of the mouse genome. *Nature* 420, 520-62.
16. Manning, G., Whyte, D. B., Martinez, R., Hunter, T. & Sudarsanam, S. (2002) The protein kinase complement of the human genome. *Science* 298, 1912-34.
17. Fromont-Racine, M., Rain, J. C. & Legrain, P. (1997) Toward a functional analysis of the yeast genome through exhaustive two-hybrid screens. *Nat Genet* 16, 277-82.
18. Schwikowski, B., Uetz, P. & Fields, S. (2000) A network of protein-protein interactions in yeast. *Nat Biotechnol* 18, 1257-61.
19. Tucker, C. L., Gera, J. F. & Uetz, P. (2001) Towards an understanding of complex protein networks. *Trends Cell Biol* 11, 102-6.

20. Ideker, T. et al. (2001) Integrated genomic and proteomic analyses of a systematically perturbed metabolic network. *Science* **292**, 929-34.
21. Gavin, A. C. et al. (2002) Functional organization of the yeast proteome by systematic analysis of protein complexes. *Nature* **415**, 141-7.
22. Buchanan, M. (2002) *Nexus : small worlds and the groundbreaking science of networks* (W.W. Norton, New York).
23. Barabasi, A.-L. (2002) *Linked : the new science of networks* (Perseus Pub., Cambridge, MA,).
24. Bollobâas, B. (1985) *Random graphs* (Academic Press, London ; Orlando).
25. Barabasi, A. L. & Albert, R. (1999) Emergence of scaling in random networks. *Science* **286**, 509-12.
26. Jeong, H., Tombor, B., Albert, R., Oltvai, Z. N. & Barabasi, A. L. (2000) The large-scale organization of metabolic networks. *Nature* **407**, 651-4.
27. Jeong, H., Mason, S. P., Barabasi, A. L. & Oltvai, Z. N. (2001) Lethality and centrality in protein networks. *Nature* **411**, 41-2.
28. Kandel, E. R., Schwartz, J. H. & Jessell, T. M. (2000) *Principles of neural science* (McGraw-Hill Health Professions Division, New York).
29. Irie, M. et al. (1997) Binding of neuroligins to PSD-95. *Science* **277**, 1511-5.
30. Garcia, R. A., Vasudevan, K. & Buonanno, A. (2000) The neuregulin receptor ErbB-4 interacts with PDZ-containing proteins at neuronal synapses. *Proc Natl Acad Sci U S A* **97**, 3596-601.
31. Sun, Y., Savanenin, A., Reddy, P. H. & Liu, Y. F. (2001) Polyglutamine-expanded huntingtin promotes sensitization of N-methyl-D-aspartate receptors via post-synaptic density 95. *J Biol Chem* **276**, 24713-8.
32. Jamain, S. et al. (2003) Mutations of the X-linked genes encoding neuroligins NLGN3 and NLGN4 are associated with autism. *Nat Genet* **34**, 27-9.
33. Stefansson, H. et al. (2002) Neuregulin 1 and susceptibility to schizophrenia. *Am J Hum Genet* **71**, 877-92.
34. Ho, L. W. et al. (2001) The molecular biology of Huntington's disease. *Psychol Med* **31**, 3-14.
35. Husi, H. & Grant, S. G. (2002) in *Neuroscience Databases: A practical Guide* 51-62 (Kluwer Academic Publishers, Boston/Dordrecht/London).

7. TABLES

Table 1. Postsynaptic proteome composition
Proteins identified in the PostSynaptic Density (PSD), NMDAR-PSD-95 complexes (NRC or MASC) and AMPA receptor complexes. Categories or families of proteins are indicated. This set of proteins results extend the list of NRC proteins, particularly in the molecular weight range below 60kDa, which had not previously been analysed using mass spectrometry[11]. Since glutamate receptors and MAGUK proteins are major constituents of the PSD fraction, we also isolated the PSD fraction and identified proteins using LC/MS-MS. Similarly, we characterized the AMPA receptor complexes using immunoprecipitation of AMPA subunits. Detailed methods are described in[12] will be reported elsewhere

Database construction
Protein interaction data from any available mammalian cell type or organ systems was included for purposes of comparison of cell-type specific proteomes and for broad application and has been described elsewhere[35]. The database was entirely manually curated to ensure accuracy and non-redundancy. Mammalian protein sequences were retrieved from SwissProt (www.expasy.org) and BLAST searched against whole genomes to identify splice modifications as well as orthologues. This generated a name field list that was used to search PubMed (www.ncbi.nlm.nih.gov) for described protein interactions and functions of these molecules. Manuscripts were retrieved and where evidence for specific binary interactions was shown at the biochemical level the interaction was accepted and referenced. This facilitates the construction of networks between sets of interacting proteins.

TABLE 1 - Postsynaptic proteome composition

PPID	Abbreviated Name	Accession	PSD #	MASC #	AMPA #	PPID	Abbreviated Name	Accession	PSD #	MASC #	AMPA #
Channels and Receptors						33	A3473	CACNA2D3	Q9Z1L5	31	
						34	A3474	CACNA2D1	O08532	32	
Glutamate Receptors						35	A3476	CACNB3	P54285	33	
1	A0103	mGluR1	Q9EPV6	1	1						
2	A1353	mGluR3	Q9QYS2	2		*Voltage-gated K+-channel*					
3	A0327	mGluR5	P31424	3	2	36	A3477	KCNQ2	Q9Z343	34	
4	A0453	mGluR7	P35400	4		37	A0764	Kv1.2	P15386	35	
5	A0296	AMPA-R1	P23818	5		1	38	A3478	Kv2.2	Q64284	36	
6	A0297	AMPA-R2	P23819	6		2	*Voltage-dependent anion channels*					
7	A0299	AMPA-R3	Q9Z2W9	7		3	39	A0425	VDAC1	Q60932	37	10
8	A0300	AMPA-R4	P19493			4	40	A0426	VDAC2	Q60930	38	11
9	A0001	NR1	Q62683	8			41	A3479	VDAC3	Q60931	39	
10	A0002	NR2A	O08948	9	4	*Na+-channel*					
11	A0003	NR2B	Q62684	10	5	42	A0598	SCN2A1	Q99250	40	
12	A0022	GluR6	P39087		6	*Ca2+-release channels*					
Other ligand receptors						43	A0154	INSP3R	P11881	41	
13	A0504	GABA-A-R1	P18504	11		44	A3480	RYR2	Q9ERN6	42	
14	A0603	GABA-B-R1	Q9WV18	12		*Other Channels and Receptors*					
15	A2000	GABA-B-R2	O88871	13		45	A1169	APP	P12023	43	
16	A0038	CIRL-1	O88917	14		46	A3481	SLC4A10	Q9EST0	44	
17	A3463	CIRL-3	Q9Z173	15		47	A3482	Plexin-1	P70206	45	
18	A3464	PGRMG	O55022	16		48	A3483	Plexin-2	P70207	46	
Ca2+-ATPases						49	A3484	SLC8A2	P48768	47	
19	A3465	ATP2B1	P11505	17		50	A3485	SLC4A4	O88343	48	
20	A0508	ATP2B2	Q9R0K7	18		51	A3486	HAC-1	O88703	49	
21	A3466	ATP2B3	Q64568	19		52	A0414	ATP6N1	Q9Z1G4	50	
22	A0114	ATP2B4	Q64542	20	7	53	A0409	ATP6V1H	Q9UI12	51	
23	A3467	ATP2A2	O55143	21		54	A3488	SLC4A3	P16283	52	
NA+/K+-ATPases						55	A1175	LRP	Q61291	53	
24	A0340	ATP1A1	P06685	22	8	5	56	A3489	SLC12A5	AF332063	54	
25	A0342	ATP1A3	Q9Z1G6	23	9	57	A4034	SLC17A7	Q62634	55	
26	A0343	ATP1A4	Q9WV28	24		58	A3490	CNTFR	Q08406	56	
27	A0344	ATP1B1	P14094	25		59	A0422	SLC1A2	P43006	57	12
Voltage- and Ligand-gated Ca2+-channel						60	A3491	EDG-1	O08530	58	
28	A0592	CACNA1B	O55017	26		61	A0637	Gephyrin	Q03555	59	
29	A0770	CACNA1C	Q01815	27		62	A1608	gC1q-R	O35658	60	
30	A0593	CACNA1A	P54282	28		**MAGUKs / Adaptors / Scaffolders**					
31	A3471	CACNA1E	Q61290	29		*PDZ-domain containing scaffolders*					
32	A3472	CACNA2D2	Q9EQG2	30		63	A0086	ZO-1	P39447	61	13

PPID		Abbreviated Name	Accession	PSD #	MASC #	AMPA #	PPID		Abbreviated Name	Accession	PSD #	MASC #	AMPA #
64	A0075	Shank	Q9WU13	62	14		100	A0018	Src	P05480	115	42	
65	A0074	ProSAP2	Q9JLU4	63			101	A0019	Fyn	P39688	116		
66	A0137	Maguin	Q9R093	64			102	A3508	FES	P16879	117		
67	A0015	Sap97	Q62402	65	15		103	A0030	PYK2	Q9QVP9		43	
68	A0013	PSD-95	Q62108	66	16		**Ser/Thr Kinases**						
69	A0014	Chapsyn-110	Q63622	67	17		104	A3499	Novel CaMK	Q63092	92		
70	A0016	Sap102	P70175	68	18		105	A0011	CamKIIalpha	S04365	93	34	
71	A0436	DLGH2	Q9WV34	69	19		106	A1851	CamKIIbeta	P28652	94	35	
72	A0793	VAM-1	Q9WV37	70			107	A1852	CamKIIgamma	P11730	95		
73	A0435	DLGH3	O88910	71	20		108	A1853	CamKIIdelta	P15791	96		
74	A0104	Homer	Q9Z216	72	21		109	A0896	CK-1	P97633	97		
75	A0568	Aczonin	Q9QYX6	73			110	A4036	CSNK1E	BAB32922	98		
76	A0125	CASK	O70589	74			111	A0753	CK-2	Q60737	99		
non-PDZ-domain containg scaffolders							112	A4037	PRKACB	P05206	100	36	
77	A0143	AKAP150	P24587	75	22		113	A0306	PKA-R2b	P12369	101	37	
78	A0107	GKAP	P97841		23		114	A0466	AKT2	Q60823	102	38	
79	A0009	Yotiao	Q99P24		24		115	A0168	PKCbeta	P04411	103	39	
80	A0126	Grb2	Q60631		25		116	A0259	PKCgamma	P05697	104	40	
81	A0188	APPL	AF169797		26		117	A0144	PKCepsilon	P16054	105	41	
82	A0124	RACK1	P25388		27		118	A0813	GRK2	P26817	106		
83	A0162	JIP-1	O35145		28		119	A3500	PRKG2	Q61410	107		
84	A3492	Neurobeachin	Q9EPM9	76			120	A3501	ROCK-2	P70336	108		
85	A3493	EPS15R	Q60902	77			121	A3502	AAK1	Q9UPV4	109		
86	A3494	Agrin	P25304	78			122	A3503	MAP4K6	Q9JM92	110		
87	A3495	LIME	Q9EQR5	79			123	A3504	CDC42BPB	O54875	111		
88	A0545	NAP-1	AAH03962	80			124	A3505	EMK	Q05512	112		
89	A3496	Flotillin-1	O08917	81			125	A3506	DCAMKL1	Q9JLM8	113		
90	A3497	Flotillin-2	Q60634	82			126	A3507	TBK1	Q9WUN2	114		
91	A1015	BAIAP2	Q60437	83			**Protein Phosphatases**						
92	A4033	RC3	O94938	84			127	A4042	PPP1CC	P37139	118	44	
93	A3990	CASKIN1	AAL49756	85			128	A0147	PPP1CA	Q9Z1G2	119		
14-3-3							129	A4043	PPP1CB	AK004686	120		
94	A0467	14-3-3beta	P35213	86			130	A0132	PPP2CA	P13353	121	45	
95	A0362	14-3-3gamma	P35214	87	29		131	A0295	PPP2R1A	Q96DH3	122	46	
96	A0280	14-3-3epsilon	U53882	88	30		132	A0123	PP2B	P20652	123	47	
97	A0633	14-3-3eta	P11576	89	31		133	A0289	PP5	O35299		48	
98	A0361	14-3-3zeta	P35215	90	32		134	A1198	PTPRD	Q64487	124		
99	A0907	14-3-3theta	P35216	91	33		135	A2624	PTPRS	Q9QW00	125		
Tyr Kinase							136	A1202	PTPzeta	Q9WUT8	126		

ORGANIZATION AND FUNCTION OF THE POST-SYNAPTIC PROTEOME

PPID	Abbreviated Name	Accession	PSD #	MASC #	AMPA #	PPID	Abbreviated Name	Accession	PSD #	MASC #	AMPA #
137	A3509	SBF-1	O60228	127		174	A3519	SEPT6	Q9R1T4	152	
138	A0223	PTPRF	M60103	128		175	A3520	SEPT7	O55131	153	
139	A0020	PTP1D	P35235		49	176	A3521	SEPT9	Q9QZR6	154	
140	A0118	PTPN5	P54830		50	*Modulators*					
141	A4016	FLJ90311	Q8VEL2	129		177	A0024	SynGAP	Q9ESK6	155	64
G-proteins and Modulators						178	A0843	ARHGEF2	Q9ESG7	156	
G-proteins						179	A3528	CGEF2	Q9Z1P0	157	
142	A3512	M-Ras	O08989	130		180	A0276	Citron	O88938	158	65
143	A0200	H-Ras	Q61411		51	181	A0196	NF1	Q04690	159	66
144	A0194	Rac1	P15154		52	182	A0214	Kalirin	Q9JIF2	160	67
145	A0033	Rap2	Q9D3D5		53	183	A0932	RGS7	O54829	161	
146	A0203	Ral-A	P05810	131	54	184	A3530	KIAA1688	BAB21779	162	
147	A0152	Ran	P28746		55	185	A4046	FMNL	Q9JL26	163	
148	A0363	Rab2	P53994		56	186	A0646	ARHGEF7	O08757	164	
149	A0037	Rab3	P05713	132	57	187	A3532	GIT1	Q9Z272	165	
150	A0359	Rab-5	O88565	133		188	A4047	RANBP9	O94764	166	
151	A0364	Rab6A	Q9JJD4		58	189	A0215	GAP43	P06837	167	68
152	A0620	Rab-8	P55258	134		190	A2428	KIAA0763	AAH10267	168	
153	A0619	Rab11	Q9JLX1			191	A0470	Frabin	Q91ZT5		69
154	A3513	Rab12	P35284	135		**Signalling molecules and Enzymes**					
155	A0365	Rab37	Q9JKM7		59	*Heat shock / Chaperones / Chaperonins*					
156	A3974	MIRO1	CAB66863	136		192	A3533	Hsp105	Q61699	169	
157	A4017	DI-RAS2	Q96HU8	137		193	A2003	Hsp90	P11499	170	
158	A0052	GNAZ	O70443	138		194	A2004	Endoplasmin	P08113	171	
159	A0051	GNAS1	P04894	139	60	195	A2007	HSPA1	P17879	172	70
160	A0043	GNAO1	P18872	140		196	A0146	Hsc-71	P08109	173	
161	A0055	GNAI2	P08752	141		197	A3534	KIAA0417	O43301	174	
162	A0042	GNAQ	P21279	142		198	A0451	BiP	P20029	175	
163	A0070	GNB2	P54312	143	61	199	A1578	Mrj	O54946	176	
164	A0072	GNB4	P29387		62	200	A3535	DNAJA2	AAH03420	177	
165	A0073	GNB5	P54314	144		201	A3536	DNAJA1	P54102	178	
166	A0069	GNB1	Q9QWG8		63	202	A3537	Novel DnaJ domain protein	Q9NVH1	179	
167	A3514	CENTG1	Q9JHW8	145		203	A3538	CCT4	P80315	180	
168	A3515	SEPT2	P42208	146		204	A3539	CCT7	P80313	181	
169	A3525	SEPT8	Q9ESF7	147		205	A3540	CCT2	P80314	182	
170	A3524	SEPT	Q9NVA2	148		206	A3541	CCT1	P11983	183	
171	A3516	SEPT3	Q9Z1S5	149		207	A3542	CCT8	P42932	184	
172	A3517	SEPT4	P28661	150		208	A3543	CCT5	P80316	185	
173	A3518	SEPT5	Q9Z2Q6	151		209	A3544	CCT3	P80318	186	

PPID		Abbreviated Name	Accession	PSD #	MASC #	AMPA #	PPID		Abbreviated Name	Accession	PSD #	MASC #	AMPA #
210	A3545	CCT6A	P80317	187			242	A3567	PSMA-4	Q9R1P0	219		
211	A3548	PDIA6	P38660	188			243	A3568	PSMD-1	O88761	220		
212	A3549	SSR1	AAK16151	189			244	A3569	PSMD-8	AAH04075	221		
213	A2010	Hsp74	P38647	190			245	A3570	COPS-3	O88543	222		
214	A3550	Hsp60	P19226	191			246	A1731	Synaptojanin-2	AK019677	223		
Phosphodiesterases							247	A3571	Kell	Q9EQF2	224		
215	A3551	PDE2A	Q01062	192			*Development*						
216	A3552	PDE10A	Q9QYJ5	193			248	A3572	LSAMP	Q62813	225		
217	A3553	CNP-1	P16330	194			249	A3573	BSG	P18572	226		
Adenylate/Guanylate cyclase							250	A3574	BASP1	Q05175	227		
218	A1979	GUCY1A2	Q9WVI4	195			251	A3575	CBCP1	AK017493	228		
secreted / secretory / signalling							252	A3998	CNNM1	Q9JIQ6	229		
219	A3555	CSPG-2	Q62059	196			253	A3576	GDAP1	O88741	230		
220	A1304	SEC-8	O35382	197			254	A3577	Dystrobrevin	Q9D2N4	231		
221	A3556	BCAN	Q61361	198			255	A3578	DBCCR-1	Q9QXL0	232		
222	A3557	OLFM1	O88998	199			256	A3579	CYFIP1	O88558	233		
ATP synthases							257	A3581	GPRIN1	Q9QZY2	234		
223	A0378	ATP5B	P56480	200			258	A2635	NUMBL	O08919	235		
224	A0377	ATP5A1	Q03265	201	71		259	A3980	GDAP1L1	Q8VE33	236		
225	A0379	ATP5C	P35435	202	72		260	A4001	DAAM1	Q9CQQ2	237		
226	A0397	ATP6A1	P50516	203			*Other signalling molecules*						
227	A0403	ATP6D	P51863	204			261	A0008	Calmodulin	P02593		74	
228	A0400	ATP6B2	P50517	205			262	A0262	nNOS	Q9Z0J4		75	
229	A0401	ATP6C	Q9Z1G3	206			263	A0330	PLCβ-1	Q9Z1B3		76	
230	A0402	ATP6M	P57746	207			264	A0012	PLC γ	Q62077		77	
231	A0404	ATP6E	P50518	208			265	A0106	cPLA2	P47713		78	
232	A3559	TOB3	Q925I1	209			266	A0265	ERK1	Q91YW5		79	
NADH-Ubiquinone Oxidoreductase							267	A0181	ERK2	P27703		80	
233	A3560	NDUFA10	Q99LC3	210			268	A0266	MEK1	P31938		81	
234	A3561	NDUFS1	P28331	211			269	A0267	MEK2	Q63932		82	
235	A3562	NDUFS2	Q91WD5	212			270	A0287	MKP2	Q62767		83	
236	A3563	NDUFA9	Q16795	213			272	A0138	RAF1	Q99N57		85	
237	A3564	NDUFS3	Q9DCT2	214			273	A0164	MKK7	O35406		86	
238	A0431	NDUFV2	Q9D6J6	215	73		274	A0489	MAPKp49	Q61831		87	
Polyubiquitin							275	A0491	MAP2K3	O09110		88	
239	A1257	UBC	Q62317	216			276	A0222	GSK3 beta	Q9WV60		89	
Proteases							277	A0476	Bad	Q61337		90	
240	A3565	DPP-6	P46101	217			278	A0277	Arg3.1	Q9WV31		91	
241	A3566	PAD-1	O35593	218			279	A0473	Stargazin	O88602		92	

ORGANIZATION AND FUNCTION OF THE POST-SYNAPTIC PROTEOME

PPID	Abbreviated Name	Accession	PSD #	MASC #	AMPA #	PPID	Abbreviated Name	Accession	PSD #	MASC #	AMPA #
280	A0488	HIF-1α	P53762		93	317	A3611	RPN2	P25235	265	
281	A0260	PDK-1	Q9Z2A0		94	318	A3612	PI3-K	P17105	266	107
282	A0484	IRS-1	P35569		95	319	A3615	ENO1	P17182	267	
283	A0095	Dynamin	Q61358	238	96	320	A3619	DDOST	O54734	268	
284	A0217	FUS	P35637		97	321	A3620	FLJ10842	Q9ESW4	269	
285	A0286	p53BP1	Q12888		98	322	A3621	SUCLA2	Q9Z2I9	270	
286	A0166	HPK1	Q9Z1W9		99	323	A3622	PIGK	Q9CXY9	271	
Other signalling molecules (Cont)						324	A0424	GLNS	P15105	272	108
287	A0434	EGF-164	Q00731		100	325	A3629	IDH3A	Q9D6R2	273	
288	A0421	ARF3	P16587		101	326	A0331	GAPDH	P16858	274	109
289	A0437	Calretinin	P47728		102	327	A3630	AGPAT5	Q9D1E8	275	
290	A3978	STUB1	Q9WUD1	239		328	A3631	PRPS1	Q9D7G0	276	
291	A4005	CDA08	Q99KW9	240		329	A3632	PPAP2B	Q99JY8	277	
292	A4012	PRNP	P04925	241		330	A3633	HSD17B12	O70503	278	
293	A4013	NRG1	O70465	242		331	A3634	FLJ23841	Q9D8V1	279	
294	A4014	SPINK5	Q9D6R9		103	332	A3635	NRDR	Q9EQU4	280	
Other enzymes						333	A3638	DPM1	Q9WU83	281	
295	A3582	ACC1	Q925C4	243		334	A0433	TPI1	P17751	282	110
296	A3584	FASN	Q9EQR0	344		335	A3639	RPN1	P07153	283	
297	A3585	ALDH1A1	P24549	245		336	A3640	KIAA0455	O75043	284	
298	A3586	ACLY	Q91V92	246		337	A3641	DKFZP761F069	Q9BQF9	285	
299	A3587	INPP4A	Q62784	247		338	A3651	DKFZP566O084	AAH03479	286	
300	A3588	PYGB	P53534	248		339	A3652	D1PAS1	P16381	287	
301	A3591	PIP5K1C	O70161	249		340	A0814	DDX5	Q61656	288	
302	A1926	PFK-C	Q9WUA3	250		*Other enzymes (Cont)*					
303	A0261	PFK-M	P47858	251		341	A3653	DDX3	Q62167	289	
304	A1925	PFK-L	P12382	252	104	342	A1245	NEDD4	P46935	290	
305	A3592	Glucosidase-1	Q9Z2W5	253		343	A3656	DDX1	Q91VR5	291	
306	A3593	BG1	Q99PU5	254		344	A3657	PYCR1	Q922W5	292	
307	A3594	FTHFD	P28037	255		345	A3658	PI4-K	O08662	293	
308	A3601	PCCA	Q91ZA3	256		346	A3659	FACL6	Q63835	294	
309	A3604	MCCC1	Q99MR8	257		347	A3660	IDH2	Q9EQK1	295	
310	A1923	ALDOA	P05064	258		348	A3662	PC	Q05920	296	
311	A0428	ALDOC	P05063	259	105	349	A3663	HK1	P17710	297	
312	A0442	DPYSL2	O08553	260	106	350	A3664	OGDH	Q60597	298	
313	A3605	CRMP1	P97427	261		351	A3667	GPD2	Q64521	299	
314	A3606	DPYSL4	O35098	262		352	A3668	IDH3B	CAC01442	300	
315	A3610	DPYSL5	Q9JMG8	263		353	A3669	IDH3G	P70404	301	
316	A0635	MGAD	P48318	264		354	A3670	GA	P28492	302	

PPID	Abbreviated Name	Accession	PSD #	MASC #	AMPA #	PPID	Abbreviated Name	Accession	PSD #	MASC #	AMPA #
355	A3672	MAOA	P21396	303			393	A0444	G9A	Q9Z148	334	120	
356	A0605	DLAT	P08461	304	111		**Transcription and Translation**					
357	A3674	DLD	O08749	305			*Transcription Elements*					
358	A3675	DBT	P53395	306			394	A3788	SND1	Q9R0S1	335		
359	A3676	UQCRC1	Q9CZ13	307			395	A3789	ATF7IP	Q9JK31	336		
360	A3677	UQCRC2	P32551	308			396	A3790	ORC1L	Q9Z1N2	337		
361	A3680	GLUD	P26443	309			397	A3792	HNRPM	Q9D0E1	338		
362	A3682	DLST	Q01205	310			398	A0439	D-Prohibitin	Q61336	339	121	
363	A3683	HADHA	Q64428	311			399	A3793	PHB	P24142	340		
364	A3684	HADHB	Q60587	312			400	A3794	NRBF1	Q99L39	341		
365	A3685	PDHA1	P35486	313			401	A3795	PURA	P42669	342		
366	A3687	PDHB	P49432	314			402	A3796	PURB	O35295	343		
367	A3688	CS	Q9CZU6	315			403	A3798	FIBP	Q9JI19	344		
368	A3690	CKMT1	P30275	316			404	A3799	TRIM32	Q8K055	345		
369	A3694	ACAT1	P17764	317			405	A3580	CYFIP2	Q924D3	346		
370	A3696	MDH2	P08249	318			406	A0298	GCFC	P58501		122	
371	A3697	SUCLG1	Q9WUM5	319			*Elongation Factors*					
372	A3698	CYC1	Q9D0M3	320			407	A1820	EF1A	P10126	347		
373	A3699	DECR1	Q64591	321			408	A3801	EEF1A2	P27706	348		
374	A3702	BDH	P29147	322			409	A3802	EEF1B	O70251	349		
375	A3731	SLC25A11	P97700	323			410	A1968	EF1G	Q9D8N0	350		
376	A3778	MTCO2	P00405	324			411	A3804	EEF2	P58252	351		
377	A3779	ETFDH	Q9DCT9	325			412	A3805	EIF3S7	O70194	352		
378	A3780	ILES	Q8R2M5	326			413	A3806	EIF4B	Q922K6	253		
379	A3665	OGDH-L1	Q9ULD0	327			414	A2508	PABPC1	P29341	354		
380	A0429	ACO2	Q99KI0	328	112		*Ribosomal Proteins*					
381	A3781	PDCD8	Q9Z0X1	329			415	A3807	RPLP0	P14869	355		
382	A3783	ENDOG	O08600	330			416	A3808	RPL4	P50878	356		
383	A0432	AOP2	Q9QWW0		113		417	A1222	RPL5	P47962	357		
384	A0427	PKLR	P53657		114		418	A3810	RPL6	P47911	358		
385	A0423	Asp amino-transfersase	P05202		115		419	A0366	RPL7	P14148	359	123	
386	A0369	LDHB	P16125		116		420	A3811	RPL7A	P12970	360		
387	A0370	LDH1	P06151		117		421	A3812	RPL8	Q9Z237	361		
388	A0371	TDPX2	P35700		118		422	A3813	RPL10A	P53026	362		
389	A0372	TPX	Q61171		119		423	A0367	RPL13	P47963	363	124	
Other enzymes (Cont)							424	A0368	RPL13A	P19253		125	
390	A3625	RP2	Q9EPK2	331			425	A3814	RPL14	Q63507	364		
391	A0685	Auxilin	O75061	332			426	A3815	RPS2	P25444	365		
392	A2429	KIAA0522	O60275	333			427	A3816	RPS3	P17073	366		

ORGANIZATION AND FUNCTION OF THE POST-SYNAPTIC PROTEOME

PPID	Abbreviated Name	Accession	PSD #	MASC #	AMPA #	PPID	Abbreviated Name	Accession	PSD #	MASC #	AMPA #
428	A3817	RPS3A	P97351	367			459	A0350	MBP	P04370		139
429	A3818	RPS4X	P47961	368			*Cell recognition molecules*					
430	A3823	RPS8	P09058	369			460	A1074	Tenascin-R	Q05546	394	
431	A3286	RBP34	Q63742	370			461	A0663	NFASC	P97685	395	
Cytoskeletal and Cell Adhesion Molecules							462	A3607	SDFR1	P97547	396	
Actin / ARP							*Keratins*					
432	A0091	Actin	O35247	371	126		463	A1669	KRT5	Q920F2	397	
433	A0235	ACTR2	P53488	372			464	A3741	2410039E07 RIK	Q9CWC9	398	
434	A0234	ACTR3	Q99JY9	373			465	A3846	KRT6A	Q9Z332	399	
435	A3287	ARPC1A	Q9R0Q6	374			*Other Cytoskeletal Proteins*					
436	A0231	ARPC4	O15509		127		466	A0081	CortBP-1	Q9QX74	400	140
437	A0232	ARPC3	O15145		128		467	A3865	EPB41L3	Q9WV92	401	
438	A0233	ARPC2	O15144		129		468	A3867	EPB41L1	Q9Z2H5	402	
Tubulin							469	A3869	PPFIA3	O75145	403	
439	A0017	Tubulin	P02551	375	130	6	470	A3870	CLMN	Q91XT7	404	
actinin							471	A3871	SYNPO	Q9Z327	405	
440	A2341	ACTN1	Q9Z1P2	376			472	A0417	CAPZ alpha	P47754	406	141
441	A0007	ACTN	Q9JI91	377	131		473	A0418	CAPZ beta	P47757	407	142
442	A2343	ACTN3	O88990		132		474	A0419	Gelsolin	P13020	408	143
443	A2344	ACTN4	P57780	378	133		475	A0273	Spinophilin	O35274	409	
Spectrin							476	A3872	WBP3	Q9NSL0	410	
444	A0010	Spectrin	Q62261	379	134		477	A1941	Plectin	P30427	411	
MAPs							478	A3873	ANK2	Q01484	412	
445	A2342	ACF7	Q9QXZ0	380			479	A3874	ANK3	Q61307	413	
446	A0085	MAP1A	P34926	381			480	A0376	Neurofilament Triplet M	P08553	414	144
447	A0625	MAP1B	P14873	382			481	A0375	Neurofilament triplet L protein	P08551	415	145
448	A0134	MTAP2	P20357	383	135		482	A3875	KIDINS220	Q9ERD4	416	
449	A0135	MAPT	P10638	384			483	A0438	DBN1	Q9QXS6	417	146
450	A3376	MTAP6	O55129	385			484	A3876	ADD1	Q9QYC0	418	
Catenins							486	A3877	ADD3	Q9QYB5	420	
451	A1914	a-catenin 2	Q61301	386			487	A3878	ELKS	Q99MI2	421	
452	A0112	b-catenin	Q02248	387	136		488	A3879	CAST	O15083	422	
453	A0469	d-catenin	O35927	388			489	A1027	WISH	Q9ESJ4	423	
454	A0218	CTNND1	P30999	389			490	A0373	ADAM22	Q9R1V6	424	147
Myelin							491	A3881	ADAM23	Q9R1V7	425	
455	A0352	PLP1	Q62079	390	137	7	492	A3880	ADAM11	Q9R1V4	426	
456	A3590	OMG	Q63912	391			493	A0148	WASF1	Q9ERQ9	427	
457	A0351	MOG	Q61885	392	138		494	A1218	Thy-1 membrane glycoprotein	P01831	428	
458	A3425	GPM6A	P35802	393			495	A3882	LASP1	Q61792	429	

	PPID	Abbreviated Name	Accession	PSD #	MASC #	AMPA #		PPID	Abbreviated Name	Accession	PSD #	MASC #	AMPA #
496	A0292	Internexin	P46660	430	148		533	A3898	MAG	P20917	461		
497	A3883	Paralemmin	Q9Z0P4	431			534	A1163	Norbin	Q9Z0E0	462		
498	A1111	E3B1	AAH04657	432			535	A1276	apoJ	Q06890	463		
499	A0285	Bassoon	O88737	433	149		536	A0493	Connexin-43	P23242	464		
500	A0654	EB2	Q61167	434			537	A3888	KILON	Q9Z0J8	465		
501	A0291	Desmoglein	Q61495		150		538	A3887	HNT	Q62718	466		
502	A0446	CSE1L	Q9UP98		151		539	A0440	Leucine-rich Glioma-inactivated 1 protein	Q9JIA1	467	159	
503	A0083	Cortactin	Q60598		152		540	A3899	VELI2	Q9Z252	468	160	
504	A0216	SLMAP	Q28623		153		541	A0141	Veli	Q9Z251	469		
505	A3976	ABLIM1	Q9EPW6	435	154		542	A3893	CRTL1	Q9QUP5	470		
506	A0204	Filamin	Q9JJ38		155		543	A0454	Desmoplakin	P15924			8
507	A0430	LMNB1	P14733		156		**Synaptic Vesicles and Protein Transport**						
508	A0660	ACTR1A	P42024	436			*Clathrin*						
509	A3975	CRTAC1	Q8R555	437			544	A0119	CLTC	P11442	471	161	
510	A3979	FLJ90835	Q96ID5	438			545	A0122	CLTA	P08081	472		
511	A3211	CKAP4	Q8R3F2	439			546	A3901	CLTB	P08082	473		
512	A3989	NG5	O35449	440			*synaptic vesicle*						
513	A3991	LYRIC	Q9D052	441			547	A3902	SV2	Q02563	474		
514	A4009	KIAA0097	Q14008	442			548	A1324	AP3D	O54774	475		
Other Cell Adhesion Molecules							549	A0684	AP180	Q61548	476		
515	A3884	NCAM1	P13595	443			550	A0560	Munc-13-1	Q62768	477		
516	A3885	NCAM2	O35136	444			551	A0035	RIM-1	O35168	478		
517	A0290	L1CAM	P11627	445	157		552	A0120	SYNJ1	Q62910	479		
518	A0219	N-cadherin	P15116	446	158		553	A2102	Coatomer alpha subunit	P53621	480		
519	A3886	OPCML	P32736	447			554	A0237	COPB2	O55029	481		
520	A0129	NLGN	Q62765	448			555	A3903	SNIP	Q9QXY3	482		
521	A3889	NLGN2	Q62888	449			556	A0562	Tomosyn	Q9Z152	483		
522	A3890	NLGN3	Q62889	450			557	A2346	CYLN2	P30622	484		
523	A3891	NRXN3	Q07314	451			558	A0161	Synapsin-I	O88935	485		
524	A0758	Caspr	O54991	452			559	A3904	Synapsin-II	Q9QWV7	486		
525	A3892	CSPG3	P55066	453			560	A3905	SYT7	Q9R0N7	487		
526	A3894	NEO1	P97798	454			561	A0121	AMPH	O08838	488		
527	A3895	CSPG5	Q9QY32	455			562	A1109	Amphiphysin2	O08839	489		
528	A0150	Densin-180	P70587	456			563	A3906	HIP1R	Q9JKY5	490		
529	A1073	Contactin-2	Q61330	457			564	A0251	AP2A1	P17426	491		
530	A0757	Contactin-1	P12960	458			565	A0252	AP2A2	P17427	492		
531	A3896	ASTN	Q61137	459			566	A0686	AP1B1	P52303	493		
532	A3897	ASTN2	Q9UHW6	460			567	A0253	AP2B1	AK004975	494		

ORGANIZATION AND FUNCTION OF THE POST-SYNAPTIC PROTEOME

PPID	Abbreviated Name	Accession	PSD #	MASC #	AMPA #	PPID	Abbreviated Name	Accession	PSD #	MASC #	AMPA #
568	A0254	AP2M1	P20172	495		606	A3942	SLP2	Q99JB2	533	
569	A3907	AP3M2	P53678	496		607	A0039	SYT1	P46096	534	167
570	A3908	VPS16	Q920Q4	497		608	A1331	EHD3	Q9QXY6	535	
571	A3909	VPS35	Q61123	498		609	A3929	VPS45A	P97390	536	
572	A1326	VPS41	Q9H348	499		610	A0445	S-REX	Q64548	537	168
573	A0596	NAPA	P54921	500		611	A4027	TJP4	Q9DCD5	538	
574	A3910	NAPB	P28663	501		612	A0441	E-FABP	Q01469	539	169
575	A3911	NAPG	AK010275	502		613	A4008	MG87	O88994	540	
576	A0333	SNAP25	P70558	503	162	614	A3863	SYCP1	Q15431	541	
577	A3913	GBAS	O55126	504		615	A3531	KTN1	Q61595	542	
578	A3914	NIPSNAP1	O55125	505		*Transporters*					
579	A3918	VCP	Q01853	506		616	A3943	SLC30A3	P97441	543	
580	A0452	Calnexin	P35564	507		617	A3944	VPS26	P40336	544	
581	A0212	RPH3A	P47708	508		618	A3945	SEC23A	Q01405	545	
582	A0117	NSF	P46460	509	163	619	A3946	ABCD3	P55096	546	
583	A3920	SEC5L1	Q9D4H1	510		620	A3947	ABCB8	Q9CXJ4	547	
584	A3921	EXO70	O54922	511		621	A3948	MTCH1	Q9QZP4	548	
585	A3922	SEC6L1	Q62825	512		622	A3949	MTCH2	AAH02152	549	
586	A3923	EXO84	O54924	513		623	A3750	SLC25A1	P32089	550	
587	A3924	SEC15L2	Q9D4R7	514		624	A3776	SLC25A3	P16036	551	
588	A3925	SEC10L1	P97878	515		625	A0395	SLC25A4	P48962	552	170
589	A3926	EPN2	Q9Z1Z3	516		626	A0396	SLC25A5	Q09073	553	171
590	A3927	PICALM	O55011	517		627	A0443	SLC25A12	O75746	554	172
591	A0238	STXBP1	O08599	518	164	9	628	A3747	SLC25A13	Q9QXX4	555	
592	A3928	RA410	Q62991	519		629	A3773	GC1	Q9H936	556	173
593	A3932	PXR2	Q9JMB9	520		630	A3950	TOMM70A	Q9CZW5	557	
Synaptic vesicle (Cont)						631	A3951	SFXN1	Q63965	558	
594	A3930	VPS33A	Q63615	521		632	A3952	SFXN3	Q9JHY2	559	
595	A3933	SYNGR3	Q9WVG8	522		633	A3954	SFXN5	Q925N0	560	
596	A0288	Synaptogyrin	O55100	523	165	634	A3956	MTX1	P47802	561	
597	A3934	VTI1L1	O88384	524		635	A3957	MTX2	O88441	562	
598	A3935	MOB3	Q9Y3A3	525		*Motor Proteins*					
599	A3936	GOSR1	O88630	526		636	A3958	MYO1B	P46735	563	174
600	A3937	VAPA	Q9Z270	527		637	A3959	MYO1D	Q63357	564	
601	A3938	VAPB	Q9QY76	528		638	A0149	Myosin (V)	Q9QYF3	565	175
602	A0040	STX	Q9QXG3	529	166	639	A3960	MYO6	Q64331	566	
603	A3939	STX12	Q9ER00	530		640	A3166	MYH9	Q8VDD5	567	176
604	A3940	STX7	O70439	531		641	A3961	MYH10	Q9JLT0	568	177
605	A3941	SYP	Q62277	532		642	A3962	FLJ13881	Q9H882	569	

	PPID	Abbreviated Name	Accession	PSD #	MASC #	AMPA #		PPID	Abbreviated Name	Accession	PSD #	MASC #	AMPA #
643	A2331	MYH11	O08638	570	178		679	A3988	TASP	Q9JJK2	603		
644	A3202	MYH7	O61355		179		680	A3992	KIAA1136	Q9ULT3	604		
645	A4048	MRLC2	Q63781		180		681	A3993	SVH	Q9CUN3	605		
646	A4049	MYL6	Q60605		181		682	A3994	0710001P09RIK	Q9D5I5	606		
647	A3964	TIAF1	Q9JMH9	571			683	A3995	2610528A17RIK	Q9CZU2	607		
648	A3965	TPM1	Q63608	572			684	A3996	1700054N08RIK	Q8QZT2	608		
649	A3205	TPM3	P21107	573			685	A3997	TTYH1	Q9EQN7	609		
650	A3966	KIF2	P28740	574			686	A3999	LRPPRC	Q8K4V0	610		
651	A3967	KIF5B	Q61768	575			687	A4000	KIAA0090	Q14700	611		
652	A3968	KLC2	O88448	576	182		688	A4002	G2	Q12914	612		
653	A0133	DNCH1	P38650	577			689	A4003	A330102H22RIK	Q9D217	613		
654	A0661	DyneinIC	O88485	578			690	A4004	LETM1	Q9Z2I0	614		
655	A2127	Dynein intermediate chain 2, cytosolic	O88487	579			691	A4006	KIAA0143	Q922I2	615		
656	A2345	Restin	P30622	580			692	A4007	CDK5RAP3	Q9JLH7	616		
657	A0658	p150Glued	O08788	581			693	A4010	WDR7	Q920I9	617		
658	A3969	DCTN4	Q9QUR2	582			694	A4011	KIAA0941	Q9Y2F0	618		
659	A0656	Dynamitin	Q99KJ8	583			695	A0281	DKFZP564A026	Q8K2R1		183	
660	A3970	CLASP2	Q99JI3	584			696	A2084	Hypothetical protein FLJ14948	Q9DB63		184	
661	A3971	OPA1	O60313	585			697	A4015	BA342L8.1	Q9D5J9		185	
662	A1178	BERP	Q9R1R2	586			698	A4023	LOC90550	Q8NE86	619		
663	A0571	Intersectin-1	Q9Z0R4	587			699	A3983	KIAA1170	Q9ULQ0	620		
664	A3972	TMOD2	P70566	588									
665	A3973	IMMT	Q9D9F6	589									
unknown													
666	A4035	KIAA0953	Q9Y2G0	590									
667	A3498	LOC145567	AL832848	591									
668	A3510	4833411B01Rik	Q9Z2D2	592									
669	A3511	PTPIP51	BAB46923	593									
670	A4045	MIC1	O35606	594									
671	A3554	KIAA1549	Q9HCM3	595									
672	A3558	FLJ20420	Q9CRB9	596									
673	A3800	DJ403A15.3	Q9CU45	597									
674	A3912	KIAA0555	O60302	598									
675	A2325	Hypothetical protein	Q9HCQ3	599									
676	A3977	KIAA0103	Q9CRD2	600									
677	A3986	KIAA1771	Q8R1A4	601									
678	A3987	DKFZP761D221	Q8VD37	602									

TABLE 3 - Biological roles of MASC proteins

	Name	PPID#	Plasticity			Rodent Behaviour			Human			
			Mutation	Drug	Other	Spatial Learning	Cued/Contextual Learning	Other	Schizophrenia	Mental Retardation	BiPolar	Depression
1	NR1	A0001	[1]	[2]	[3]	[1], [4]	[5], [6], [7], [8], [9]	[10],[11] [12]	[13], [14], [15]			
2	NR2A	A0002	[16], [17]			[16]	[17-20]	[21], [22]	[13], [14]			
3	NR2B	A0003	[23]	[6]	[24]	[24],[23], [25]	[26],[27]	[21], [22]	[14], [28]			
4	CALM	A0008		[29], [30], [31]			[32-34]					
5	PRKA9	A0009										
6	SPNB	A0010			[35],[36]							
7	CAMK2A	A0011	[37], [38]	[31], [39]		[40]	[41, 42]				[43]	
8	DLG4	A0013	[44]			[44]		[45-47]	[13, 48]	[49]		
9	DLG1	A0015			[50]				[51]			
10	DLG3	A0016							[51]			
11	Tubulin	A0017			[52]							
12	Src	A0018	[53], [54]		[55]	[56]		[57]				
13	PTP1D	A0020	[58]			[58]				[59]		
14	GRIK2	A0022	[60]						[61]			
15	SynGAP	A0024	[62], [63]			[62]						
16	FAK2	A0030		[64]	[65]							
17	Rap2	A0033			[66]				[67]			
18	Rab3	A0037	[68]		[69]				[70]			
19	SYT1	A0039			[71]		[72]					
20	STX	A0040			[73], [74]	[75],[76]			[77-80]			
21	GNAS1	A0051							[81], [82]			
22	ACT	A0091		[83]	[84], [85]		[85], [86]		[87], [88]			
23	DNM1	A0095										
24	MGluR1a	A0103	[89, 90]		[91]	[92]	[93]	[94]				
25	HOMER1	A0104			[95], [96], [97], [98]							
26	Cpla2	A0106			[99], [100], [101]	[102]	[103]		[104-106]			
27	NSF	A0117		[107]					[108]			
28	PTPN5	A0118		[109]								
29	CLTC	A0119								[110]		
30	PP2B	A0123	[111] [112] [113] [114]	[115] [116] [117]		[118] [112] [113]	[115] [119] [120]					
31	RACK-1	A0124					[121]				[122]	
32	MTAP2	A0134			[123] [124] [125]		[126]	[127]				

	Name	PPID#	Plasticity			Rodent Behaviour			Human			
			Mutation	Drug	Other	Spatial Learning	Cued/Contexual Learning	Other	Schizophrenia	Mental Retardation	BiPolar	Depression
33	AKAP5	A0143			[128]		[129]					
34	PKC epsilon	A0144		[130] [131]			[132], [133]	[134], [135]				
35	Erk2	A0181			[62], [136]	[137]			[138]			[139]
36	Rac 1	A0194								[140], [141]		
37	NF-1	A0196	[142]					[143]		[144-146]		
38	H-Ras	A0200	[62],[147]									
39	Filamin	A0204							[148]			
40	GAP43	A0215	[149]					[150],[151]	[152]			
41	N-cadherin	A0219		[153]	[154], [155]							
42	GSK3 beta	A0222				[156]					[157]	
43	PKCgama	A0259	[158-160]			[161]	[132, 162-165]	[166]				
44	Nnos	A0262	[167-170]	[171-173]			[174]	[175-181]	[182]			
45	Erk1	A0265	[183], [184]		[136]	[184]		[185]	[138]			[139]
46	MEK1	A0266		[186-188]		[189]	[188]					
47	Rsk-2	A0268								[190-192]		
48	Arg3.1	A0277		[193]	[98], [194]	[193], [195]		[196]				
49	MKP2	A0287							[138]			[139]
50	Synaptogyrin	A0288	[197]									
51	L1CAM	A0290	[198], [199]			[200]		[201]	[202]	[203], [204]		
52	MgluR5	A0327	[205]			[92,206,207]	[208-211]		[212], [213]			
53	PLCb	A0330	[160], [214-216]			[207]	[214,] [217]	[215]				[218]
54	GAPDH	A0331						[219]				
55	SNAP25	A0333			[123], [220]			[221]	[78, 222-226]		[226]	
56	Atp1A3	A0342									[227]	
57	MOG	A0351						[228]				
58	PLP1	A0352								[229], [230]		
59	LDHB	A0369							[231]			
60	Neurofilament Triplet L protein	A0375			[232], [233]				[234]			
61	Neurofilament Triplet M	A0376				[235]						
62	CAPZ alpha	A0417								[236]		
63	SLC1A2	A0422		[237], [238]			[238]	[239]	[212],[213],[240] [241]			
64	Gln synthetase	A0424						[242]				

| | Name | PPID# | Plasticity | | | Rodent Behaviour | | | Human | | | |
			Mutation	Drug	Other	Spatial Learning	Cued/Contextual Learning	Other	Schizophrenia	Mental Retardation	BiPolar	Depression
65	PKLR	A0427							[243]			
66	Triosephosphate Isomerase	A0433							[244]			
67	Calretinin	A0437	[245], [246]									
68	DPYSL2	A0442							[247]			
69	14-3-3eta	A0633							[248]			
70	Phosphofructokinase B	A1925								[249]		
71	HSPA1	A2007							[250], [251]			

Table 3. Biological roles of MASC proteins in plasticity, behaviour and disease.
All proteins in MASC were annotated according to their reported roles in synaptic plasticity (left section), rodent behaviour (middle section) and human psychiatric diseases (right section). Proteins are listed and specific references reported within the table. The Plasticity section reports molecules involved with synaptic plasticity (long-term potentiation or depression) and is divided into mutation, drugs and other. Mutation cites papers where mutation in the gene encoding the specific protein is reported; Drug cites papers where a specific pharmacological antagonist was reported; Other refers to reports where there is no causal link to phenotype, but reports changes in the protein or gene. The rodent Behaviour section is divided into Spatial learning, Cued/Contextual Learning, Other. Drugs, mutations and other results implicating the proteins in these 3 sections are annotated. Major psychiatric disorders are annotated; schizophrenia, mental retardation, Bipolar disorder and Depression. This table is not exhaustive and additional reports for these proteins are available. Apologies are made to authors whose work was not cited.

1. Tsien, J.Z., Huerta, P.T. and Tonegawa, S. (1996) The essential role of hippocampal CA1 NMDA receptor-dependent synaptic plasticity in spatial memory. *Cell*, **87**(7), 1327-1338.
2. Cammarota, M., et al. (2000) Rapid and transient learning-associated increase in NMDA NR1 subunit in the rat hippocampus. *Neurochem Res* **25**(5), 567-572.
3. Collingridge, G.L., Kehl, S.J. and McLennan, H. (1983) Excitatory amino acids in synaptic transmission in the Schaffer collateral-commissural pathway of the rat hippocampus. *J Physiol* **334**, 33-46.
4. Morris, R.G., et al. (1986) Selective impairment of learning and blockade of long-term potentiation by an N-methyl-D-aspartate receptor antagonist, AP5. *Nature* **319**(6056), 774-776.
5. Fanselow, M.S., et al. (1994) Differential effects of the N-methyl-D-aspartate antagonist DL-2-amino-5-phosphonovalerate on acquisition of fear of auditory and contextual cues. *Behav Neurosci* **108**(2), 235-240.
6. Bauer, E.P., Schafe G.E., and LeDoux, J.E. (2002) NMDA receptors and L-type voltage-gated calcium channels contribute to long-term potentiation and different components of fear memory formation in the lateral amygdala. *J Neurosci* **22**(12), 5239-5249.
7. Gould, T.J., McCarthy, M.M. and Keith, R.A. (2002) MK-801 disrupts acquisition of contextual fear conditioning but enhances memory consolidation of cued fear conditioning. *Behav Pharmacol* **13**(4), 287-294.
8. Lu, Y. and Wehner, J.M. (1997) Enhancement of contextual fear-conditioning by putative (+/-)-alpha-amino-3-hydroxy-5-methylisoxazole-4-propionic acid (AMPA) receptor modulators and N-methyl-D-aspartate (NMDA) receptor antagonists in DBA/2J mice. *Brain Res* **768**(1-2), 197-207.
9. Davis, M., et al. (1993) Fear-potentiated startle: a neural and pharmacological analysis. *Behav Brain Res* **58**(1-2), 175-98.
10. Miyamoto, Y., et al. (2002) Lower sensitivity to stress and altered monoaminergic neuronal function in mice lacking the NMDA receptor epsilon 4 subunit. *J Neurosci* **22**(6), 2335-2342.
11. Mohn, A.R., et al. (1999) Mice with reduced NMDA receptor expression display behaviors related to schizophrenia. *Cell* **98**(4), 427-436.
12. Rondi-Reig, L., et al. (2001) CA1-specific N-methyl-D-aspartate receptor knockout mice are deficient in solving a nonspatial transverse patterning task. *Proc Natl Acad Sci U S A* **98**(6), 3543-3548.
13. Dracheva, S., et al. (2001) N-methyl-D-aspartic acid receptor expression in the dorsolateral prefrontal cortex of elderly patients with schizophrenia. *Am J Psychiatry* **158**(9), 1400-1410.
14. Akbarian, S., et al. (1996) Selective alterations in gene expression for NMDA receptor subunits in prefrontal cortex of schizophrenics. *J Neurosci* **16**(1), 19-30.

15. Gao, X.M., et al. (2000) Ionotropic glutamate receptors and expression of N-methyl-D-aspartate receptor subunits in subregions of human hippocampus: effects of schizophrenia. *Am J Psychiatry* **157**(7), 1141-1149.
16. Sakimura, K., et al. (1995) Reduced hippocampal LTP and spatial learning in mice lacking NMDA receptor epsilon 1 subunit. *Nature* **373**(6510),. 151-155.
17. Sprengel, R., et al. (1998) Importance of the intracellular domain of NR2 subunits for NMDA receptor function in vivo. *Cell* **92**(2): p. 279-289.
18. Kiyama, Y., et al. (1998) Increased thresholds for long-term potentiation and contextual learning in mice lacking the NMDA-type glutamate receptor epsilon1 subunit. *J Neurosci* **18**(17), 6704-6712.
19. Moriya, T., et al. (2000) Close linkage between calcium/calmodulin kinase II alpha/beta and NMDA-2A receptors in the lateral amygdala and significance for retrieval of auditory fear conditioning. *Eur J Neurosci* **12**(9), 3307-3314.
20. Kishimoto, Y., et al. (1997) Conditioned eyeblink response is impaired in mutant mice lacking NMDA receptor subunit NR2A. *Neuroreport* **8**(17), 3717-3721.
21. Doyle, K.M., et al. (1998) Comparison of various N-methyl-D-aspartate receptor antagonists in a model of short-term memory and on overt behaviour. *Behav Pharmacol* **9**(8), 671-681.
22. Khan, A.M., et al. (1999) Lateral hypothalamic NMDA receptor subunits NR2A and/or NR2B mediate eating: immunochemical/behavioral evidence. *Am J Physiol* **276**(3 Pt 2), R880-891.
23. Tang, Y.P., et al. (1999) Genetic enhancement of learning and memory in mice. *Nature* **401**(6748), 63-69.
24. Clayton, D.A., et al. (2002) A hippocampal NR2B deficit can mimic age-related changes in long-term potentiation and spatial learning in the Fischer 344 rat. *J Neurosci* **22**(9), 3628-3637.
25. Wong, R.W., et al. (2002) Overexpression of motor protein KIF17 enhances spatial and working memory in transgenic mice. *Proc Natl Acad Sci U S A* **99**(22), 14500-14505.
26. Tang, Y.P., et al. (2001) Differential effects of enrichment on learning and memory function in NR2B transgenic mice. *Neuropharmacology* **41**(6), 779-790.
27. Rodrigues, S.M., Schafe, G.E. and LeDoux, J.E. (2001) Intra-amygdala blockade of the NR2B subunit of the NMDA receptor disrupts the acquisition but not the expression of fear conditioning. *J Neurosci* **21**(17), 6889-6896.
28. Grimwood, S., et al. (1999) NR2B-containing NMDA receptors are up-regulated in temporal cortex in schizophrenia. *Neuroreport* **10**(3), 461-465.
29. Liu, J., et al. (1999) Differential roles of Ca(2+)/calmodulin-dependent protein kinase II and mitogen-activated protein kinase activation in hippocampal long-term potentiation. *J Neurosci* **19**(19), 8292-8299.
30. Fukunaga, K., et al. (2000) Decreased protein phosphatase 2A activity in hippocampal long-term potentiation. *J Neurochem* **74**(2), 807-817.
31. Malenka, R.C., et al. (1989) An essential role for postsynaptic calmodulin and protein kinase activity in long-term potentiation. *Nature* **340**(6234), 554-557.
32. Menendez, L., Hidalgo, A. and Baamonde, A. (1997) Spinal calmodulin inhibitors reduce N-methyl-D-aspartate- and septide-induced nociceptive behavior. *Eur J Pharmacol* **335**(1), 9-14.
33. Alvarez-Vega, M., et al. (1998) Comparison of the effects of calmidazolium, morphine and bupivacaine on N-methyl-D-aspartate- and septide-induced nociceptive behaviour. *Naunyn Schmiedebergs Arch Pharmacol* **358**(6), 628-634.
34. Alvarez-Vega, M., et al. (2000) Intrathecal N-methyl-D-aspartate (NMDA) induces paradoxical analgesia in the tail-flick test in rats. *Pharmacol Biochem Behav* **65**(4), 621-625.
35. Tomimatsu, Y., et al. (2002) Proteases involved in long-term potentiation. *Life Sci* **72**(4-5), 355-361.
36. Massicotte, G., et al. (1991) Modulation of DL-alpha-amino-3-hydroxy-5-methyl-4-isoxazolepropionic acid/quisqualate receptors by phospholipase A2: a necessary step in long-term potentiation? *Proc Natl Acad Sci U S A* **88**(5), 1893-1897.
37. Silva, A.J., et al. (1992) Deficient hippocampal long-term potentiation in alpha-calcium-calmodulin kinase II mutant mice. *Science* **257**(5067), 201-206.
38. Hinds, H.L., Tonegawa, S. and Malinow, R. (1998) CA1 long-term potentiation is diminished but present in hippocampal slices from alpha-CaMKII mutant mice. *Learn Mem* **5**(4-5), 344-354.
39. Malinow, R., Madison, D.V. and Tsien, R.W. (1988) Persistent protein kinase activity underlying long-term potentiation. *Nature* **335**(6193), 820-824.
40. Silva, A.J., et al. (1992) Impaired spatial learning in alpha-calcium-calmodulin kinase II mutant mice. *Science* **257**(5067), 206-211.
41. Szapiro, G., et al. (2003) The role of NMDA glutamate receptors, PKA, MAPK, and CAMKII in the hippocampus in extinction of conditioned fear. *Hippocampus* **13**(1), 53-58.
42. Chen, C., et al. (1994) Abnormal fear response and aggressive behavior in mutant mice deficient for alpha-calcium-calmodulin kinase II. *Science* **266**(5183), 291-294.
43. Xing, G., et al. (2002) Decreased prefrontal CaMKII alpha mRNA in bipolar illness. *Neuroreport* **13**(4), 501-505.
44. Migaud, M., et al. (1998) Enhanced long-term potentiation and impaired learning in mice with mutant postsynaptic density-95 protein. *Nature* **396**(6710), 433-439.
45. Skibinska, A., Lech, M. and Kossut, M. (2001) PSD95 protein level rises in murine somatosensory cortex after sensory training. *Neuroreport* **12**(13), 2907-2910.
46. Tao, F., et al. (2001) Knockdown of PSD-95/SAP90 delays the development of neuropathic pain in rats. *Neuroreport* **12**(15), 3251-3255.
47. Garry, E.M., et al. (2003) Neuropathic Sensitization of Behavioral Reflexes and Spinal NMDA Receptor/CaM Kinase II Interactions Are Disrupted in PSD-95 Mutant Mice. *Curr Biol* **13**(4), 321-328.
48. Ohnuma, T., et al. (2000) Gene expression of PSD95 in prefrontal cortex and hippocampus in schizophrenia. *Neuroreport* **11**(14), 3133-3137.
49. Ranta, S., et al. (2000) Positional cloning and characterisation of the human DLGAP2 gene and its exclusion in progressive epilepsy with mental retardation. *Eur J Hum Genet* **8**(5), 381-384.
50. Tavalin, S.J., et al. (2002) Regulation of GluR1 by the A-kinase anchoring protein 79 (AKAP79) signaling complex shares properties with long-term depression. *J Neurosci* **22**(8), 3044-3051.
51. Toyooka, K., et al. (2002) Selective reduction of a PDZ protein, SAP-97, in the prefrontal cortex of patients with chronic schizophrenia. *J Neurochem* **83**(4), 797-806.
52. Roberts, L.A., et al. (1996) Changes in hippocampal gene expression associated with the induction of long-term potentiation. *Brain Res Mol Brain Res* **42**(1), 123-127.

53. Grant, S.G., et al, (1992) Impaired long-term potentiation, spatial learning, and hippocampal development in fyn mutant mice. *Science* **258**(5090), 1903-1910.
54. Korte, M., et al. (2000) Shc-binding site in the TrkB receptor is not required for hippocampal long-term potentiation. *Neuropharmacology* **39**(5), 717-724.
55. Salter, M.W. (1998) Src, N-methyl-D-aspartate (NMDA) receptors, and synaptic plasticity. *Biochem Pharmacol* **56**(7), 789-798.
56. Zhao, W., et al. (2000) Nonreceptor tyrosine protein kinase pp60c-src in spatial learning: synapse-specific changes in its gene expression, tyrosine phosphorylation, and protein-protein interactions. *Proc Natl Acad Sci U S A* **97**(14), 8098-8103.
57. Nishihara, E., et al. (2003) SRC-1 null mice exhibit moderate motor dysfunction and delayed development of cerebellar Purkinje cells. *J Neurosci* **23**(1), 213-222.
58. Uetani, N., et al. (2000) Impaired learning with enhanced hippocampal long-term potentiation in PTPdelta-deficient mice. *Embo J* **19**(12), 2775-2785.
59. Tartaglia, M., et al. (2001) Mutations in PTPN11, encoding the protein tyrosine phosphatase SHP-2, cause Noonan syndrome. *Nat Genet* **29**(4), 465-468.
60. Contractor, A., Swanson, G. and Heinemann, S.F. (2001) Kainate receptors are involved in short- and long-term plasticity at mossy fiber synapses in the hippocampus. *Neuron* **29**(1), 209-216.
61. Porter, R.H., Eastwood, S.L. and Harrison, P.J. (1997) Distribution of kainate receptor subunit mRNAs in human hippocampus, neocortex and cerebellum, and bilateral reduction of hippocampal GluR6 and KA2 transcripts in schizophrenia. *Brain Res* **751**(2), 217-231.
62. Komiyama, N.H., et al. (2002) SynGAP regulates ERK/MAPK signaling, synaptic plasticity, and learning in the complex with postsynaptic density 95 and NMDA receptor. *J Neurosci* **22**(22), 9721-9732.
63. Kim, J.H., et al. (2003) The role of synaptic GTPase-activating protein in neuronal development and synaptic plasticity. *J Neurosci* **23**(4), 1119-1124.
64. Huang, Y., et al. (2001) CAKbeta/Pyk2 kinase is a signaling link for induction of long-term potentiation in CA1 hippocampus. *Neuron* **29**(2), 485-496.
65. Lauri, S.E., Taira, T. and Rauvala, H. (2000) High-frequency synaptic stimulation induces association of fyn and c-src to distinct phosphorylated components. *Neuroreport* **11**(5), 997-1000.
66. Yamagata, K., et al. (1994) rheb, a growth factor- and synaptic activity-regulated gene, encodes a novel Ras-related protein. *J Biol Chem* **269**(23), 16333-16339.
67. Geist, R.T., et al. (1996) Expression of the tuberous sclerosis 2 gene product, tuberin, in adult and developing nervous system tissues. *Neurobiol Dis* **3**(2), 111-20.
68. Castillo, P.E., et al. (1997) Rab3A is essential for mossy fibre long-term potentiation in the hippocampus. *Nature* **388**(6642), 590-3.
69. Lonart, G., et al. (1998) Mechanism of action of rab3A in mossy fiber LTP. *Neuron* **21**(5), 1141-1150.
70. D'Adamo, P. et al. (1998) Mutations in GDI1 are responsible for X-linked non-specific mental retardation. *Nat Genet* **19**(2), 134-139.
71. Lynch, M.A., et al. (1994) Increase in synaptic vesicle proteins accompanies long-term potentiation in the dentate gyrus. *Neuroscience* **60**(1), 1-5.
72. Ferguson, G.D. et al. (2000) Deficits in memory and motor performance in synaptotagmin IV mutant mice. *Proc Natl Acad Sci USA* **97**(10), 5598-5603.
73. Rodger, J., et al. (1998) Induction of long-term potentiation in vivo regulates alternate splicing to alter syntaxin 3 isoform expression in rat dentate gyrus. *J Neurochem* **71**(2), 666-675.
74. Helme-Guizon, A., et al. (1998) Increase in syntaxin 1B and glutamate release in mossy fibre terminals following induction of LTP in the dentate gyrus: a candidate molecular mechanism underlying transsynaptic plasticity. *Eur J Neurosci* **10**(7), 2231-2237.
75. Davis, S. et al. (1998) Increase in syntaxin 1B mRNA in hippocampal and cortical circuits during spatial learning reflects a mechanism of trans-synaptic plasticity involved in establishing a memory trace. *Learn Mem* **5**(4-5), 375-390.
76. Hu, J.Y., Meng, X. and Schacher, S. (2003) Redistribution of syntaxin mRNA in neuronal cell bodies regulates protein expression and transport during synapse formation and long-term synaptic plasticity. *J Neurosci* **23**(5), 1804-1815.
77. Honer, W.G. et al. (1997) Cingulate cortex synaptic terminal proteins and neural cell adhesion molecule in schizophrenia. *Neuroscience* **78**(1), 99-110.
78. Honer, W.G. et al. (2002) Abnormalities of SNARE mechanism proteins in anterior frontal cortex in severe mental illness. *Cereb Cortex* **12**(4), 349-356.
79. Gabriel, S.M. et al. (1997) Increased concentrations of presynaptic proteins in the cingulate cortex of subjects with schizophrenia. *Arch Gen Psychiatry* **54**(6), 559-566.
80. Sokolov, B.P. et al. (2000) Levels of mRNAs encoding synaptic vesicle and synaptic plasma membrane proteins in the temporal cortex of elderly schizophrenic patients. *Biol Psychiatry* **48**(3), 184-196.
81. Aldred, M.A. et al. (2002) Constitutional deletion of chromosome 20q in two patients affected with albright hereditary osteodystrophy. *Am J Med Genet* **113**(2), 167-172.
82. Freson, K. et al. (2001) Genetic variation of the extra-large stimulatory G protein alpha-subunit leads to Gs hyperfunction in platelets and is a risk factor for bleeding. *Thromb Haemost* **86**(3), 733-738.
83. Kim, C.H. and Lisman, J.E. (1999) A role of actin filament in synaptic transmission and long-term potentiation. *J Neurosci* **19**(11), 4314-4324.
84. Raymond, C.R., Redman, S.J. and Crouch, M.F. (2002) *The phosphoinositide 3-kinase and p70 S6 kinase regulate long-term potentiation in hippocampal neurons.* Neuroscience, **109**(3), 531-536.
85. Meng, Y. et al. (2002) Abnormal spine morphology and enhanced LTP in LIMK-1 knockout mice. *Neuron* **35**(1), 121-133.
86. Stork, O. et al. (2001) Identification of genes expressed in the amygdala during the formation of fear memory. *Learn Mem* **8**(4), 209-219.
87. Suchy, S.F. and Nussbaum, R.L. (2002) The deficiency of PIP2 5-phosphatase in Lowe syndrome affects actin polymerization. *Am J Hum Genet* **71**(6), 1420-1427.
88. Nunoi, H. et al. (1999) A heterozygous mutation of beta-actin associated with neutrophil dysfunction and recurrent infection. *Proc Natl Acad Sci USA* **96**(15), 8693-8698.
89. Conquet, F. et al. (1994) Motor deficit and impairment of synaptic plasticity in mice lacking mGluR1. *Nature* **372**(6503), 237-243.
90. Aiba, A. et al. (1994) Deficient cerebellar long-term depression and impaired motor learning in mGluR1 mutant mice. *Cell* **79**(2), 377-388.

91. Thomas, K.L. et al. (1996) Alterations in the expression of specific glutamate receptor subunits following hippocampal LTP in vivo. *Learn Mem* 3(2-3), 197-208.
92. Petersen, S. et al. (2002) Differential effects of mGluR1 and mGlur5 antagonism on spatial learning in rats. *Pharmacol Biochem Behav* 73(2), 381-389.
93. Riedel, G., Sandager-Nielsen, K. and Macphail, E.M. (2002) Impairment of contextual fear conditioning in rats by Group I mGluRs: reversal by the nootropic nefiracetam. *Pharmacol Biochem Behav* 73(2), 391-399.
94. Neugebauer, V. et al. (2003) Synaptic plasticity in the amygdala in a model of arthritic pain: differential roles of metabotropic glutamate receptors 1 and 5. *J Neurosci* 23(1), 52-63.
95. Kato, A. et al. (1997) vesl, a gene encoding VASP/Ena family related protein, is upregulated during seizure, long-term potentiation and synaptogenesis. *FEBS Lett* 412(1), 183-189.
96. Kato, A. et al. (1998) Novel members of the Vesl/Homer family of PDZ proteins that bind metabotropic glutamate receptors. *J Biol Chem* 273(37), 23969-23975.
97. Matsuo, R. et al. (2000) Identification and cataloging of genes induced by long-lasting long-term potentiation in awake rats. *J Neurochem* 74(6), 2239-2349.
98. French, P.J. et al. (2001) Subfield-specific immediate early gene expression associated with hippocampal long-term potentiation *in vivo*. *Eur J Neurosci* 13(5), 968-976.
99. Massicotte, G. (2000) Modification of glutamate receptors by phospholipase A2: its role in adaptive neural plasticity. *Cell Mol Life Sci* 57(11), 1542-1550.
100. Chabot, C. et al. (1998) Bidirectional modulation of AMPA receptor properties by exogenous phospholipase A2 in the hippocampus. *Hippocampus* 8(3), 299-309.
101. Normandin, M. et al. (1996) Involvement of the 12-lipoxygenase pathway of arachidonic acid metabolism in homosynaptic long-term depression of the rat hippocampus. *Brain Res* 730(1-2), 40-46.
102. Fujita, S. et al.(2000) Ca2+-independent phospholipase A2 inhibitor impairs spatial memory of mice. *Jpn J Pharmacol* 83(3), 277-278.
103. Holscher, C. and Rose, S.P. (1994) Inhibitors of phospholipase A2 produce amnesia for a passive avoidance task in the chick. *Behav Neural Biol* 61(3), 225-232.
104. Peet, M. et al. (1998) Association of the Ban I dimorphic site at the human cytosolic phospholipase A2 gene with schizophrenia. *Psychiatr Genet* 8(3), 191-192.
105. Hudson, C.J. et al. (1996) Genetic variant near cytosolic phospholipase A2 associated with schizophrenia. *Schizophr Res* 21(2), 111-116.
106. Gattaz, W.F. and Brunner, J. (1996) Phospholipase A2 and the hypofrontality hypothesis of schizophrenia. *Prostaglandins Leukot Essent Fatty Acids* 55(1-2), 109-113.
107. Luthi, A. et al. (1999) Hippocampal LTD expression involves a pool of AMPARs regulated by the NSF-GluR2 interaction. *Neuron* 24(2), 389-399.
108. Mirnics, K. et al. (2000) Molecular characterization of schizophrenia viewed by microarray analysis of gene expression in prefrontal cortex. *Neuron* 28(1), 53-67.
109. Pelkey, K.A. et al. (2002) Tyrosine phosphatase STEP is a tonic brake on induction of long-term potentiation. *Neuron* 34(1), 127-138.
110. Holmes, S.E. et al. (1997) Disruption of the clathrin heavy chain-like gene (CLTCL) associated with features of DGS/VCFS: a balanced (21;22)(p12;q11) translocation. *Hum Mol Genet* 6(3), 357-367.
111. Zhuo, M. et al. (1999) A selective role of calcineurin aalpha in synaptic depotentiation in hippocampus. *Proc Natl Acad Sci USA* 96(8), 4650-4655.
112. Zeng, H. et al. (2001) Forebrain-specific calcineurin knockout selectively impairs bidirectional synaptic plasticity and working/episodic-like memory. *Cell* 107(5), 617-629.
113. Malleret, G. et al. (2001) Inducible and reversible enhancement of learning, memory, and long-term potentiation by genetic inhibition of calcineurin. *Cell* 104(5), 675-586.
114. Winder, D.G. et al. (1998) Genetic and pharmacological evidence for a novel, intermediate phase of long-term potentiation suppressed by calcineurin. *Cell* 92(1), 25-37.
115. Lin, C.H., Lee, C.C. and Gean, P.W. (2003) Involvement of a calcineurin cascade in amygdala depotentiation and quenching of fear memory. *Mol Pharmacol* 63(1), 44-52.
116. Kang-Park, M.H. et al. (2000) Protein phosphatases mediate depotentiation induced by high-intensity theta-burst stimulation. *J Neurophysiol* 89(2), 684-690.
117. Mulkey, R.M. et al. (1994) Involvement of a calcineurin/inhibitor-1 phosphatase cascade in hippocampal long-term depression. *Nature* 369(6480), 486-488.
118. Mansuy, I.M. et al. (1998) Restricted and regulated overexpression reveals calcineurin as a key component in the transition from short-term to long-term memory. *Cell* 92(1), 39-49.
119. Ikegami, S. and Inokuchi, K. (2000) Antisense DNA against calcineurin facilitates memory in contextual fear conditioning by lowering the threshold for hippocampal long-term potentiation induction. *Neuroscience* 98(4), 637-646.
120. Lin, C.H. et al. (2003) Identification of calcineurin as a key signal in the extinction of fear memory. *J Neurosci* 23(5), 1574-1579.
121. Birikh, K.R. et al. (2003) Interaction of "readthrough" acetylcholinesterase with RACK1 and PKCbeta II correlates with intensified fear-induced conflict behavior. *Proc Natl Acad SciU SA* 100(1), 283-288.
122. Wang, H. and Friedman, E. (2001) Increased association of brain protein kinase C with the receptor for activated C kinase-1 (RACK1) in bipolar affective disorder. *Biol Psychiatry* 50(5), 364-370.
123. Roberts, L.A. et al. (1998) Increased expression of dendritic mRNA following the induction of long-term potentiation. *Brain Res Mol Brain Res* 56(1-2), 38-44.
124. Fukunaga, K. (1993) [The role of Ca2+/calmodulin-dependent protein kinase II in the cellular signal transduction]. *Nippon Yakurigaku Zasshi* 102(6), 355-369.
125. Fukunaga, K., Muller, D. and Miyamoto, E. (1996) CaM kinase II in long-term potentiation. *Neurochem Int* 28(4), 343-358.
126. Woolf, N.J. et al. (1994) Pavlovian conditioning alters cortical microtubule-associated protein-2. *Neuroreport* 5(9), 1045-1048.
127. Bury, S.D. and Jones, T.A. (2002) Unilateral sensorimotor cortex lesions in adult rats facilitate motor skill learning with the "unaffected" forelimb and training-induced dendritic structural plasticity in the motor cortex. *J Neurosci* 22(19),. 8597-8606.
128. Genin, A. et al. (2003) LTP but not seizure is associated with up-regulation of AKAP-150. *Eur J Neurosci* 17(2), 331-340.

129. Moita, M.A. et al. (2002) A-kinase anchoring proteins in amygdala are involved in auditory fear memory. *Nat Neurosci* **5**(9), 837-838.
130. Terrian, D.M., Ways, D.K. and Gannon, R.L. (1991) A presynaptic role for protein kinase C in hippocampal mossy fiber synaptic transmission. *Hippocampus* **1**(3), 303-314.
131. Roisin, M.P., Leinekugel, X. and Tremblay, E. (1997) Implication of protein kinase C in mechanisms of potassium-induced long-term potentiation in rat hippocampal slices. *Brain Res* **745**(1-2), 222-230.
132. Young, E. et al. (2002) Changes in protein kinase C (PKC) activity, isozyme translocation, and GAP-43 phosphorylation in the rat hippocampal formation after a single-trial contextual fear conditioning paradigm. *Hippocampus* **12**(4), 457-464.
133. Hodge, C.W., et al. (2002) Decreased anxiety-like behavior, reduced stress hormones, and neurosteroid supersensitivity in mice lacking protein kinase Cepsilon. *J Clin Invest* **110**(7), 1003-1010.
134. Choi, D.S. et al. (2002) Conditional rescue of protein kinase C epsilon regulates ethanol preference and hypnotic sensitivity in adult mice. *J Neurosci* **22**(22), 9905-9911.
135. Hodge, C.W. et al. (1999) Supersensitivity to allosteric GABA(A) receptor modulators and alcohol in mice lacking PKCepsilon. *Nat Neurosci* **2**(11), 997-1002.
136. Norman, E.D. et al. (2000) Long-term depression in the hippocampus in vivo is associated with protein phosphatase-dependent alterations in extracellular signal-regulated kinase. *J Neurochem* **74**(1), 192-198.
137. Akirav, I., Sandi, C. and Richter-Levin, G. (2001) Differential activation of hippocampus and amygdala following spatial learning under stress. *Eur J Neurosci* **14**(4), 719-725.
138. Kyosseva, S.V., et al. (1999) Mitogen-activated protein kinases in schizophrenia. *Biol Psychiatry* **46**(5), 689-696.
139. Dwivedi, Y. et al. (2001) Reduced activation and expression of ERK1/2 MAP kinase in the post-mortem brain of depressed suicide subjects. *J Neurochem* **77**(3), 916-928.
140. Schenck, A. et al. (2001) A highly conserved protein family interacting with the fragile X mental retardation protein (FMRP) and displaying selective interactions with FMRP-related proteins FXR1P and FXR2P. *Proc Natl Acad Sci USA* **98**(15), 8844-8849.
141. Bardoni, B. and Mandel, J.L. (2002) Advances in understanding of fragile X pathogenesis and FMRP function, and in identification of X linked mental retardation genes. *Curr Opin Genet Dev* **12**(3), 284-293.
142. Costa, R.M. et al. (2002) Mechanism for the learning deficits in a mouse model of neurofibromatosis type 1. *Nature* **415**(6871), 526-530.
143. Silva, A.J. et al. (1997) A mouse model for the learning and memory deficits associated with neurofibromatosis type I. *Nat Genet* **15**(3), 281-284.
144. Samuelsson, B. and Samuelsson, S. (1989) Neurofibromatosis in Gothenburg, Sweden. I. Background, study design and epidemiology. *Neurofibromatosis* **2**(1), 6-22.
145. Samuelsson, B. and Riccardi, V.M. (1989) Neurofibromatosis in Gothenburg, Sweden. II. Intellectual compromise. *Neurofibromatosis* **2**(2), 78-83.
146. von Deimling, A., Krone, W. and Menon, A.G. (1995) Neurofibromatosis type 1: pathology, clinical features and molecular genetics. *Brain Pathol* **5**(2), 153-162.
147. Manabe, T. et al. (2000) Regulation of long-term potentiation by H-Ras through NMDA receptor phosphorylation. *J Neurosci* **20**(7), 2504-2511.
148. Moro, F. et al. (2002) Familial periventricular heterotopia: missense and distal truncating mutations of the FLN1 gene. *Neurology* **58**(6), 916-921.
149. Hulo, S. et al. (2002) A point mutant of GAP-43 induces enhanced short-term and long-term hippocampal plasticity. *Eur J Neurosci* **15**(12), 1976-1982.
150. Wong, K.L., Murakami, K. and Routtenberg, A. (1989) *Dietary cis-fatty acids that increase protein F1 phosphorylation enhance spatial memory*. *Brain Res* **505**(2), 302-305.
151. Zhao, W., Ng, K.T. and Sedman, G.L. (1995) Passive avoidance learning induced change in GAP43 phosphorylation in day-old chicks. *Brain Res Bull* **36**(1), 11-17.
152. Perrone-Bizzozero, N.I. et al. (1996) Levels of the growth-associated protein GAP-43 are selectively increased in association cortices in schizophrenia. *Proc Natl Acad Sci USA* **93**(24), 14182-14187.
153. Tang, L., Hung, C.P. and Schuman, E.M. (1998) A role for the cadherin family of cell adhesion molecules in hippocampal long-term potentiation. *Neuron* **20**(6), 1165-1175.
154. Huntley, G.W., Gil, O. and Bozdagi, O. (2002) The cadherin family of cell adhesion molecules: multiple roles in synaptic plasticity. *Neuroscientist* **8**(3), 221-233.
155. Bozdagi, O. et al. (2000) Increasing numbers of synaptic puncta during late-phase LTP: N-cadherin is synthesized, recruited to synaptic sites, and required for potentiation. *Neuron* **28**(1), 245-259.
156. Hernandez, F. et al. (2002) Spatial learning deficit in transgenic mice that conditionally over-express GSK-3beta in the brain but do not form tau filaments. *J Neurochem* **83**(6), 1529-1533.
157. Li, X., Bijur, G.N. and Jope, R.S. (2002) Glycogen synthase kinase-3beta, mood stabilizers, and neuroprotection. *Bipolar Disord* **4**(2), 137-144.
158. Shahraki, A. and Stone, T.W. (2002) Long-term potentiation and adenosine sensitivity are unchanged in the AS/AGU protein kinase Cgamma-deficient rat. *Neurosci Lett* **327**(3), 165-168.
159. Abeliovich, A. et al. (1993) Modified hippocampal long-term potentiation in PKC gamma-mutant mice. *Cell* **75**(7), 1253-1262.
160. Hirono, M. et al. (2001) Phospholipase Cbeta4 and protein kinase Calpha and/or protein kinase Cbeta1 are involved in the induction of long term depression in cerebellar Purkinje cells. *J Biol Chem* **276**(48), 45236-45242.
161. Abeliovich, A. et al. (1993) PKC gamma mutant mice exhibit mild deficits in spatial and contextual learning. *Cell* **75**(7), 1263-1271.
162. Colombo, P.J. and Gallagher, M. (2002) Individual differences in spatial memory among aged rats are related to hippocampal PKCgamma immunoreactivity. *Hippocampus* **12**(2), 285-289.
163. Colombo, P.J., Wetsel, W.C. and Gallagher, M. (1997) Spatial memory is related to hippocampal subcellular concentrations of calcium-dependent protein kinase C isoforms in young and aged rats. *Proc Natl Acad Sci USA* **94**(25), 14195-14199.
164. Douma, B.R., Van der Zee, E.A. and Luiten, P.G. (1998) Translocation of protein kinase Cgamma occurs during the early phase of acquisition of food rewarded spatial learning. *Behav Neurosci* **112**(3), 496-501.
165. Krugers, H.J. et al. (1997) Exposure to chronic psychosocial stress and corticosterone in the rat: effects on spatial discrimination learning and hippocampal protein kinase Cgamma immunoreactivity. *Hippocampus* **7**(4), 427-436.
166. Malmberg, A.B. et al. (1997) Preserved acute pain and reduced neuropathic pain in mice lacking PKCgamma. *Science* **278**(5336), 279-283.

167. O'Dell, T.J. et al. (1994) Endothelial NOS and the blockade of LTP by NOS inhibitors in mice lacking neuronal NOS. *Science* **265**(5171), 542-526.
168. Linden, D.J., Dawson, T.M. and Dawson, V.L. (1995) An evaluation of the nitric oxide/cGMP/cGMP-dependent protein kinase cascade in the induction of cerebellar long-term depression in culture. *J Neurosci* **15**(7 Pt 2), 5098-5105.
169. Doreulee, N. et al. (2003) Cortico-striatal synaptic plasticity in endothelial nitric oxide synthase deficient mice. *Brain Res* **964**(1), 159-163.
170. Son, H. et al. (1996) Long-term potentiation is reduced in mice that are doubly mutant in endothelial and neuronal nitric oxide synthase. *Cell* **87**(6), 1015-1023.
171. Doyle, C. et al. (1996) The selective neuronal NO synthase inhibitor 7-nitro-indazole blocks both long-term potentiation and depotentiation of field EPSPs in rat hippocampal CA1 in vivo. *J Neurosci* **16**(1), 418-424.
172. Malen, P.L. and Chapman, P.F. (1997) Nitric oxide facilitates long-term potentiation, but not long-term depression. *J Neurosci* **17**(7), 2645-2651.
173. Haley, J.E., Malen, P.L. and Chapman, P.F. (1993) Nitric oxide synthase inhibitors block long-term potentiation induced by weak but not strong tetanic stimulation at physiological brain temperatures in rat hippocampal slices. *Neurosci Lett* **160**(1), 85-88.
174. Maren, S. (1998) Effects of 7-nitroindazole, a neuronal nitric oxide synthase (nNOS) inhibitor, on locomotor activity and contextual fear conditioning in rats. *Brain Res* **804**(1), 155-158.
175. Nelson, R.J. et al. (1995) Behavioural abnormalities in male mice lacking neuronal nitric oxide synthase. *Nature* **378**(6555), 383-386.
176. Kriegsfeld, L.J. et al. (1999) Nocturnal motor coordination deficits in neuronal nitric oxide synthase knock-out mice. Neuroscience **89**(2), 311-315.
177. Le Roy, I. et al. (2000) Loss of aggression, after transfer onto a C57BL/6J background, in mice carrying a targeted disruption of the neuronal nitric oxide synthase gene. *Behav Genet* **30**(5), 367-373.
178. Gammie, S.C. and Nelson, R.J. (1999) Maternal aggression is reduced in neuronal nitric oxide synthase-deficient mice. *J Neurosci* **19**(18), 8027-8035.
179. Kriegsfeld, L.J. et al. (1997) Aggressive behavior in male mice lacking the gene for neuronal nitric oxide synthase requires testosterone. *Brain Res* **769**(1), 66-70.
180. Demas, G.E. et al. (1997) Inhibition of neuronal nitric oxide synthase increases aggressive behavior in mice. Mol Med **3**(9), 610-616.
181. Araki, T. et al. (2001) Nitric oxide synthase inhibitors cause motor deficits in mice. *Eur Neuropsychopharmacol* **11**(2), 125-133.
182. Shinkai, T. et al. (2002) Allelic association of the neuronal nitric oxide synthase (NOS1) gene with schizophrenia. *Mol Psychiatry* **7**(6), 560-563.
183. Selcher, J.C. et al. (2001) Mice lacking the ERK1 isoform of MAP kinase are unimpaired in emotional learning. *Learn Mem* **8**(1), 11-19.
184. Mazzucchelli, C. et al. (2002) Knockout of ERK1 MAP kinase enhances synaptic plasticity in the striatum and facilitates striatal-mediated learning and memory. *Neuron* **34**(5), 807-820.
185. Jones, M.W. et al. (1999) Molecular mechanisms of long-term potentiation in the insular cortex in vivo. *J Neurosci* **19**(21), RC36.
186. Watabe, A.M., Zaki, P.A. and O'Dell, T.J. (2000) Coactivation of beta-adrenergic and cholinergic receptors enhances the induction of long-term potentiation and synergistically activates mitogen-activated protein kinase in the hippocampal CA1 region. *J Neurosci* **20**(16), 5924-5931.
187. Winder, D.G. et al. (1999) ERK plays a regulatory role in induction of LTP by theta frequency stimulation and its modulation by beta-adrenergic receptors. *Neuron* **24**(3), 715-726.
188. Atkins, C.M. et al. (1998) The MAPK cascade is required for mammalian associative learning. *Nat Neurosci* **1**(7), 602-609.
189. Kahn, L. et al. (2001) Group 2 metabotropic glutamate receptors induced long term depression in mouse striatal slices. *Neurosci Lett* **316**(3), 178-182.
190. Trivier, E. et al. (1996) Mutations in the kinase Rsk-2 associated with Coffin-Lowry syndrome. *Nature* **384**(6609), 567-570.
191. Abidi, F. et al. (1999) Novel mutations in Rsk-2, the gene for Coffin-Lowry syndrome (CLS). *Eur J Hum Genet* **7**(1), 20-26.
192. McCandless, S.E. et al. (2000) Adult with an interstitial deletion of chromosome 10 [del(10)(q25. 1q25.3)]: overlap with Coffin-Lowry syndrome. *Am J Med Genet* **95**(2), 93-98.
193. Guzowski, J.F. et al. (2000) Inhibition of activity-dependent arc protein expression in the rat hippocampus impairs the maintenance of long-term potentiation and the consolidation of long-term memory. *J Neurosci* **20**(11), 3993-4001.
194. Waltereit, R. et al. (2001) Arg3.1/Arc mRNA induction by Ca2+ and cAMP requires protein kinase A and mitogen-activated protein kinase/extracellular regulated kinase activation. *J Neurosci* **21**(15), 5484-5493.
195. Guzowski, J.F. et al. (2001) Experience-dependent gene expression in the rat hippocampus after spatial learning: a comparison of the immediate-early genes Arc, c-fos, and zif268. *J Neurosci* **21**(14), 5089-5098.
196. Kelly, M.P. and Deadwyler, S.A. (2002) Acquisition of a novel behavior induces higher levels of Arc mRNA than does overtrained performance. *Neuroscience* **110**(4), 617-626.
197. Janz, R. et al.(1999) Essential roles in synaptic plasticity for synaptogyrin I and synaptophysin I. *Neuron* **24**(3), 687-700.
198. Luthi, A. et al. (1996) Reduction of hippocampal long-term potentiation in transgenic mice ectopically expressing the neural cell adhesion molecule L1 in astrocytes. *J Neurosci Res* **46**(1), 1-6.
199. Bliss, T. et al. (2000) Long-term potentiation in mice lacking the neural cell adhesion molecule L1. *Curr Biol* **10**(24), 1607-1610.
200. Wolfer, D.P. et al. (1998) Increased flexibility and selectivity in spatial learning of transgenic mice ectopically expressing the neural cell adhesion molecule L1 in astrocytes. *Eur J Neurosci* **10**(2), 708-717.
201. Montag-Sallaz, M., Schachner, M. and Montag, D. (2002) Misguided axonal projections, neural cell adhesion molecule 180 mRNA upregulation, and altered behavior in mice deficient for the close homolog of L1. *Mol Cell Biol* **22**(22), 7967-7981.
202. Kurumaji, A. et al. (2001) An association study between polymorphism of L1CAM gene and schizophrenia in a Japanese sample. *Am J Med Genet* **105**(1), 99-104.
203. Wong, E.V. et al. (1995) Mutations in the cell adhesion molecule L1 cause mental retardation. *Trends Neurosci* **18**(4), 168-172.
204. Rosenthal, A., Jouet, M. and Kenwrick, S. (1992) Aberrant splicing of neural cell adhesion molecule L1 mRNA in a family with X-linked hydrocephalus. *Nat Genet* **2**(2), 107-112.
205. Lu, Y.M. et al. (1997) Mice lacking metabotropic glutamate receptor 5 show impaired learning and reduced CA1 long-term potentiation (LTP) but normal CA3 LTP. *J Neurosci* **17**(13), 5196-5205.
206. Balschun, D. and Wetzel, W. (2002) Inhibition of mGluR5 blocks hippocampal LTP in vivo and spatial learning in rats. *Pharmacol Biochem Behav* **73**(2), 375-380.

207. Nicolle, M.M. et al. (1999) Metabotropic glutamate receptor-mediated hippocampal phosphoinositide turnover is blunted in spatial learning-impaired aged rats. *J Neurosci* 19(21), 9604-9610.
208. Fendt, M. and Schmid, S. (2002) Metabotropic glutamate receptors are involved in amygdaloid plasticity. *Eur J Neurosci* 15(9), 1535-1541.
209. Rodrigues, S.M. et al. (2002)The group I metabotropic glutamate receptor mGluR5 is required for fear memory formation and long-term potentiation in the lateral amygdala. *J Neurosci* 22(12), 5219-5229.
210. Riedel, G. et al. (2000) Fear conditioning-induced time- and subregion-specific increase in expression of mGlu5 receptor protein in rat hippocampus. *Neuropharmacology* 39(11), 1943-1951.
211. Schulz, B. et al. (2001) The metabotropic glutamate receptor antagonist 2-methyl-6-(phenylethynyl)-pyridine (MPEP) blocks fear conditioning in rats. *Neuropharmacology* 41(1), 1-7.
212. Ohnuma, T. et al. (1998) Expression of the human excitatory amino acid transporter 2 and metabotropic glutamate receptors 3 and 5 in the prefrontal cortex from normal individuals and patients with schizophrenia. *Brain Res Mol Brain Res* 56(1-2), 207-217.
213. Ohnuma, T. et al. (2000) Gene expression of metabotropic glutamate receptor 5 and excitatory amino acid transporter 2 in the schizophrenic hippocampus. *Brain Res Mol Brain Res* 85(1-2), 24-31.
214. Kishimoto, Y. et al. (2001) Impaired delay but normal trace eyeblink conditioning in PLCbeta4 mutant mice. *Neuroreport* 12(13), 2919-2922.
215. Miyata, M. et al. (2001) Deficient long-term synaptic depression in the rostral cerebellum correlated with impaired motor learning in phospholipase C beta4 mutant mice. *Eur J Neurosci* 13(10), 1945-1954.
216. Hashimoto, K. et al. (2001) Roles of phospholipase Cbeta4 in synapse elimination and plasticity in developing and mature cerebellum. *Mol Neurobiol* 23(1), 69-82.
217. Weeber, E.J. et al. (2001) Fear conditioning-induced alterations of phospholipase C-beta1a protein level and enzyme activity in rat hippocampal formation and medial frontal cortex. *Neurobiol Learn Mem* 76(2), 151-182.
218. Pacheco, M.A. et al. (1996) Alterations in phosphoinositide signaling and G-protein levels in depressed suicide brain. *Brain Res* 723(1-2), 37-45.
219. Eravci, M. et al. (1999) Gene expression of glucose transporters and glycolytic enzymes in the CNS of rats behaviorally dependent on ethanol. *Brain Res Mol Brain Res* 65(1), 103-111.
220. Roberts, L.A., Morris, B.J. and O'Shaughnessy, C.T. (1998) Involvement of two isoforms of SNAP-25 in the expression of long-term potentiation in the rat hippocampus. Neuroreport 9(1), 33-36.
221. Hess, E.J., Collins, K.A. and Wilson, M.C. (1996) Mouse model of hyperkinesis implicates SNAP-25 in behavioral regulation. *J Neurosci* 16(9), 3104-3111.
222. Thompson, P.M., Sower, A.C. and Perrone-Bizzozero, N.I. (1998) Altered levels of the synaptosomal associated protein SNAP-25 in schizophrenia. *Biol Psychiatry* 43(4), 239-243.
223. Thompson, P.M., Rosenberger, C. and Qualls, C. (1999) CSF SNAP-25 in schizophrenia and bipolar illness. A pilot study. *Neuropsychopharmacology* 21(6), 717-722.
224. Saito, T. et al. (2001) Polymorphism in SNAP29 gene promoter region associated with schizophrenia. *Mol Psychiatry* 6(2), 193-201.
225. Mukaetova-Ladinska, E.B. et al. (2002) Loss of synaptic but not cytoskeletal proteins in the cerebellum of chronic schizophrenics. *Neurosci Lett* 317(3), 161-165.
226. Fatemi, S.H. et al. (2001) Altered levels of the synaptosomal associated protein SNAP-25 in hippocampus of subjects with mood disorders and schizophrenia. *Neuroreport* 12(15), 3257-3262.
227. Mynett-Johnson, L. et al. (1998) Evidence for an allelic association between bipolar disorder and a Na+, K+ adenosine triphosphatase alpha subunit gene (ATP1A3). *Biol Psychiatry* 44(1), 47-51.
228. Varshavskaia, V.M., Ivanova, O.N. and Iakimovskii, A.F. (2002) [Locomotor behavior in rats after separate and simultaneous intrastriatal microinjections of GABA-ergic drugs]. *Ross Fiziol Zh Im I M Sechenova* 88(10), 1317-1323.
229. Hodes, M.E. et al.(1997) Nonsense mutation in exon 3 of the proteolipid protein gene (PLP) in a family with an unusual form of Pelizaeus-Merzbacher disease. *Am J Med Genet* 69(2), 121-125.
230. Saito-Ohara, F. et al. (2002) The Xq22 inversion breakpoint interrupted a novel Ras-like GTPase gene in a patient with Duchenne muscular dystrophy and profound mental retardation. *Am J Hum Genet* 71(3), 637-645.
231. Magenis, E. et al. (1981) Resolution of breakpoints in a complex rearrangement by use of multiple staining techniques: confirmation of suspected 12p12.3 intraband by deletion dosage effect of LDHB. *Am J Med Genet* 9(2), 95-103.
232. Hashimoto, R. et al. (2000) Site-specific phosphorylation of neurofilament-L is mediated by calcium/calmodulin-dependent protein kinase II in the apical dendrites during long-term potentiation. *J Neurochem* 75(1), 373-382.
233. Hashimoto, R. et al. (2000) Phosphorylation of neurofilament-L during LTD. *Neuroreport* 11(12), 2739-2742.
234. Gemignani, F. and Marbini A. (2001) Charcot-Marie-Tooth disease (CMT): distinctive phenotypic and genotypic features in CMT type 2. *J Neurol Sci* 184(1), 1-9.
235. Haroutunian, V. et al. (1996) Age-dependent spatial memory deficits in transgenic mice expressing the human mid-sized neurofilament gene: I. *Brain Res Mol Brain Res* 42(1), 62-70.
236. Gulesserian, T. et al. (2002) Aberrant expression of centractin and capping proteins, integral constituents of the dynactin complex, in fetal down syndrome brain. *Biochem Biophys Res Commun* 291(1), 62-7.
237. Katagiri, H., Tanaka, K. and Manabe, T. (2001) Requirement of appropriate glutamate concentrations in the synaptic cleft for hippocampal LTP induction. *Eur J Neurosci* 14(3), 547-553.
238. Levenson, J. et al. (2002) Long-term potentiation and contextual fear conditioning increase neuronal glutamate uptake. *Nat Neurosci* 5(2), 155-161.
239. Tsuru, N., Ueda, Y. and Doi, T. (2002) Amygdaloid kindling in glutamate transporter (GLAST) knockout mice. *Epilepsia* 43(8), 805-811.
240. McCullumsmith, R.E. and Meador-Woodruff, J.H. (2002) Striatal excitatory amino acid transporter transcript expression in schizophrenia, bipolar disorder, and major depressive disorder. *Neuropsychopharmacology* 26(3), 368-375.
241. Smith, R.E. et al. (2001) Expression of excitatory amino acid transporter transcripts in the thalamus of subjects with schizophrenia. *Am J Psychiatry* 158(9), 1393-1399.
242. Burbaeva, G. et al. (2001) [Impaired cerebral glutamate metabolism in mental diseases (Alzheimer's disease, schizophrenia). *Vestn Ross Akad Med Nauk* (7), 34-37.

243. Indo, Y. et al. (2001) Congenital insensitivity to pain with anhidrosis (CIPA): novel mutations of the TRKA (NTRK1) gene, a putative uniparental disomy, and a linkage of the mutant TRKA and PKLR genes in a family with CIPA and pyruvate kinase deficiency. *Hum Mutat* **18**(4), 308-318.
244. Eber, S.W. et al. (1991) Triosephosphate isomerase deficiency: haemolytic anaemia, myopathy with altered mitochondria and mental retardation due to a new variant with accelerated enzyme catabolism and diminished specific activity. *Eur J Pediatr* **150**(11), 761-766.
245. Schurmans, S. et al. (1997) Impaired long-term potentiation induction in dentate gyrus of calretinin-deficient mice. *Proc Natl Acad Sci USA* **94**(19), 10415-10420.
246. Gurden, H. et al. (1998) Calretinin expression as a critical component in the control of dentate gyrus long-term potentiation induction in mice. *Eur J Neurosci* **10**(9), 3029-3033.
247. Edgar, P.F. et al. (2000) Comparative proteome analysis of the hippocampus implicates chromosome 6q in schizophrenia. *Mol Psychiatry* **5**(1), 85-90.
248. Toyooka, K. et al. (1999) 14-3-3 protein eta chain gene (YWHAH) polymorphism and its genetic association with schizophrenia. *Am J Med Genet* **88**(2), 164-167.
249. Pangalos, C. et al. (1992) No significant effect of monosomy for distal 21q22.3 on the Down syndrome phenotype in "mirror" duplications of chromosome 21. *Am J Hum Genet* **51**(6), 1240-1250.
250. Schwarz, M.J. et al. (1999) Antibodies to heat shock proteins in schizophrenic patients: implications for the mechanism of the disease. *Am J Psychiatry* **156**(7), 1103-1104.
251. Kim, J.J. et al. (2001) Identification of antibodies to heat shock proteins 90 kDa and 70 kDa in patients with schizophrenia. *Schizophr Res* **52**(1-2), 127-135.

2

DYNAMISM OF POSTSYNAPTIC PROTEINS AS THE MECHANISM OF SYNAPTIC PLASTICITY

Kensuke Futai and Yasunori Hayashi*

1. OVERVIEW

At excitatory synapses of the vertebrate central nervous system, the neuronal information is transmitted via glutamate which is released from presynaptic terminals and opens postsynaptic cation channels integrated in glutamate receptor proteins. This transmission is not static, but rather is dynamically regulated in both a positive and negative manner by its own activity level as well as by interactions with other synaptic inputs. Such regulation is collectively called synaptic plasticity. After the discovery of synaptic plasticity, intensive work over the past three decades has shown that excitatory synapses in the central nervous system exhibit a remarkable degree of plasticity. Most importantly, growing evidence suggests that synaptic plasticity is the molecular/cellular basis of learning and memory.

The most thoroughly studied form of synaptic plasticity is long-term potentiation (LTP) in the hippocampal CA1 region. LTP was first described in 1970 by Bliss and co-workers[1]. In the 1980s, the requirement for postsynaptic depolarization coupled with presynaptic glutamate release and a resultant influx of Ca^{2+} through N-methyl-D-aspartate (NMDA) receptors was established. Thereafter, debate about whether LTP is expressed as an increase in transmitter release or as an increase in the 'sensitivity' of the postsynapse to released glutamate continued almost the entire decade of the 1990s. In the late 1990s, the new concept of a postsynaptically silent synapse and dynamic α-amino-3-hydroxy-5-methyl-4-isoxazole propionic acid (AMPA) receptors mostly put an end to this long-standing debate. These evidences concluded changes occurring postsynaptically, although presynaptic changes have not been entirely ruled out. Much research along this line has been conducted, boosted by the isolation of cDNAs for the glutamate receptors and other proteins as well as by advances in imaging technology. Such studies have drastically changed the view of postsynaptic receptor proteins as being rather static to being highly dynamic and regulated by synaptic activity.

In this chapter, we will briefly overview the history of LTP research and then summarize the current understanding of the activity-dependent regulation of postsynaptic protein localization in the context of hippocampal LTP.

*K. Futai[1,2] and Y. Hayashi[1], 1. RIKEN-MIT Neuroscience Research Center, The Picower Center for Learning and Memory, Department of Brain and Cognitive Science, Massachusetts Institute of Technology E18-270, 77 Massachusetts Avenue Cambridge MA 02139 U.S.A., 2. Laboratory for Neural Architecture, Brain Science Institute, RIKEN, Wako, Saitama 351-0198, Japan

2. MOLECULAR MECHANISMS OF HIPPOCAMPAL CA1 LTP

2.1. LTP Induction

Glutamatergic excitatory synaptic transmission is mediated through two types of ionotropic glutamate receptors: AMPA receptors and NMDA receptors[2,3]. AMPA receptors are permeable only to monovalent cations (Na^+, K^+) and mediate synaptic transmission under low frequency synaptic inputs. In contrast, NMDA receptors are regulated by membrane potential as well as ligands and are permeable to Ca^{2+} as well as Na^+ and K^+. Because their dependence on membrane potentials is caused by an extracellular Mg^{2+} block of the channel pore to just under the resting membrane potential, the contribution of NMDA receptors to basal synaptic transmission is minimal. Thus, synaptic plasticity is expressed as AMPA receptor response. When the cell is depolarized by high frequency synaptic input stimulation, Mg^{2+} dissociates from the NMDA receptor allowing Ca^{2+} influx into the postsynapse. This increase in intracellular Ca^{2+} through NMDA receptors is essential for the induction of LTP[4]. Thus, it is well accepted that LTP is induced by postsynaptic machinery.

Intracellular Ca^{2+} influx causes the activation of multiple signaling molecules that triggers an increase in AMPA receptor-mediated transmission. The most well-characterized and critical molecule for triggering LTP is calcium/calmodulin-dependent protein kinase II (CaMKII). CaMKII is a serine/threonine protein kinase highly enriched in the brain (1-2% of total protein)[5]. Although Ca^{2+} is required for the initial activation of CaMKII, this kinase no longer requires Ca^{2+} once it is autophosphorylated at threonine 286[5]. From this feature, CaMKII has been suggested as a memory molecule that keeps enhanced transmission after the transient increase in synaptic activity. Indeed the autophosphorylated status of CaMKII increases and is sustained after the induction of LTP[6]. Drugs that inactivate CaMKII or genetic ablation of CaMKII block LTP[7-9]. Furthermore, the constitutively active form of CaMKII can potentiate AMPA receptor-mediated excitatory postsynaptic current (EPSC) without changing NMDA receptor mediated EPSC in a manner occluding electrophysiologically induced LTP[10,11]. The functional significance of CaMKII autophosphorylation has been also confirmed by genetic methods. Mice carrying a CaMKII point mutation at threonine 286 exhibit a loss of plasticity and are deficient in learning[12].

2.2. LTP Expression

How does the increased activity of CaMKII and likely other signal transduction molecules lead to the expression of LTP? Over the past few decades, significant attention was dedicated to one conceptually simple question: Is the expression of LTP due to presynaptic or postsynaptic modifications? Postsynaptic modification would be attributed to the change in postsynaptic AMPA receptor properties or number, whereas presynaptic modulation would be the enhancement of the glutamate release probability from the presynaptic nerve terminus. Much evidence in support of postsynaptic modification and against presynaptic modification has been generated. Most investigators report that AMPA receptor-mediated EPSCs increase significantly more than NMDA receptor-mediated EPSCs[13], and the sensitivity against exogenous AMPA is increased following LTP[14]. If LTP is caused by presynaptic modifications, the effect of these two components of EPSCs should be the same, although other explanations are also possible[15]. Experiments to determine release probability by measurement of the paired-pulse facilitation (a presynaptically originated short-term plasticity)[16-18], glutamate receptors in out-side-out patch excised from neurons[19] or glial cell glutamate transporter currents as a indirect measure of synaptically released glutamate[20,21], and the use-dependent antagonists of NMDA or AMPA receptors[22,23] all showed no change in the release of glutamate from the presynaptic terminus before and after LTP.

On the other hand, the presynaptic theory relied heavily on results from quantal analysis that the failure rate of synaptic transmission changes after LTP, because the transmission failure was thought to result from a failure in neurotransmitter release[24]. However, this is not the case for glutamatergic synapses in the central nervous system. Several groups have reported that synaptic failures can be caused by a postsynaptic mechanism[25-27]. They identified synapses that have only NMDA receptors. Because NMDA receptors are blocked at a hyperpolarized membrane potential by extracellular Mg^{2+}, these synapses are functionally silent under basal synaptic activity. However, application of LTP induction procedures to such "silent" synapses causes rapid appearance of AMPA receptor-mediated EPSCs. Thus synaptic failure is attributable not only to transmitter release failure but also to postsynaptic failure in the response to released glutamate. Immunohistochemical studies at both the light and electromicroscopic levels also support the existence of silent synapses[28-32].

2.3. Activity-Dependent AMPA Receptor Synaptic Insertion

How are silent synapses 'unsilenced' by the induction of LTP? The simplest hypothesis is that there exists a delivery/exocytotic mechanism that introduces AMPA receptors into silent synapses. Evidence favoring this concept has accumulated over the last few years. LTP induction increases the surface binding of ^{3}H-glutamate[33]. Postsynaptic infusion of toxin that selectively inhibits the exocytotic process blocks LTP[34]. Prolonged FM1-43 staining revealed dendritic organella that cause exocytosis in a Ca^{2+} and CaMKII activity-dependent manner[35,36]. In these reports, however, the identity of the substance(s) subject to exocytosis was not known.

The first evidence for activity-dependent synaptic delivery of the AMPA receptor came from a study using the AMPA receptor subtype GluR1 tagged with green fluorescence protein (GFP) expressed in organotypic hippocampal slice cultures through a viral expression system[37]. Newly synthesized recombinant GluR1-GFP is diffusely distributed throughout and retained within the dendrite. However, high frequency stimulation induces the movement of GluR1-GFP into the spine. This movement is dependent on the activation of NMDA receptors and is sustained in the spine for at least 50 min. This effect can be mimicked by coexpression of GluR1-GFP with truncated CaMKII, a constitutively active form of the enzyme[38].

Furthermore, use of an electrophysiological tagging technique supported this conclusion[39]. Most endogenous AMPA receptors in hippocampal CA1 pyramidal cells contain the GluR2 subunit, and the current-voltage relationship is linear. However, GluR1 overexpressed from a strong viral promoter makes a homomeric receptor lacking GluR2. Such exogenous receptors exhibit inward rectification, allowing them to be distinguished from endogenous receptors. Using this approach, an insertion of exogenous AMPA receptors into the spine was monitored by determination of the ratio of EPSCs at negative and positive membrane potentials. This assay confirmed that LTP induction or coexpression of truncated CaMKII drives GluR1-containing receptors into the synapse in an organotypic

```
Long Forms
              M4         4.1N            CaMKII,PKC   PKC   PKA                                           RIL,SAP97
              ||||||||||||             |            |                                                    |||
GluR1         LIEFCYKSRSESKRMKGFCLIPQQSINEAIRTSELPRNSGAGASGGGGSGENGRVVSQDFPKSMQSIPCMSHSSGMPLGATGL
GluR2L        LIEFCYKSRAEAKRMK-------MTLSAATRNK-------ARLSITGSTGENGRVMTPEFFPKAVHAVPYVSPGMGMNVSVTDLS
GluR4         LIEFCYKSRAEAKRMK-------LTFSEAIRNK-------ARLSITGSVGENGRVLTPDCPKVHTGTAIRQSSGLAVIASDLP
Homology      **********.*.****       ...*  *         *   * *.******....  **....   .  *. ..*
                                         |                |                                              |||
                                        PKC          CaMKII,PKC,PKA                                     PDZ

Short Forms
              M4         NSF&AP2            GRIP/ABP(GRIP2)/Pick1/rDLG6/afadin
              __         ||||||||||         ||||
GluR2         LIEFCYKSRAEAKRMKVAKNPQNINPSSSQNSQNFATYKEGYNVYGIESVKI
GluR3         LIEFCYKSRAESKRMKLTKNTQNFKPAPATNTQNYATYREGYNVYGTESVKI
GluR4c        LIEFCYKSRAEAKRMKVAKSAQTFNPTSSQNTHNLATYREGYNVYGTESIKI
Homology      ************.****..*. *  ..*... *..* ***.******.**.**
                                                                         |
                                                                   PKC/CKII(GluR2)
```

Fig. 1. Homology alignment of intracellular carboxyl tails of AMPA receptor subunits. Identified interacting proteins and corresponding regions on receptor (shaded) are shown. See text for detail. M4, 4th membrane associated region; 4.1N, band 4.1 neuronal subtype; CaMKII, Ca^{2+}/calmodulin dependent protein kinase type II; PKC, protein kinase C; PKA protein kinase A; RIL, reversion-induced LIM domain gene product; SAP97, synapse associated protein 97; NSF, N-ethylmaleimide sensitive factor; AP2, adapter protein 2; GRIP, glutamate receptor interacting protein; ABP, AMPA receptor binding protein; Pick1, protein interacting with C kinase 1; rDLG6, rat disc large.

hippocampal slice culture[39]. This activity-dependent synaptic insertion of GluR1-GFP is also observed *in vivo* in the rodent barrel cortex, indicating that the same mechanism works at *in vivo* cortical synapses[40]. In support of this hypothesis, mice lacking GluR1 cannot induce LTP[41], an effect that can be rescued by transgenic expression of GluR1-GFP[42].

2.4. Interplay between AMPA Receptor Phosphorylation and Synaptic Delivery

What triggers the synaptic delivery of AMPA receptors? There are three phosphorylation sites on the carboxyl terminus of GluR1[43,44] and two of them are well-characterized (Figure 1). Serine 831 of GluR1 is the CaMKII phosphorylation site[43,45,46]. Phosphorylation at this site increases the single channel conductance[47]. LTP induction increases this phosphorylation[45,48] as well as the AMPA receptor conductance[49]. Thus, CaMKII phosphorylation of GluR1 may be a mechanism for the enhancement of AMPA receptor responsiveness through the modulation of channel conductance. However, this phosphorylation does not appear to be necessary for delivery of GluR1 to the synapse since a mutant receptor that cannot be phosphorylated by CaMKII at this site was still delivered to the synapse by coexpression of a truncated CaMKII[39].

In contrast, phosphorylation at serine 845 is necessary for CaMKII-induced synaptic delivery of GluR1. A mutant GluR1 which has mutation at this site is not delivered to the synapse by the coexpression with truncated CaMKII[50]. This site is phosphorylated by protein kinase A (PKA) but not by CaMKII itself. Interestingly, the application of forskolin, an activator of adenylate cyclase, cannot induce the AMPA receptor delivery although it can drastically enhance the serine 845 phosphorylation. It has been considered that the cAMP pathway works as a 'gating mechanism' that can modulate LTP through the modulation of protein phosphatase and inhibitor-1[51,52]. Thus this PKA phosphorylation of GluR1 may act as the second gating mechanism. Consistent with this importance of GluR1 phosphorylation in the regulation of AMPA receptor delivery, mutant knock-in mice lacking serines 831 and 845 exhibited impaired synaptic plasticity[53].

Neither of these phosphorylation sites on GluR1 is the direct substrate of CaMKII necessary for synaptic delivery of GluR1. This implies that there must be another CaMKII substrate that acts for the synaptic delivery of GluR1. Although there are many CaMKII substrates in the postsynaptic density (PSD) fraction[54,55], only two proteins, other than CaMKII itself, are known to contribute to the activity-dependent synaptic delivery of GluR1. SynGAP [synaptic Ras GTPase-activating protein (GAP)] is a negative regulator of Ras and interacts with PSD-95 and SAP102[56,57]. SynGAP phosphorylation by CaMKII inactivates Ras GAP activity thereby relieving Ras from inhibition and increasing its activity. Activation of Ras is important for inducing LTP and synaptic delivery of GluR1 through a CaMKII-dependent signal cascade[58].

Another candidate is PSD-95, a major PSD protein that binds directly to the NMDA receptor[59]. Overexpression of PSD-95 in hippocampal neurons enhanced AMPA receptor-mediated EPSC without changing NMDA receptor mediated EPSC[60,61] in a manner occluding with LTP. This enhancement is mediated by the synaptic delivery of GluR1 receptor[60,61] likely through an interaction with stargazin, an AMPA receptor binding transmembrane protein[62-64]. Interestingly, the neuronal activity induced by eye opening can increase the amount of PSD-95 in the synaptoneurosome fraction prepared from the superior colliculus and visual cortex without changing the total amount of PSD-95[65]. The Drosophila homologue of PSD-95, *dlg*, changes its localization at the synapse by CaMKII phosphorylation[66]. Thus similar activity-dependent translocation of PSD-95 may take place in CA1 neurons.

In addition to a role as a signal transduction molecule, CaMKII may have structural role as well. It has been shown that CaMKII is translocated to the synapse by synaptic activity (see section 3.2. for detail). Inspired by this, Lisman established a model in which CaMKII has a structural role[5]. In his model, CaMKII, when activated, moves to the post-synapse thereby triggering the assembly of other postsynaptic proteins through direct or indirect interactions with molecules such as actinin-2, band 4.1, SAP97, and eventually AMPA receptors[5]. In fact, the amount of CaMKII is very high in the postsynapse, almost comparable with actin and well beyond the levels of other signal transduction components, which may be consistent with a structural role of CaMKII in the postsynapse.

In addition, the PDZ-protein binding motif on the very end of GluR1 is also important for GluR1 synaptic insertion[39,67]. SAP-97, a PDZ domain containing membrane associated guanylate kinase (MAGUKs) family protein similar to PSD-95, is known to bind to this site[68]. However, mutant mice lacking the last 7 amino acids of GluR1 including the PDZ protein binding domain showed normal LTP[69]. Therefore, the contribution of the PDZ binding domain to the delivery is still controversial. These results imply that the GluR1 delivery mechanism cannot be explained by single protein-protein interaction. Multiple protein interactions taking place in parallel are likely to be necessary for this delivery.

2.5. Subunit-Specific Rules of AMPA Receptor Delivery to the Synapse

Two-hybrid screening boosted the identification of glutamate receptor-interacting proteins. Several independent groups noticed that GluR2 binds to *N*-ethylmaleimid-sensitive factor (NSF), a protein involved in the membrane fusion process[70-72]. Intracellular infusion of partial peptide corresponding to the NSF-binding site of GluR2 induced a rapid but partial suppression of AMPA receptor-mediated current. Viral expression of this peptide reduced the surface amount of the receptor as detected by surface immunostaining[73]. A recent study showed that clathrin adapter protein type 2 (AP2) binds to an overlapping sequence, an event that is necessary for NMDA-receptor-induced internalization of AMPA receptor and essential for long-term depression (LTD)[74].

The rapid reduction in synaptic response and surface GluR2 by the infusion of the peptide indicates that constitutive recycling of GluR2 protein occurs between the synaptic and extrasynaptic pools. In this way, GluR2 is thought to play a maintenance role in keeping a certain synaptic strength through its own recycling. At a glance, this sounds contradictory to results indicating that synaptic delivery of the AMPA receptor requires synaptic activity. However, the one critical difference is that these studies used GluR2 whereas activity-dependent delivery was studied with the GluR1 subunit.

When we compare amino acid sequences among different subunits of the AMPA receptor, extracellular and transmembrane regions are relatively conserved, whereas the intracellular cytoplasmic tails of each subunit, which bind to intracellular scaffolding proteins and are involved in determining receptor trafficking, diverge into two groups[75,76]. GluR1, GluR4, and an alternative splice form of GluR2 (GluR2L) have longer cytoplasmic tails whereas GluR2, GluR3, and an alternative splice form of GluR4 (GluR4c) have shorter cytoplasmic tails. A side-by-side comparison of GluR1-GFP and GluR2-GFP revealed that these two forms show very different behaviors: Whereas GluR1 was retained within dendritic shafts, GluR2 was constitutively delivered to the synapses in a manner not requiring synaptic activity as determined by both imaging and electrophysiological tagging[76]. These results are consistent with the data obtained for the peptide interfering with NSF interaction. Surface immunostaining in dissociated cultures of GluR1 and GluR2 epitope tagged at the extracellular domain led to an essentially similar conclusion[67].

What, then, will happen if these two subunits are combined? In fact, most AMPA receptors in hippocampal CA1 synapses contain GluR2. They make two distinct populations: those composed of GluR1 and GluR2 or those composed of GluR2 and GluR3[77]. Coexpression experiments indicated that a heteromeric AMPA receptor including GluR1 and GluR2 behaved like GluR1 whereas a GluR2 and GluR3 complex had properties similar to GluR2. Thus GluR1/2 acts as activity-dependent receptor, and GluR2/3 maintains normal synaptic transmission.

2.6. Developmental Switch of Activity-Dependent AMPA Receptor Subunits

In both GluR1 knock-out mice and GluR1 phosphorylation site mutant knock-in mice, whereas adult mice exhibit impairments in LTP, juvenile mice show normal LTP[42,53]. These results suggest that different AMPA receptor subunits may participate in activity-dependent delivery throughout development, although other explanations such as presynaptic mechanisms are still possible. GluR4 and GluR2L, which have sequences similar to GluR1, may contribute to activity-dependent AMPA receptor trafficking at different developmental stages. The immature hippocampus (<P10) expresses GluR4 that complexes with GluR2[78]. Studies of activity-dependent GluR4 delivery using electrophysiological tagging indicate that this subunit mediates activity-dependent delivery during the immature period. As the expression of GluR4 disappears by postnatal day 10, GluR2L may contribute to the activity-dependent delivery of AMPA receptor during the following juvenile period[79]. GluR2L is expressed mainly in pyramidal neurons and dentate gyrus granule cells in the hippocampus. At both the mRNA and protein level, GluR2L expression in the hippocampus peaks approximately 2 weeks after birth. At this stage, GluR2L is assembled with ~20% of the GluR1 and GluR3 subunit populations.

How is the delivery of these receptors controlled? GluR4 is phosphorylated by PKA at serine 842, which is also conserved in GluR2L, and this phosphorylation is sufficient for delivery of GluR4 to the synapse. Unlike for GluR1, CaMKII activity is not necessary[50,78]. Consistent with this, the kinase requirement of LTP developmentally switches from PKA at younger age to CaMKII at older age[80].

3. ACTIVITY-DEPENDENT DYNAMICS OF POSTSYNAPTIC PROTEINS

AMPA receptors do not exist by themselves. They bind to various interacting proteins that can regulate the AMPA receptor. Such proteins often have multiple protein-interaction interfaces and, in turn, bind to other binding partners, eventually making a dense network of protein interactions at the postsynapse that include NMDA receptor and other signaling and structural components. This network of proteins comprises the electron-dense structure, which has been recognized as postsynaptic density. Therefore, it would not be hard to imagine that such postsynaptic proteins also respond to neuronal activity and change their properties such as conformation, posttranslational modifications, and interacting partners, thereby changing their localization. Now, accumulating evidence suggests that acute activity-dependent delivery to the synapse or removal from the synapse occurs not only for the AMPA receptor, but also for many other postsynaptic proteins.

3.1. Activity-Dependent Delivery of NMDA Receptors

Although the synaptic insertion or removal of the NMDA receptor by chronic activity modulation has been reported[81], generally, the mobility of the NMDA receptor during acute synaptic plasticity, such as LTP, has been considered to be lower than that of the AMPA receptor. Only a few studies have noticed that the NMDA receptor also rapidly changes its location under certain conditions of activity.

In slices obtained from adult animals, 20%-40% of the NMDA receptor resides within the cells. LTP induction induces surface delivery of the NMDA receptor as determined by biotinylation and chymotrypsinization of receptors on the surface[82]. During synaptogenesis, a membranous packet containing the NMDA receptor is rapidly recruited to nascent synaptic contacts[83]. Activation of type I metabotropic glutamate receptors (mGluRs) in cortical slices induces internalization of NMDA receptors[84].

Phosphorylation, at least in part, may regulate these processes. Application of phorbol ester, an activator of protein kinase C or intracellular infusion of the active form of protein kinase C increased NMDA-evoked current in dissociated hippocampal neurons[85]. In contrast, application of purified serine/threonine protein phosphatases decreased NMDA receptor-mediated current[86]. Phorbol ester treatment changes subcellular distribution and increases the surface amount of the NMDA receptor in both heterologous systems (Xenopus oocytes) and hippocampal neurons[82,87-89], though it may also explained by modulation of channel activity. The NMDA receptor has been shown to be phosphorylated by CaMKII and PKC at the carboxyl tail, and the actual sites of phosphorylation have been identified[90,91]. The NR1 subunit of the NMDA receptor has an ER retention signal regulated by an adjacent PKC phosphorylation site and PDZ domain protein binding[92,93]. However, the phorbol ester still showed an effect, even in a mutant where all known PKC phosphorylation sites carboxyl termini are eliminated[94], leaving much ambiguity as to the precise mechanism of this process.

3.2. Ca^{2+}/Calmodulin-Dependent Protein Kinase Type II (CaMKII)

It had been noticed that the amount of CaMKII in the postsynaptic density differed depending on the way animal brain tissue is prepared[95]. This observation was the first indication that CaMKII acutely change its localization depending on cellular conditions. The effect was reproduced in vitro with isolated crude PSD-like fractions. This in vitro study found that CaMKII translocation is dependent on its autophosphorylation[96]. An elegant series of experiments by Shen and Meyer using GFP-labeled CaMKII demonstrated activity-dependent translocation in living neurons. This delivery of α-CaMKII requires its autophosphorylation but those at threonine 286, which renders α-CaMKII constitutively active, does not define the binding[97]. Rather, autophosphorylation of the regulatory domain at threonine 305/threonine 306, which regulates calmodulin binding, also regulates localization. β-CaMKII has a distinct mechanism for its subcellular localization. It has a unique filamentous actin (F-actin)-binding domain, and, upon activation, is liberated and relocated to the PSD[98]. Furthermore, recent in vivo studies also showed that the translocation of endogenous CaMKII is driven by synaptic activity[99].

What provides the binding sites for α-CaMKII at the synapse? It has been suggested that the carboxyl tail of the NR2B subtype of the NMDA receptor binds to α-CaMKII[100,101]. This binding keeps α-CaMKII in the constitutively active form independent of calmodulin binding[102]. This process may act as another mechanism for keeping the constitutive activity of this enzyme independent of phosphorylation at threonine 286. However, NR2B cannot be the only binding site for α-CaMKII. Whereas α-CaMKII is a highly abundant protein constituting up to 10% of the protein at the postsynaptic density, the number of NMDA receptors on a single synaptic contact will be much less considering single channel conductance and the size of the synaptic response. In fact, pharmacological activation of protein kinase C with phorbol ester delivers α-CaMKII to synapses while rapidly dispersing NMDA receptor from the synapse[89]. These mismatches suggest the presence of another mechanism that provides a binding site for α-CaMKII. Another CaMKII-binding protein enriched in the postsynapse, densin-180[103,104] and F-actin[98], may underlie such a mechanism.

Other kinases such as protein kinase C family members have also been shown to rapidly change their distribution in a nonneuronal cells in studies with GFP-fusion proteins. It would be interesting to know whether activity-dependent redistribution is a general phenomenon among various kinase molecules.

3.3. Other Postsynaptic Scaffolding and Cytoskeletal Proteins

Activity-dependent delivery of PSD-95 and homer1C (PSD-Zip45), binding proteins for the NMDA receptor and mGluR, respectively, were determined by time-lapse imaging of GFP-fusion proteins[105]. This study first indicated that, whereas PSD-95 is relatively stable, Homer1C rapidly redistributes in response to high extracellular potassium and glutamate application. Further, the turnover rates of these proteins were estimated by use of a fluorescent recovery after photobleaching (FRAP) assay. In this assay, GFP signals in the spine head were photobleached by repeated local scanning. The recovery time course of the fluorescence in the spine head by influx from the dendritic shaft is measured as the turnover rate. The fluorescence level of homer1C returns to 50% of the original within 5 minutes[105]. A protein with an even faster rate is actin, which recovers within 1 minutes, while GFP itself takes only 1 second[106]. PSD-95 showed slower kinetics with recovery of only 20% of PSD-95-GFP taking more than 30 minutes. This assay demonstrated that postsynaptic scaffolding proteins show unexpectedly fast turnover rates even in mature hippocampal neurons.

Differential stimulation protocols were found to alter the direction of homer1C assembly-disassembly[105]. Transient increases in intracellular Ca^{2+} by voltage-dependent Ca^{2+}

channel activation induced homer1C clustering. In contrast, NMDA receptor-dependent Ca^{2+} influx resulted in the disassembly of PSD-Zip45 clusters. In contrast, PSD-95 distribution was relatively stable under these stimulation[105]. Thus, neuronal activity differentially redistributes a specific subset of PSD proteins, which are important for localization of both surface receptors and intracellular signaling complexes.

4. A GENERAL MECHANISM FOR ACTIVITY-DEPENDENT DELIVERY OF POSTSYNAPTIC PROTEINS

Are there any general mechanisms that regulate protein delivery to the postsynapse? We may be able to classify such mechanisms into three general categories (Figure 2).

For the first mechanism, there may be a structure that actively delivers proteins to the postsynapse, analogous to transport of proteins along dendrites (Figure 2A). The efficiency of this transport mechanism may be regulated by synaptic activity.

For the second mechanism, molecules may not undergo active transport, rather they move in and out of the postsynaptic site by simple passive diffusion, at a rate largely determined by shape and molecular weight. However, once they enter the postsynapse, they are trapped by putative binding sites. The capacity of these binding sites may be modulated by synaptic activity (Figure 2B). Binding of β-CaMKII to F-actin or post-translational modifications, such as myristoylation of PSD-95, would also fall into this category. Consistent with this notion, prevention of PSD-95 myristoylation blocks its delivery to the synapse[107].

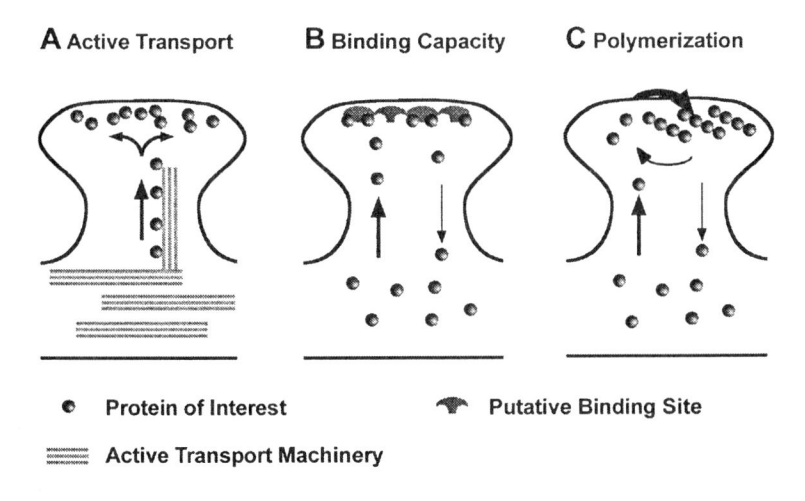

Fig. 2. Possible mechanisms for activity dependent delivery of synaptic proteins to postsynapse. **A** A modulation of active transport machinery. **B** A modulation of binding capacity at the postsynapse. **C** A modulation of polymerization status at the postsynapse. In **B** and **C**, the molecules enter the postsynaptic site by passive diffusion and do not require active transport mechanism.

For the third mechanism, the molecule can form a polymer, and the equilibration between polymerization and depolymerization may be regulated by synaptic activity (Figure 2C). Once the monomers diffuse to the synaptic site, they will be trapped as polymer. One example of a protein that is regulated in this manner is actin. Actin exist in an equilibration between monomer (globular or G-actin) and polymer (F-actin) forms and F-actin predominantly exists in the postsynapse[105,106]. Its polymerization is indispensable for the formation of dendritic spines as well as for synaptic plasticity[105,108]. Importantly, various signal transduction mechanisms regulate the polymerization/depolymerization of actin. This increases or decreases in the F-actin may modulate binding sites for other actin-binding proteins such as cortactin, β-CaMKII, and various types of myosins. In this sense, F-actin would serve as the binding site described in the second mechanism of delivery and can be a "master regulator" of synaptic delivery for various postsynaptic proteins. Other proteins that exist as both monomers and polymers may also be regulated in this fashion. These include homer1B/C, PSD-95, and glutamate receptor-interacting proteins (GRIPs).

5. CONCLUDING REMARKS

Within the last few years, the view of postsynaptic proteins has completely changed from being associated with each other in a stable and fixed protein complexes to being highly dynamic components of the synaptic architecture. Not only that, this dynamism of postsynaptic proteins is the one of the major site of synaptic plasticity. Many questions still need to be addressed: How is it maintained at constant level? What is the exact mechanism of its regulation? How is the specificity to individual synapses are maintained? Fortunately, the field is rapidly advancing, promising imminent answers to these and other important issues in learning and memory.

6. ACKNOWLEDGMENT

We thank Dr. Michel Cleary for her comments on English.

7. REFERENCES

1. T. V. Bliss and T. Lømo (1970) Plasticity in a monosynaptic cortical pathway, *J Physiol (Lond)*, **207**, 61P.
2. M. Hollmann and S. Heinemann (1994) Cloned glutamate receptors, *Annu Rev Neurosci*, **17**, 31-108.
3. R. Dingledine, K. Borges, D. Bowie and S. F. Traynelis (1999) The glutamate receptor ion channels, *Pharmacol Rev*, **51**, 7-61.
4. T. V. Bliss and G. L. Collingridge (1993) A synaptic model of memory: long-term potentiation in the hippocampus, *Nature*, **361**, 31-39.
5. J. Lisman, H. Schulman and H. Cline (2002) The molecular basis of CaMKII function in synaptic and behavioural memory, *Nat Rev Neurosci*, **3**, 175-190.
6. K. Fukunaga, D. Muller and E. Miyamoto (1995) Increased phosphorylation of Ca^{2+}/calmodulin-dependent protein kinase II and its endogenous substrates in the induction of long-term potentiation, *J Biol Chem*, **270**, 6119-6124.
7. R. C. Malenka, J. A. Kauer, D. J. Perkel, M. D. Mauk, P. T. Kelly, R. A. Nicoll and M. N. Waxham (1989) An essential role for postsynaptic calmodulin and protein kinase activity in long-term potentiation, *Nature*, **340**, 554-557.
8. R. Malinow, H. Schulman and R. W. Tsien (1989) Inhibition of postsynaptic PKC or CaMKII blocks induction but not expression of LTP, *Science*, **245**, 862-866.
9. A. J. Silva, C. F. Stevens, S. Tonegawa and Y. Wang (1992) Deficient hippocampal long-term potentiation in

alpha-calcium-calmodulin kinase II mutant mice, *Science*, **257**, 201-206.
10. D. L. Pettit, S. Perlman and R. Malinow (1994) Potentiated transmission and prevention of further LTP by increased CaMKII activity in postsynaptic hippocampal slice neurons, *Science*, **266**, 1881-1885
11. P. M. Lledo, G. O. Hjelmstad, S. Mukherji, T. R. Soderling, R. C. Malenka and R. A. Nicoll (1995) Calcium/calmodulin-dependent kinase II and long-term potentiation enhance synaptic transmission by the same mechanism, *Proc Natl Acad Sci U S A*, **92**, 11175-11179.
12. K. P. Giese, N. B. Fedorov, R. K. Filipkowski and A. J. Silva (1998) Autophosphorylation at Thr286 of the α-calcium-calmodulin kinase II in LTP and learning, *Science*, **279**, 870-873.
13. D. M. Kullmann, G. Erdemli and F. Asztely (1996) LTP of AMPA and NMDA receptor-mediated signals: evidence for presynaptic expression and extrasynaptic glutamate spill-over, *Neuron*, **17**, 461-474.
14. S. N. Davies, R. A. Lester, K. G. Reymann and G. L. Collingridge (1989) Temporally distinct pre- and post-synaptic mechanisms maintain long-term potentiation, *Nature*, **338**, 500-503.
15. J. J. Renger, C. Egles and G. Liu (2001) A developmental switch in neurotransmitter flux enhances synaptic efficacy by affecting AMPA receptor activation, *Neuron*, **29**, 469-484.
16. D. Muller and G. Lynch (1989) Evidence that changes in presynaptic calcium currents are not responsible for long-term potentiation in hippocampus, *Brain Res*, **479**, 290-299.
17. T. Manabe, D. J. Wyllie, D. J. Perkel and R. A. Nicoll (1993) Modulation of synaptic transmission and long-term potentiation: effects on paired pulse facilitation and EPSC variance in the CA1 region of the hippocampus, *J Neurophysiol*, **70**, 1451-1459.
18. F. Asztely, M. Y. Xiao and B. Gustafsson (1996) Long-term potentiation and paired-pulse facilitation in the hippocampal CA1 region, *Neuroreport*, **7**, 1609-1612.
19. T. Maeda, S. Kaneko, A. Akaike and M. Satoh (1997) Direct evidence for increase in excitatory amino acids release during mossy fiber LTP in rat hippocampal slices as revealed by the patch sensor methods, *Neurosci Lett*, **224**, 103-106.
20. J. S. Diamond, D. E. Bergles and C. E. Jahr (1998) Glutamate release monitored with astrocyte transporter currents during LTP, *Neuron*, **21**, 425-433.
21. C. Lüscher, R. C. Malenka and R. A. Nicoll (1998) Monitoring glutamate release during LTP with glial transporter currents, *Neuron*, **21**, 435-441.
22. T. Manabe and R. A. Nicoll (1994) Long-term potentiation: evidence against an increase in transmitter release probability in the CA1 region of the hippocampus, *Science*, **265**, 1888-1892.
23. Z. F. Mainen, Z. Jia, J. Roder and R. Malinow (1998) Use-dependent AMPA receptor block in mice lacking GluR2 suggests postsynaptic site for LTP expression, *Nat Neurosci*, **1**, 579-586.
24. R. Malinow, Z. F. Mainen and Y. Hayashi (2000) LTP mechanisms: from silence to four-lane traffic, *Curr Opin Neurobiol*, **10**, 352-357.
25. J. T. Isaac, R. A. Nicoll and R. C. Malenka (1995) Evidence for silent synapses: implications for the expression of LTP, *Neuron*, **15**, 427-434.
26. D. Liao, N. A. Hessler and R. Malinow (1995) Activation of postsynaptically silent synapses during pairing-induced LTP in CA1 region of hippocampal slice, *Nature*, **375**, 400-404.
27. G. M. Durand, Y. Kovalchuk and A. Konnerth (1996) Long-term potentiation and functional synapse induction in developing hippocampus, *Nature*, **381**, 71-75.
28. S. N. Gomperts, A. Rao, A. M. Craig, R. C. Malenka and R. A. Nicoll (1998) Postsynaptically silent synapses in single neuron cultures, *Neuron*, **21**, 1443-1451.
29. Z. Nusser, R. Lujan, G. Laube, J. D. Roberts, E. Molnar and P. Somogyi (1998) Cell type and pathway dependence of synaptic AMPA receptor number and variability in the hippocampus, *Neuron*, **21**, 545-559.
30. D. Liao, X. Zhang, R. O'Brien, M. D. Ehlers and R. L. Huganir (1999) Regulation of morphological postsynaptic silent synapses in developing hippocampal neurons, *Nat Neurosci*, **2**, 37-43.
31. R. S. Petralia, J. A. Esteban, Y. X. Wang, J. G. Partridge, H. M. Zhao, R. J. Wenthold and R. Malinow (1999) Selective acquisition of AMPA receptors over postnatal development suggests a molecular basis for silent synapses, *Nat Neurosci*, **2**, 31-36.
32. Y. Takumi, V. Ramirez-Leon, P. Laake, E. Rinvik and O. P. Ottersen (1999) Different modes of expression of AMPA and NMDA receptors in hippocampal synapses, *Nat Neurosci*, **2**, 618-624.
33. S. Maren, G. Tocco, S. Standley, M. Baudry and R. F. Thompson (1993) Postsynaptic factors in the expression of long-term potentiation (LTP): increased glutamate receptor binding following LTP induction in vivo, *Proc Natl Acad Sci U S A*, **90**, 9654-9658.
34. P. M. Lledo, X. Zhang, T. C. Sudhof, R. C. Malenka and R. A. Nicoll (1998) Postsynaptic membrane fusion and long-term potentiation, *Science*, **279**, 399-403.
35. M. Maletic-Savatic, T. Koothan and R. Malinow (1998) Calcium-evoked dendritic exocytosis in cultured hippocampal neurons. Part II: mediation by calcium/calmodulin-dependent protein kinase II, *J Neurosci*, **18**, 6814-6821.

36. M. Maletic-Savatic and R. Malinow (1998) Calcium-evoked dendritic exocytosis in cultured hippocampal neurons. Part I: trans-Golgi network-derived organelles undergo regulated exocytosis, *J Neurosci*, **18**, 6803-6813.
37. S. H. Shi, Y. Hayashi, R. Petralia, S. Zaman, R. Wenthold, K. Svoboda and R. Malinow (1999) Rapid Spine Delivery and Redistribution of AMPA Receptors after Synaptic NMDA Receptor Activation, *Science*, **284**, 1811-1816.
38. A. Piccini and R. Malinow (2002) Critical postsynaptic density 95/disc large/zonula occludens-1 interactions by glutamate receptor 1 (GluR1) and GluR2 required at different subcellular sites, *J Neurosci*, **22**, 5387-5392.
39. Y. Hayashi, S. H. Shi, J. A. Esteban, A. Piccini, J. C. Poncer and R. Malinow (2000) Driving AMPA receptors into synapses by LTP and CaMKII: requirement for GluR1 and PDZ domain interaction, *Science*, **287**, 2262-2267.
40. T. Takahashi, K. Svoboda and R. Malinow (2002) Experience strengthening transmission by driving AMPA receptors into synapses. *Science* **299**, 1585-1588.
41. D. Zamanillo, R. Sprengel, O. Hvalby, V. Jensen, N. Burnashev, A. Rozov, K. M. Kaiser, H. J. Koster, T. Borchardt, P. Worley, J. Lubke, M. Frotscher, P. H. Kelly, B. Sommer, P. Andersen, P. H. Seeburg and B. Sakmann (1999) Importance of AMPA receptors for hippocampal synaptic plasticity but not for spatial learning, *Science*, **284**, 1805-1811.
42. V. Mack, N. Burnashev, K. M. Kaiser, A. Rozov, V. Jensen, O. Hvalby, P. H. Seeburg, B. Sakmann and R. Sprengel (2001) Conditional restoration of hippocampal synaptic potentiation in GluR-A-deficient mice, *Science*, **292**, 2501-2504.
43. K. W. Roche, R. J. O'Brien, A. L. Mammen, J. Bernhardt and R. L. Huganir (1996) Characterization of multiple phosphorylation sites on the AMPA receptor GluR1 subunit, *Neuron*, **16**, 1179-1188.
44. H.-K. Lee, K. Takamiya, K. Kameyama, S. Yu, L. Rossetti C. He, D. Wilen and R. L. Huganir (2002) Identification and characterization of a novel phosphorylation site on the GluR1 subunit of AMPA receptors. *US Soc. Neurosci. Abstr.*
45. A. Barria, D. Muller, V. Derkach, L. C. Griffith and T. R. Soderling (1997) Regulatory phosphorylation of AMPA-type glutamate receptors by CaM-KII during long-term potentiation, *Science*, **276**, 2042-2045.
46. A. L. Mammen, K. Kameyama, K. W. Roche and R. L. Huganir (1997) Phosphorylation of the α-amino-3-hydroxy-5-methylisoxazole-4-propionic acid receptor GluR1 subunit by calcium/calmodulin-dependent kinase II, *J Biol Chem*, **272**, 32528-32533.
47. V. Derkach, A. Barria and T. R. Soderling (1999) Ca^{2+}/calmodulin-kinase II enhances channel conductance of α-amino-3-hydroxy-5-methyl-4-isoxazolepropionate type glutamate receptors, *Proc Natl Acad Sci U S A*, **96**, 3269-3274.
48. H. K. Lee, M. Barbarosie, K. Kameyama, M. F. Bear and R. L. Huganir (2000) Regulation of distinct AMPA receptor phosphorylation sites during bidirectional synaptic plasticity, *Nature*, **405**, 955-959.
49. T. A. Benke, A. Luthi, J. T. Isaac and G. L. Collingridge (1998) Modulation of AMPA receptor unitary conductance by synaptic activity, *Nature*, **393**, 793-797.
50. J. A. Esteban, S. H. Shi, C. Wilson, M. Nuriya, R. L. Huganir and R. Malinow (2003) PKA phosphorylation of AMPA receptor subunits controls synaptic trafficking underlying plasticity, *Nat Neurosci*, **6**, 136-143.
51. R. D. Blitzer, J. H. Connor, G. P. Brown, T. Wong, S. Shenolikar, R. Iyengar and E. M. Landau (1998) Gating of CaMKII by cAMP-regulated protein phosphatase activity during LTP, *Science*, **280**, 1940-1942.
52. M. J. Thomas, T. D. Moody, M. Makhinson and T. J. O'Dell (1996) Activity-dependent beta-adrenergic modulation of low frequency stimulation induced LTP in the hippocampal CA1 region, *Neuron*, **17**, 475-482.
53. H. K. Lee, K. Takamiya, J. S. Han, H. Man, C. H. Kim, G. Rumbaugh, S. Yu, L. Ding, C. He, R. S. Petralia, R. J. Wenthold, M. Gallagher and R. L. Huganir (2003) Phosphorylation of the AMPA Receptor GluR1 Subunit Is Required for Synaptic Plasticity and Retention of Spatial Memory, *Cell*, **112**, 631-643.
54. Y. Yoshimura, C. Aoi and T. Yamauchi (2000) Investigation of protein substrates of Ca^{2+}/calmodulin-dependent protein kinase II translocated to the postsynaptic density, *Brain Res Mol Brain Res*, **81**, 118-128.
55. Y. Yoshimura, T. Shinkawa, M. Taoka, K. Kobayashi, T. Isobe and T. Yamauchi (2002) Identification of protein substrates of Ca^{2+}/calmodulin-dependent protein kinase II in the postsynaptic density by protein sequencing and mass spectrometry, *Biochem Biophys Res Commun*, **290**, 948-954.
56. H. J. Chen, M. Rojas-Soto, A. Oguni and M. B. Kennedy (1998) A synaptic Ras-GTPase activating protein (p135 SynGAP) inhibited by CaM kinase II, *Neuron*, **20**, 895-904.
57. J. H. Kim, D. Liao, L. F. Lau and R. L. Huganir (1998) SynGAP: a synaptic RasGAP that associates with the PSD-95/SAP90 protein family, *Neuron*, **20**, 683-691.
58. J. J. Zhu, Y. Qin, M. Zhao, L. Van Aelst and R. Malinow (2002) Ras and Rap control AMPA receptor trafficking during synaptic plasticity, *Cell*, **110**, 443-455.
59. H. C. Kornau, L. T. Schenker, M. B. Kennedy and P. H. Seeburg (1995) Domain interaction between NMDA

receptor subunits and the postsynaptic density protein PSD-95, *Science*, **269**, 1737-1740.
60. A. E. El-Husseini, E. Schnell, D. M. Chetkovich, R. A. Nicoll and D. S. Bredt (2000) PSD-95 involvement in maturation of excitatory synapses, *Science*, **290**, 1364-1368.
61. I. D. Ehrlich and R. Malinow (2002) PSD-95 mimics, occludes and dominant negative forms block LTP in hippocampal slice cultures. *US Soc. Neurosc. Abstr.*
62. E. Schnell, M. Sizemore, S. Karimzadegan, L. Chen, D. S. Bredt and R. A. Nicoll (2002) Direct interactions between PSD-95 and stargazin control synaptic AMPA receptor number, *Proc Natl Acad Sci U S A*, **99**, 13902-13907.
63. L. Chen, D. M. Chetkovich, R. S. Petralia, N. T. Sweeney, Y. Kawasaki, R. J. Wenthold, D. S. Bredt and R. A. Nicoll (2000) Stargazin regulates synaptic targeting of AMPA receptors by two distinct mechanisms, *Nature*, **408**, 936-943.
64. D. M. Chetkovich, L. Chen, T. J. Stocker, R. A. Nicoll and D. S. Bredt (2002) Phosphorylation of the postsynaptic density-95 (PSD-95)/discs large/zona occludens-1 binding site of stargazin regulates binding to PSD-95 and synaptic targeting of AMPA receptors, *J Neurosci*, **22**, 5791-5796.
65. A. Yoshii, M. H. Sheng and M. Constantine-Paton (2003) Eye opening induces a rapid dendritic localization of PSD-95 in central visual neurons, *Proc Natl Acad Sci U S A*, **100**, 1334-1339.
66. Y. H. Koh, E. Popova, U. Thomas, L. C. Griffith and V. Budnik (1999) Regulation of DLG localization at synapses by CaMKII-dependent phosphorylation, *Cell*, **98**, 353-363.
67. M. Passafaro, V. Piech and M. Sheng (2001) Subunit-specific temporal and spatial patterns of AMPA receptor exocytosis in hippocampal neurons, *Nat Neurosci*, **4**, 917-926.
68. A. S. Leonard, M. A. Davare, M. C. Horne, C. C. Garner and J. W. Hell (1998) SAP97 is associated with the α-amino-3-hydroxy-5-methylisoxazole-4-propionic acid receptor GluR1 subunit, *J Biol Chem*, **273**, 19518-19524.
69. C.-H. Kim, K. Takamiya, R. Sattler, R. S. Petralia, S. Yu, L. Ding, R. Wenthold and R. Huganir (2002) Role of the AMPA receptor GluR1 subunit carboxyl-terminal PDZ ligand in synaptic plasticity and receptor distribution. *Neuroscience Meeting*.
70. I. Song, S. Kamboj, J. Xia, H. Dong, D. Liao and R. L. Huganir (1998) Interaction of the N-ethylmaleimide-sensitive factor with AMPA receptors, *Neuron*, **21**, 393-400.
71. P. Osten, S. Srivastava, G. J. Inman, F. S. Vilim, L. Khatri, L. M. Lee, B. A. States, S. Einheber, T. A. Milner, P. I. Hanson and E. B. Ziff (1998) The AMPA receptor GluR2 C terminus can mediate a reversible, ATP-dependent interaction with NSF and α- and β-SNAPs, *Neuron*, **21**, 99-110.
72. A. Nishimune, J. T. Isaac, E. Molnar, J. Noel, S. R. Nash, M. Tagaya, G. L. Collingridge, S. Nakanishi and J. M. Henley (1998) NSF binding to GluR2 regulates synaptic transmission, *Neuron*, **21**, 87-97.
73. J. Noel, G. S. Ralph, L. Pickard, J. Williams, E. Molnar, J. B. Uney, G. L. Collingridge and J. M. Henley (1999) Surface expression of AMPA receptors in hippocampal neurons is regulated by an NSF-dependent mechanism, *Neuron*, **23**, 365-376.
74. S. H. Lee, L. Liu, Y. T. Wang and M. Sheng (2002) Clathrin adaptor AP2 and NSF interact with overlapping sites of GluR2 and play distinct roles in AMPA receptor trafficking and hippocampal LTD, *Neuron*, **36**, 661-674.
75. R. Malinow and R. C. Malenka (2002) AMPA receptor trafficking and synaptic plasticity, *Annu Rev Neurosci*, **25**, 103-126.
76. S. Shi, Y. Hayashi, J. A. Esteban and R. Malinow (2001) Subunit-specific rules governing AMPA receptor trafficking to synapses in hippocampal pyramidal neurons, *Cell*, **105**, 331-343.
77. R. J. Wenthold, R. S. Petralia, J. Blahos, II and A. S. Niedzielski (1996) Evidence for multiple AMPA receptor complexes in hippocampal CA1/CA2 neurons, *J Neurosci*, **16**, 1982-1989.
78. J. J. Zhu, J. A. Esteban, Y. Hayashi and R. Malinow (2000) Postnatal synaptic potentiation: delivery of GluR4-containing AMPA receptors by spontaneous activity, *Nat Neurosci*, **3**, 1098-1106.
79. A. Kolleker, B. J. Schupp, V. Mack, G. Köhr, P. H. Seeburg and P. Osten (2002) GluR-Blong AMPA receptor subunit mediates spontaneous activity-driven receptor insertion in young principal neurons. *Neuroscience Meeting*, Orland.
80. H. Yasuda, A. L. Barth, D. Stellwagen and R. C. Malenka (2003) A developmental switch in the signaling cascades for LTP induction, *Nat Neurosci*, **6**, 15-16.
81. A. M. Craig (1998) Activity and synaptic receptor targeting: the long view, *Neuron*, **21**, 459-462.
82. D. R. Grosshans, D. A. Clayton, S. J. Coultrap and M. D. Browning (2002) LTP leads to rapid surface expression of NMDA but not AMPA receptors in adult rat CA1, *Nat Neurosci*, **5**, 27-33.
83. P. Washbourne, J. E. Bennett and A. K. McAllister (2002) Rapid recruitment of NMDA receptor transport packets to nascent synapses, *Nat Neurosci*, **5**, 751-759.
84. E. M. Snyder, B. D. Philpot, K. M. Huber, X. Dong, J. R. Fallon and M. F. Bear (2001) Internalization of ionotropic glutamate receptors in response to mGluR activation, *Nat Neurosci*, **4**, 1079-1085.

85. L. Y. Wang, E. M. Dudek, M. D. Browning and J. F. MacDonald (1994) Modulation of AMPA/kainate receptors in cultured murine hippocampal neurones by protein kinase C, *J Physiol*, **475**, 431-437.
86. D. N. Lieberman and I. Mody (1994) Regulation of NMDA channel function by endogenous Ca(2+)-dependent phosphatase, *Nature*, **369**, 235-239.
87. J. Y. Lan, V. A. Skeberdis, T. Jover, S. Y. Grooms, Y. Lin, R. C. Araneda, X. Zheng, M. V. Bennett and R. S. Zukin (2001) Protein kinase C modulates NMDA receptor trafficking and gating, *Nat Neurosci*, **4**, 382-390.
88. M. D. Ehlers, W. G. Tingley and R. L. Huganir (1995) Regulated subcellular distribution of the NR1 subunit of the NMDA receptor, *Science*, **269**, 1734-1737.
89. D. K. Fong, A. Rao, F. T. Crump and A. M. Craig (2002) Rapid synaptic remodeling by protein kinase C: reciprocal translocation of NMDA receptors and calcium/calmodulin-dependent kinase II, *J Neurosci*, **22**, 2153-2164.
90. W. G. Tingley, M. D. Ehlers, K. Kameyama, C. Doherty, J. B. Ptak, C. T. Riley and R. L. Huganir (1997) Characterization of protein kinase A and protein kinase C phosphorylation of the N-methyl-D-aspartate receptor NR1 subunit using phosphorylation site-specific antibodies, *J Biol Chem*, **272**, 5157-5166.
91. A. S. Leonard and J. W. Hell (1997) Cyclic AMP-dependent protein kinase and protein kinase C phosphorylate N-methyl-D-aspartate receptors at different sites, *J Biol Chem*, **272**, 12107-12115.
92. D. B. Scott, T. A. Blanpied, G. T. Swanson, C. Zhang and M. D. Ehlers (2001) An NMDA receptor ER retention signal regulated by phosphorylation and alternative splicing, *J Neurosci*, **21**, 3063-3072.
93. S. Standley, K. W. Roche, J. McCallum, N. Sans and R. J. Wenthold (2000) PDZ domain suppression of an ER retention signal in NMDA receptor NR1 splice variants, *Neuron*, **28**, 887-898.
94. X. Zheng, L. Zhang, A. P. Wang, M. V. Bennett and R. S. Zukin (1999) Protein kinase C potentiation of N-methyl-D-aspartate receptor activity is not mediated by phosphorylation of N-methyl-D-aspartate receptor subunits, *Proc Natl Acad Sci U S A*, **96**, 15262-15267.
95. T. Suzuki, K. Okumura-Noji, R. Tanaka and T. Tada (1994) Rapid translocation of cytosolic Ca^{2+}/calmodulin-dependent protein kinase II into postsynaptic density after decapitation, *J Neurochem*, **63**, 1529-1537.
96. S. Strack, S. Choi, D. M. Lovinger and R. J. Colbran (1997) Translocation of autophosphorylated calcium/calmodulin-dependent protein kinase II to the postsynaptic density, *J Biol Chem*, **272**, 13467-13470.
97. K. Shen, M. N. Teruel, J. H. Connor, S. Shenolikar and T. Meyer (2000) Molecular memory by reversible translocation of calcium/calmodulin-dependent protein kinase II, *Nat Neurosci*, **3**, 881-886.
98. K. Shen, M. N. Teruel, K. Subramanian and T. Meyer (1998) CaMKIIβ functions as an F-actin targeting module that localizes CaMKIIα/β heterooligomers to dendritic spines, *Neuron*, **21**, 593-606.
99. M. R. Gleason, S. Higashijima, J. Dallman, K. Liu, G. Mandel and J. R. Fetcho (2003) Translocation of CaM kinase II to synaptic sites in vivo, *Nat Neurosci*, **6**, 217-218.
100. S. Strack and R. J. Colbran (1998) Autophosphorylation-dependent targeting of calcium/ calmodulin-dependent protein kinase II by the NR2B subunit of the N-methyl-D-aspartate receptor, *J Biol Chem*, **273**, 20689-20692.
101. A. S. Leonard, I. A. Lim, D. E. Hemsworth, M. C. Horne and J. W. Hell (1999) Calcium/calmodulin-dependent protein kinase II is associated with the N-methyl-D-aspartate receptor, *Proc Natl Acad Sci U S A*, **96**, 3239-3244.
102. K. U. Bayer, P. De Koninck, A. S. Leonard, J. W. Hell and H. Schulman (2001) Interaction with the NMDA receptor locks CaMKII in an active conformation, *Nature*, **411**, 801-805.
103. R. S. Walikonis, A. Oguni, E. M. Khorosheva, C. J. Jeng, F. J. Asuncion and M. B. Kennedy (2001) Densin-180 forms a ternary complex with the α-subunit of Ca^{2+}/calmodulin-dependent protein kinase II and α-actinin, *J Neurosci*, **21**, 423-433.
104. S. Strack, A. J. Robison, M. A. Bass and R. J. Colbran (2000) Association of calcium/calmodulin-dependent kinase II with developmentally regulated splice variants of the postsynaptic density protein densin-180, *J Biol Chem*, **275**, 25061-25064.
105. S. Okabe, T. Urushido, D. Konno, H. Okado and K. Sobue (2001) Rapid redistribution of the postsynaptic density protein PSD-Zip45 (Homer 1c) and its differential regulation by NMDA receptors and calcium channels, *J Neurosci*, **21**, 9561-9571.
106. E. N. Star, D. J. Kwiatkowski and V. N. Murthy (2002) Rapid turnover of actin in dendritic spines and its regulation by activity, *Nat Neurosci*, **5**, 239-246.
107. D. El-Husseini Ael, E. Schnell, S. Dakoji, N. Sweeney, Q. Zhou, O. Prange, C. Gauthier-Campbell, A. Aguilera-Moreno, R. A. Nicoll and D. S. Bredt (2002) Synaptic strength regulated by palmitate cycling on PSD-95, *Cell*, **108**, 849-863.
108. C. H. Kim and J. E. Lisman (1999) A role of actin filament in synaptic transmission and long-term potentiation, *J Neurosci*, **19**, 4314-4324.

3

CONSTRUCTION, STABILITY AND DYNAMICS OF THE INHIBITORY POSTSYNAPTIC MEMBRANE

Christian Vannier and Antoine Triller[*]

1. INTRODUCTION

In the mature central nervous system (CNS), two amino acids, glycine and GABA, mediate inhibition by activating chloride channels. Their ionotropic receptors share a common pentameric structure delineating the channel. The subunits harbor an extracellular N-terminal domain, four transmembrane segments (M1-4), the M2 segments forming the pore walls. The ligand binding sites are associated with α subunits, while the other ones contains regulatory sites and interact through their large M3-M4 cytoplasmic loop with proteins involved in their synaptic localization in the CNS (ref. in Moss and Smart[1]). Actually, the glycine receptor (GlyR) was the first receptor in the CNS, shown to be accumulated at postsynaptic differentiations (PSD), in front of presynaptic release sites[2]. Later on, the distributions of many other receptors including the $GABA_A$ receptors (GABAR) were analyzed at the electron microscopy level, and variable subcellular organizations have been described. It was shown that the subcellular distribution of $GABA_A$ receptors varies according to the isotypes. Some GABARs are accumulated at PSDs while other spread throughout the somato-dendritic membrane independently of synapses[3]. The electron dense PSD, which is facing the pre-synaptic active zone is now partly understood and corresponds to a high density of proteins that constitute the sub-synaptic scaffold. At the electron microscopy level, and using conventional fixative procedures and staining, two types of PSDs have been described, corresponding to the so-called type I and type II synapses, depending upon the thickness of the PSD. The differences in thickness may reflect the differences in the molecular organization of the postsynaptic scaffold. Interestingly, this ultrastructural feature, together with the shape of pre-synaptic vesicles and the width of the synaptic cleft, are in most cases the signature of inhibitory and excitatory synapses, respectively (ref. in Sotelo and Triller[4]).

The presynaptic element plays a key role in the postsynaptic accumulation of receptors. (In the CNS, this was demonstrated for the excitatory and inhibitory synapses

---

[*] C. Vannier, A. Triller, Inserm UR497, Ecole Normale Supérieure, Paris, France 75005.

Excitatory-Inhibitory Balance: Synapses, Circuits, Systems
Edited by Hensch and Fagiolini, Kluwer Academic/Plenum Publishers, 2003

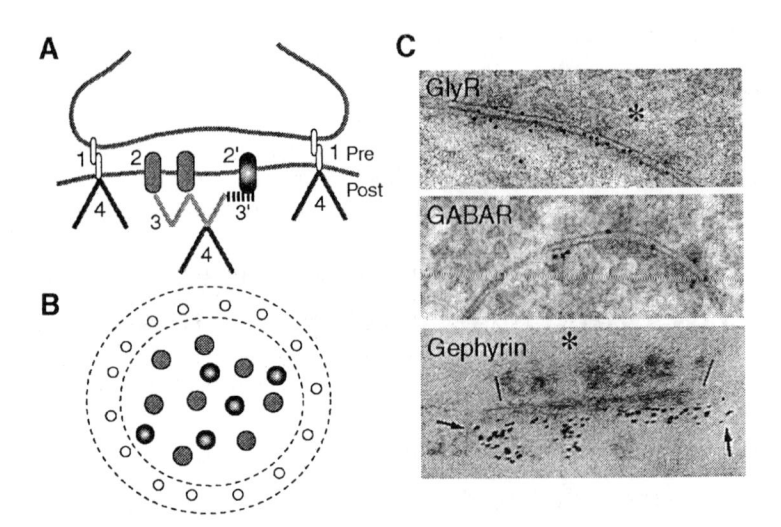

Fig. 1. Organization of a generic inhibitory synapse. **A**, transverse view of synaptic contacts with emphasis on the postsynaptic scaffold network organization. Transmembrane proteins, adhesive proteins such as cadherins (1). Glycine (2) and GABA (2') receptors bind directly and indirectly, respectively, through a non-identified intermediate molecule (3') to gephyrin (3). The cytoskeleton (4) interacts at various levels within the molecular network. **B**: top view of the synapse. Cadherins are concentrated on an annulus around both excitatory and inhibitory synapses. Transmembrane proteins are shown with the same code as in A. **C**: Ultrastructural immunogold localizations of some postsynaptic components of inhibitory synapses with GlyR (Triller et al. unpublished), GABAR (3) and from the gephyrin[2] antigenic sites. In C, the presynaptic element is indicated by an asterisk.

(ref. in Craig and Boudin[5]). The case of inhibitory synapses is particularly intriguing as both glycine and GABA activate a chloride conductance and have roughly the same effect on the postsynaptic membrane. Indeed, in the spinal cord, as well as in the brain stem, inhibitory synapses are either GABAergic, glycinergic or mixed, using the two neurotransmitters[6]. The corresponding receptors accumulate to form microdomains enriched with GABAR, GlyR, or a mixture of both, and the composition of the postsynaptic membrane is determined by the identity of the presynaptic element[7].

As GlyR was initially purified with an affinity column[8-9] an associated extrinsic membrane protein, subsequently named gephyrin was copurified. Antisense experiments[10] and knock out mice[11] have established that gephyrin, is involved in the synaptic localization of GlyR. Gephyrin is also a pivotal protein for the stabilization of GABAR at synapses. The GABAR γ2 subunit is influential in the synaptic localization since mice lacking this subunit had a significant reduction of the number of GABA receptor clusters at synapses[12]. This effect is mediated through gephyrin as shown by antisense experiments[12], or by examining gephyrin KO mice[13a]. However a direct link between GABARs and gephyrin could never be demonstrated. GABARAP (GABA-receptor-associated protein), a candidate protein for linking GABAR to the cytoskeleton binds to GABAR γ2 subunit, gephyrin, and to N-ethylmaleimide-sensitive factor (NSF)[14,15], but it is seen at synapses only occasionally[15,16]. As GABARAP and gephyrin colocalize intracellularly along membranous structures including those of the secretory pathway, it is likely that GABARAP is involved in the receptor intracellular trafficking rather than in stabilization of GABAR at synapses[16].

2. GEPHYRIN AND GEPHYRIN-BINDING PROTEINS

2.1. The gephyrin polypeptide

Gephyrin was initially identified in the spinal cord as the Mr=93,000 cytoplasmic protein associated with large complexes of purified GlyR[8,9]. Its biochemical characterization and studies using heterologous expression showed that gephyrin not only could bind cooperatively to tubulin dimers with a high affinity[17], but also to GlyR via an 18-aminoacid amphipathic helix (βgb, see below) within the M3-M4 cytoplasmic loop of the β subunit[18,19]. These remarkable properties raised the hypothesis that the protein was a linker of the receptor to microtubules and was thus functionally adapted to anchor GlyR at synapses via the cytoskeleton. The sequencing of proteolytic products from material copurifying with GlyR led to the isolation of several cDNAs partially or entirely covering gephyrin coding sequence[20]. This work revealed the highly variable structure of the molecule that exists as several splice variants differing by the presence of distinct nucleotide sequences termed cassettes. The characterization of the mouse gephyrin single gene and of the molecule exonic structure[21], together with the cloning of seven different full-length coding sequences expressed in the adult rat spinal cord[22] confirmed the mosaic structure of the gephyrin polypeptide.

The gephyrin exons (35 base pairs (bps) on the average) represent less than two percent of a gene more than 150 kilobase pairs long. Ten exons (cassettes, ranging from 39 to 120 base pairs) are known, out of a total of 30, that can be alternatively used to built up a set of at least 11 identified molecular variants. Some variants are N- or C-terminally truncated. All are differentially expressed in numerous tissues including brain, heart, skeletal muscle, and kidney, liver, lung, spleen, or testis, illustrating the ubiquitous nature of gephyrin. Most of the alternatively spliced cassettes are located in the N-terminal half of the molecule, likely making this domain a major site for the modulation of gephyrin activity(ies). The variation in gephyrin structure raises both the question of the role of the distinct cassettes, and the hypothesis that the various forms of the protein may have in synapses several functions in-, or in addition to the control of GlyR targeted insertion.

It is currently unknown which molecule(s) is associated with GlyR in synapses. However, *in vitro* binding assays of gephyrin variants using GST fusion proteins bearing cytoplasmic loop of the β subunit have recently illustrated the functional heterogeneity of gephyrin with regard to GlyR anchoring. The biological activity of gephyrin relies on the cassette composition of the polypeptide because some variants only can bind the GlyR β subunit[22] (see below). As a consequence, gephyrins of defined structure only can function in the formation of glycinergic synapses.

2.2. Tertiary and Quaternary Structures of Gephyrin

It is thought that during evolution, the gephyrin polypeptide originated from the fusion of two genes of bacterial origin respectively encoding the MogA and MoeA proteins involved in the biosynthesis of the molybdenum cofactor (MoCo) in *E. coli*[20,23,24]. These proteins are homologous to the gephyrin N- and C-terminal domains (G- and E-domains) respectively, flanking a 170-residue unique linker region not found in the homologous fusion proteins from *Arabidopsis* (Cnx1) or *Drosophila* (Cinnamon)[21] An intriguing feature of gephyrin is that these domains are enzymatically functional. Indeed, the vertebrate protein (clone p1 or Ge2,6; [20,22]) rescues MoCo biosynthesis in deficient prokaryotic and eukaryotic cells[25], and disruption of the gephyrin gene leads to molybdoenzyme activity deficiency in the mouse[11]. These results are consistent with the high conservation of the MoCo biosynthetic pathway and, together with the evidence that gephyrin mutation in human can cause Moco deficiency[26,27] they also point to the notion that gephyrin has maintained MogA and MoeA activities in numerous tissues.

Undoubtedly, delineating structure-function relationships for identified gephyrin molecular variants will benefit from cues provided by the determination of tertiary or quaternary structures, after the recently achieved crystallization of MogA, MoeA, or of Cnx1 and gephyrin G-domains[28-31]. It is now acquired that the MogA and G-domains (including rat and human gephyrin G-domains) are compact structures forming stable trimers in solution that are maintained and can be detected in crystals. The MoeA polypeptide is composed of four distinct tertiary domains forming an elongated L-shaped molecule. It is predominantly encountered as a dimer. Very likely, if they are conserved in the gephyrin molecule, these oligomerization properties would lead to the construction of a two-dimensional hexagonal gephyrin lattice, as recently proposed[30] but without binding of a connecting protein[32]. This oligomerization pattern is the key factor confering gephyrin its function as a scaffold core protein and its ability to recruit GlyR as postsynaptic clusters.

It is currently unknown whether the GlyR β subunit binds to the trimer- or the dimer-forming domain of gephyrin. It is believed that an alteration of the trimer interface might alter the formation of a hexagonal lattice[31] because the absence or presence of some cassettes (such as cassettes 1 and 5, see ref. 33) in the MogA-like domain might strongly alter its competent conformation. The N-terminal end of the molecule has been shown to have a crucial role in promoting receptor binding function since splice variants with modified N-terminal structures do not interact with GlyR. More precisely, disruption of a highly conserved α-helical structure (α-helix 4) upon cassette 5 insertion could result either in the mere destruction of GlyR binding site or in a remote effect on a key conformation in the N-terminal domain of gephyrin, affecting the overall molecule folding. In these studies it was not demonstrated that it contains the β subunit binding site of gephyrin. In fact, a direct trimer-gephyrin interaction cannot be detected[29]. The mechanism by which a precise conformation of the gephyrin G-domain is required for the whole structure function remains to be directly investigated[31,33]. Therefore it remains to assign to domains or oligomerized domains the GlyR binding activity of gephyrin, but also to measure whether a full hexagonal lattice is formed to evaluate the receptor binding capacity of a defined scaffold.

The tertiary structure of the linker domain is not known. This domain contains part of the consensus sequences for phosphorylation by either protein kinase C or cyclic nucleotide-dependent kinase[20]. Four of the alternatively spliced exons out of eight contribute to the relatively high variabilty of its structure. Among them, exon 11 creates a potential phosphorylation substrate for casein kinase II and exon 13 introduces a sequence similar to a α-helical dimerization domain of keratin. Interestingly however, constant sequences are found which might be involved in protein-protein interactions. This is the case of exons 8 and 10 or 14 which encode peptidic stretches respectively rich in hydroxylated, charged, and proline residues, or 60/80% identical to the second repeat motifs of rat tau/MAP2[34,35]. This last feature could mediate the reported ability of tubulin binding. An important issue of the biology of gephyrin will be to establish the structure-function relationships of this domain in the context of the already variable whole molecule.

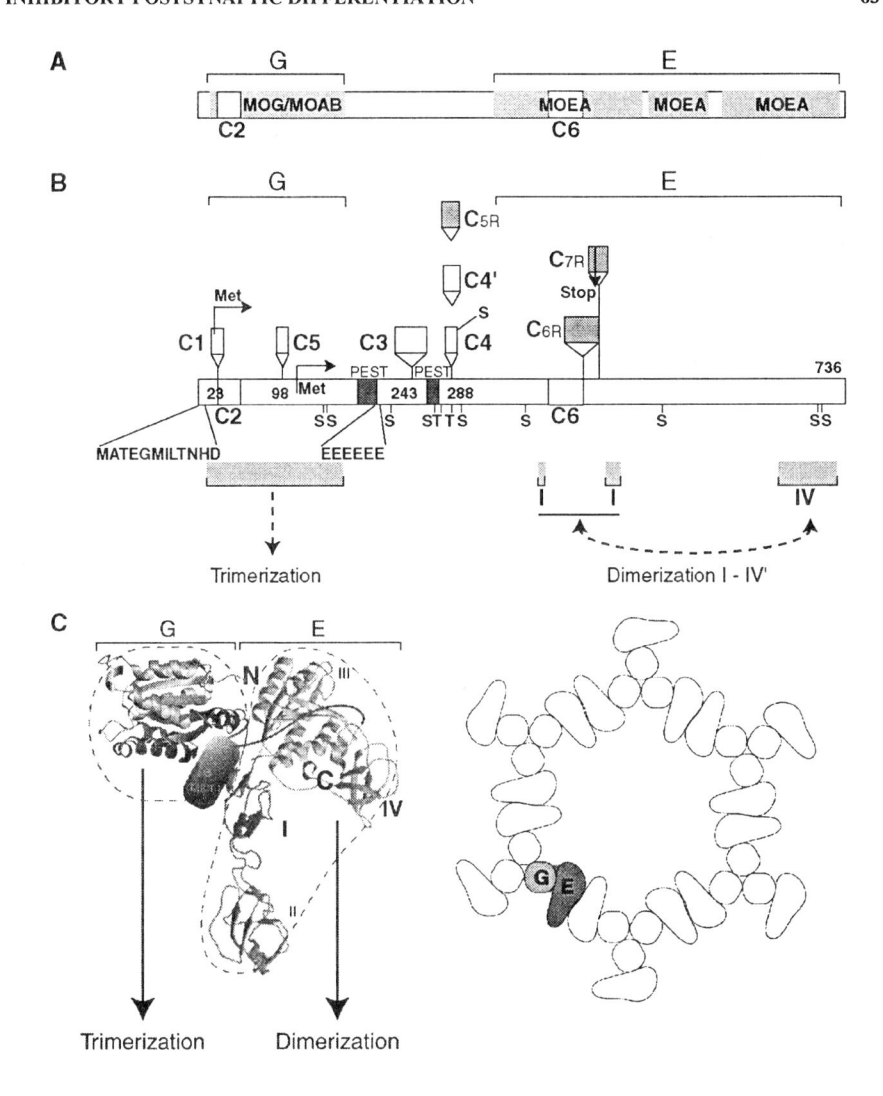

Fig. 2. Structure of gephyrin. **A**, Homology domains in mouse/rat (clone p1) gephyrin. Grey boxes indicate regions with homology to the bacterial MogA/MoaB (G) and MoeA (E) proteins. White boxes labelled C2 and C6 indicate the alternatively spliced cassettes[22]. **B**, Cassettes and oligomerization domains. Cassettes are shown in scale with their insertion site, and are labelled according to Meier and coll. (2000b) (C1 to C6, white boxes), or to Ramming and coll. (2000) (C5R to C7R, grey boxes). Dark grey boxes labelled PEST correspond to proline-rich domains. S and T denote serine and threonine residues representing potential phosphorylation sites, and the N-terminal sequence shows the cryptic myristoylation site, G[20]. Positions of alternative N- and C-termini (arrows) are indicated by Met and Stop, respectively. Shadowed bars below the rod show the position and extent of identified domains involved in the quaternary structure of the molecule; I and IV are the tertiary domains of the E region allowing dimerization[30]. **C**, Model for gephyrin oligomerization. *Left*, tentative tertiary structure of gephyrin in which the G and E domains are fused via the linker region (grey block), and the G domain is in close proximity with the E-III domain[30]. The position of the protein extremities (N, C) are indicated. *Right*, representation of the hypothetical hexagonal scaffold resulting from combined dimerization and trimerization of the E and G domains, respectively.

2.3. Gephyrin-interacting Proteins

Gephyrin is the core element of the inhibitory postsynaptic scaffold. Binding partners for gephyrin other than GlyR have recently been identified but their exact function in the receptor postsynaptic clustering is not yet delineated.

Two-hybrid screening revealed that various proteins such as collybistin[36], RAFT1[37] or dynein light chains 1 and 2 (ref. 38) bind to gephyrin. Collybistin, a nervous system specific protein existing as two splice variants in rat, collybistin I and II, is a member of the diffuse B-cell lymphoma-like (dbl) GDP/GTP-exchange factors (GEFs) family. As recently demonstrated for its human homolog hPEM-2 (ref. 39) collybistin likely activates Rac/Rho-like GTPases such as Cdc42, which plays key roles not only in submembraneous actin polymerization but also in transcriptional/translational and phosphorylation processes[40]. As other GEFs, collybistin is a modular protein possessing both a dbl homology (DH) domain responsible for the GDP/GTP-exchange reaction and a pleckstrin homology (PH) domain. The presence of the latter can be viewed as a key structural determinant with regard to membrane association of postsynaptic complexes. Some PH domains are known to bind 3-phosphoinositides (such as PIP3 and PI(3,4)P2), providing a means to recruit rapidly to the plasma membrane the target proteins of signaling pathways generated by extracellular factors[41].

Interestingly co-expression of gephyrin and collybistin II (but not collybistin I) in non-neuronal cells results in the translocation of otherwise intracellular gephyrin to sub-membraneous micro aggregates which in turn recruit a co-expressed β subunit-harboring GlyR. This is consistent with the demonstration that, *in vitro*, gephyrin can bind simultaneously both the βgb sequence and collybistin[42]. The latter interaction involves polar amino acid residues upstream to the DH domain of collybistin. In addition, collybistin I possesses a SH3 domain (highly homologous to those of cytoskeletal proteins) and a putative coiled-coil domain in its N-terminal and C-terminal region respectively, which could mediate interactions with other proteins. So far, it has not been determined whether ternary assemblies of GlyR, gephyrin and collybistin are formed at inhibitory synaptic sites in neurons. If it were so, such an association, through an action on Cdc42, might have an important role at glycinergic synapses in signaling events controlling the organization of subsynaptic microfilaments and scaffold.

Gephyrin-RAFT1 interaction was discovered during a search for RAFT1 partners, leading to the identification of a 118-aminoacid domain, GBD (Gephyrin Binding Domain) able to bind to a gephyrin sequence encompassing residues 209 to 685 (ref. 37). Within GBD a 30-amino acid polypeptide stretch shares 45% sequence similarity to the region of the same length in the GlyR β subunit that interacts with gephyrin[18]. RAFT1 (rapamycin and FKBP12 target) is a protein kinase, despite a similarity to the catalytic domains of phosphoinositide 3- and 4-kinases. There is increasing evidence that RAFT1 participates in mitogen-stimulated signaling pathways controlling translation and regulating the cell cycle. In mammals RAFT1 is a mediator of the immunosuppressant actions of rapamycin which impedes phosphorylation of the translational regulators pp70S6k and 4E-BP1. RAFT1 forms with mutations in the conserved residues leading to altered interaction with gephyrin fail to signal to downstream proteins, including pp70S6k and 4E-BP1. This observation is exciting because it suggests that a gephyrin-RAFT1 association could modulate a signal transduction pathway controlling translation. In this case, gephyrin by being able to bind GlyR and RAFT1 might function as a sensor of synaptic activity and control local, dendritic or synaptic protein synthesis from transported mRNAs. This must be put in line with the fact that GlyR α subunits mRNAs were detected within the dendrites, below synapses[43].

So far however, it has not been demonstrated that gephyrin and RAFT1 co-exist at inhibitory synapses. With regard to binding sites occupancy on gephyrin molecules within the postsynaptic scaffold, it is also necessary that the structural requirements for gephyrin-RAFT1 interaction be delineated. Considering gephyrin diversity together with the similarity of binding sites in β subunit and RAFT1, other questions arise as to whether all gephyrin variants are RAFT1 partners, and whether the interaction selectively occurs with synaptic variants.

The dynein light chains-1 and 2 (Dlc-1/2) were identified in screens using full-length gephyrin (clone p1) as bait[38]. As subunits of the cytoplasmic dynein and myosin-Va, they provide another example of likely interaction of a component of the inhibitory postsynaptic scaffold with molecular motors. This had only been shown in the case of the NMDA receptor-associated complex[44,45]. It is worth noting that the binding site of Dlc-1/2 to gephyrin involves a 63-aminoacid polypeptide stretch (residues 181-243) in the linker domain of gephyrin, suggesting that this interaction is phylogenetically recent. Dlc accumulates in dendritic spines at the edge of some PSDs in hippocampal synapses and at the periphery of Golgi apparatus. The observation that Dlc-1/2 can be detected at the edge of gephyrin-containing postsynaptic areas (but not all), indicates that the interaction can take place at synapses, even if Dlc-1/2 is not a structural element of the inhibitory scaffold. Although it is tempting to interpret such an interaction as a means by which gephyrin or gephyin-associated complexes could be routed to- or removed from synapses, Dlc-1/2 is not essential to the synaptic distribution of the protein. Further investigation is needed to understand the implications and roles of this interaction since, alternatively, it equally puts forward a function of the scaffold in a motor-based organization of the subsynaptic cytoskeleton.

Gephyrin has also been identified as being a profilin I ligand by an affinity purification approach[46]. Profilin is considered primarily as a key regulator of actin polymerization in cells via its unique mediation of the ADP to ATP exchange on actin monomer. Beside that for actin, profilin harbors two other main binding sites, respectively for phosphoinositides such as PIP2 (overlapping with the actin binding site) and for proteins with poly(L-proline) (PLP) stretches[47]. Proline-rich ligands of profilin include MENA, VASP and Arp2/3 which are found at focal contacts (but also in *listeria* actin tail) are thought to be involved in recruiting profilin-actin complexes to the site of actin polymerization in cells. It is not known whether gephyrin also utilizes this site to interact with profilin. Profilin is engaged in a complex interplay of signaling pathways in higher eucaryotes involving partners as diverse as PI3K, PLC, and downstream targets of Rho enzymes family, all controlling microfilaments dynamics and regulating functions such as motility and cell shape changes (for a review see ref. 47). Moreover, in brain extracts, dynamin I, clathrin, synapsin, NAP1, Rho-associated kinase, and a member of the NSF family have been found to be ligands of profilins I and II (ref. 48).

This suggests that profilin complexes help link the actin-based cytoskeleton and signaling molecules to the membrane flow of endocytosis and synaptic recycling. It is tempting to hypothesize that associated gephyrin and profilin might be involved (as suggested for collybistin) in the local organization of microfilament-dependent membrane traffic as well as in signaling. So far however, there is no evidence for the presence of profilin at inhibitory postsynaptic loci in neurons. It is nevertheless intriguing that some partners of gephyrin, collybistin and profilin, are capable of both membrane binding and actin polymerization control, two functions which are determinant in building up submembranous multimolecular assemblies. Gephyrin could therefore specify the sites where they occur. Whether the gephyrin and its variants function as regulators or effectors of their partners is still to be delineated. The multiplicity of the potential functions of the protein reinforces the idea that a member of the so-called moonlighting protein family[49] such as gephyrin operates at the heart of dynamic interactions in inhibitory synapses.

As a phosphoprotein, gephyrin could also regulate its own availability as a partner not only for GlyR synaptic anchoring but also for signal transduction events involving the other gephyrin-binding proteins. It cannot be excluded that a particular phosphorylation state of the molecule is required for the assembly of a particular element of the scaffold. Since it is conceivable that local phosphorylation-dephosphorylation interplays alter the various interactions within the scaffold, a major task will be to determine the causal sequences underlying the construction of the whole postsynaptic assembly to fully understand its dynamic properties.

3. RECEPTORS AND CYTOSKELETON

Because of its interaction with scaffold proteins the cytoskeleton is a component of the PSD, at least during differentiation or plastic changes. As such, it can help regulate the local concentration or activity of proteins which themselves control receptor stability by direct interactions. Therefore, it can be involved in the regulation of the number of binding sites for receptors within postsynaptic sites. In the case of excitatory synapses this can also be achieved by a changes in the shape of dendritic spines. It has not been demonstrated that microfilaments or microtubules interact directly with receptors in cells. Assessing the mechanisms by which they contribute to receptor-enriched domains will be complicated by the multiplicity of linker proteins which may explain why their selective disruption has differential effects. For example, in excitatory spines of hippocampal neurons F-actin is required for clustering of AMPA receptors (and CaMKII, α-actinin, drebrin) but not for that of NMDA receptors (and PSD-95, GKAP)[50]. In addition, microtubules are dispensable for the stabilization of protein clusters on excitatory spines[50].

The situation is still unclear for inhibitory synapses. It has been reported that, in spinal cord neurons, microtubule depolymerization results in lateral spreading of GlyR clusters of lower relative gephyrin density, whereas microfilament depolymerization generated smaller clusters with higher gephyrin content[51]. This does not seem to be the case in hippocampal inhibitory synapses since microtubules integrity is not essential for maintaining the localization of gephyrin and $GABA_AR$ (ref. 50). This differential stability of gephyrin in receptor-associated scaffold may reflects the structural heterogeneity of inhibitory synapses (see above) that likely relies on receptor-specific assemblies of cytoplasmic proteins. In line with this, stronger interactions within $GABA_AR$/gephyrin- than within GlyR/gephyrin-containing scaffolds could lead to differential solubilizing effects of detergents. Indeed it has been shown that mild detergents fail to extract $GABA_AR$ and gephyrin from hippocampal synapses likely explaining why $GABA_AR$ and gephyrin do not copurify.

4. DYNAMICS OF GLYCINE RECEPTOR

Synapses, and in particular postsynaptic membrane domains and their associated PSDs appear as stable multimolecular assemblies ensuring that a unique set of functions are performed locally. The notion however emerged these last years that they represent highly dynamic protein complexes adapted to the mechanisms underlying the construction of synaptic contacts and their plasticity. As for organelles or other cellular multimolecular assemblies, postsynaptic complexes thus correspond to the equilibrium between the exo- and endocytic delivery routes for membrane proteins on the one hand, and the interactions involving cytoplasmic scaffolding proteins on the other hand. Cell differentiation and receptor activity can modify this steady-state balance.

4.1. Activity and GlyR stability at synapses

A current view is that receptor stabilization and aggregation, mediated by interactions with scaffold elements, are coupled at synaptic loci. As mentioned above this is not necessarily the case. A distinction could also be made concerning the postsynaptic formation of scaffold assemblies and that of receptor clusters. To understand the relationships between the formation of GlyR clusters, the retention of the receptor in the postsynaptic membrane or scaffold formation, the effect of GlyR activation has been studied using an activity blockade approach. Chronic exposure of cultured spinal cord neurons to strychnine leads to a reversible intracellular redistribution of GlyRs which are lost from synaptic loci[52,53]. In contrast it is without any effect on gephyrin[53]. This unambiguously reflects the fact that aggregation of gephyrin and that of GlyR are governed by distinct mechanisms[53].

In their work[51,52], Kirsch and col. proposed that glycine, which depolarizes immature neurons, could activate a voltage sensitive calcium channel. The entry of calcium would then promote the assembly of the postsynaptic scaffold. Although seductive, this mechanism alone cannot account for the specific accumulation of GlyR in front of glycinergic terminals since GABA and glutamate also depolarize neurons at early stages of development. The strychnine experiments rather provided the experimental basis to investigate whether the binding of the antagonist disrupts the gephyrin-receptor interaction, or the exocytic transport of GlyR to the cell surface is blocked at a post-Golgi step[54]. Combination of biochemical analysis of biotinylated cells with fluorescence microscopy revealed that strychnine-induced activity blockade has no effect on either the gephyrin-GlyR interaction or the half-life of GlyR in synapses. Instead, it turns out that GlyR redistribution results from a default in biosynthetic routing which discloses a transport compartment so far unidentified. Importantly, this study demonstrated that, at the steady-state, only a minor fraction ($\approx$ 5%) of newly-synthesized GlyR is incorporated in synapses and that the receptor mainly undergoes degradation. This indicates that the stability of the glycinergic synapse is due to a tight control of the relative rates of subunit synthesis, synaptic anchoring, and degradation.

Our view of GlyR steady-state presence at synapse is that the activation of the receptor allows its turn-over at this site ($t_{1/2}$=16h), the constitutive replenishment of synapse with newly-plasma membrane inserted molecules being the only process abolished in absence of activity. Therefore, the effect of a lack of activity would be to switch GlyR molecules from a degradative to an apparent storage pathway. A similar switch had been proposed for the AMPARs upon NMDAR activation that involves a recycling process[56]. Interestingly, although degradation is prevented by two opposite effects (NMDAR activation vs GlyR inhibition), in the case of GlyR degradation is a constitutive route. The diffusion/retention mechanism that we have proposed for GlyR accumulation in postsynaptic domains is consistent with a process in which GlyR subunits are synthesized in excess relatively to synaptic anchoring sites[56,57]. In this model it is the availability/saturation of the postsynaptic binding sites (represented by gephyrin and the scaffold) that controls the rate-limiting step of postsynaptic capture.

An important mechanism, ubiquitination, was recently proposed to account for the regulation of plasma membrane GlyR α1 subunit endocytosis and degradation, suggesting that ubiquitin-conjugating enzymes could have a role in determining postsynaptic receptor density[58]. This hypothesis is based on the demonstration that ubiquitination of homopentameric GlyR occurs almost exclusively at the plasma membrane rather than during intracellular transport of the receptor. This study was carried out using heterologous GlyR α1 subunit expression in the absence of gephyrin. Therefore, it is still unknown which receptor state, free or gephyrin-associated, homo- or heteropentameric, would be the target of ubiquitination in neurons. Acting on postsynaptic GlyR this process could disrupt heteropentameric GlyR-gephyrin interaction and provoke receptor internalization.

Alternatively, gephyrin interaction might as well prevent ubiquitination, making extrasynaptic or homopentameric GlyR, or even newly-synthesized molecules the best candidates for internalization. This notion is important because it could provide an explanation of the effect a lack of GlyR activity. It could indeed be speculated that GlyR channels that are not open by glycine become targets of ubiquitination and thus can no longer associate with postsynaptic gephyrin, leading to receptor disapppearrance from synapses and intracellular accumulation. Interestingly, the absence of (indirect) gephyrin interaction is thought to be at the origin of $GABA_AR$ intracellular accumulation in neurons of gephyrin-deficient mice[13].

4.2. Routing of newly-synthesized GlyR to synapses

The details of selective accumulation of receptors at synaptic sites are unknown. As a first step to understand how GlyR was inserted in the postsynaptic membrane, the distribution of newly-synthesized molecules has been monitored in transfected neurons devoid of β subunit. This study involved tagged α1 and α2 subunits or α1 subunit harboring the gephyrin binding site (βgb) belonging to the β subunit[21]. For all three types of subunits, newly-synthesized homomeric receptors spontaneously form micro aggregates devoid of gephyrin in the somatodendritic membrane, and are able to progressively associate with endogenous gephyrin aggregates (a phenomenon accelerated by the presence of the βgb domain in the α1 subunit). The correct accumulation within synaptic loci thus indicates that gephyrin-non associated clusters represent kinetic intermediates between molecules newly-inserted in the plasma membrane and synaptic ones. This finding has important implications: 1) the gephyrin-independent formation of GlyR clusters precedes the receptor synaptic accumulation, 2) aggregation and synaptic localization of GlyR correspond to two non-coupled mechanisms, the latter only being a biological function of gephyrin[21].

The apparently spontaneous formation of GlyR clusters is not yet understood. An observation common to heterologous expression experiments is that cell surface receptors or ion channels do not form clusters unless their normally associated scaffold partners are co-expressed. Intriguingly, in neurons, GlyR gephyrin-non associated clusters are of homogenous size suggesting that they form according to a regulated process in which a still unknown linker protein would be involved. However, the results obtained might explain two sets of observations: 1) in cultured spinal cord neurons, GlyR clusters can be observed that are not associated with inhibitory synapses during the early phase of synaptogenesis[59] in cultured pure motoneurons, GlyR and GABA$_A$R form extrasynaptic cell surface clusters not associated with gephyrin. These clusters become synaptic upon reconstitution of the inhibitory innnervation but can also in part remain in the extrasynaptic space[7].

The ability of exogenous GlyR to populate synapses in cultured neurons provided the basis for a further analysis of two aspects of the insertion of the receptor: the exocytosis site of newly-synthesized molecules and the formation of receptor clusters[56]. A strategy suitable for the selective visualization of the insertion of newly-synthesized GlyR in the plasma membrane was developed. A transient block of intracellular transport in the trans-Golgi Network (TGN) was combined with the expression of a GlyR α1 subunit bearing a thrombin-cleavable N-terminal c-myc epitope. The myc tags of cell surface GlyR could thus be trimmed off from transfected neurons prior to the analysis of the synchronous restoration of exocytosis upon released of the block. This showed that GlyR was inserted in the plasma membrane as stable clusters which initially appeared in the soma and dendritic proximal segments. These clusters subsequently redistributed over the dendritic tree during the restoration phase, by diffusion of cell surface molecules but not as a result of the exocytosis of vesicles routed to dendrites.

These results thus provided the first demonstration that GlyR postsynaptic accumulation results from a diffusion/retention mechanism. In this case, the addition of receptor clusters (and not necessarily that of isolated receptors) and their postsynaptic stabilization by gephyrin is consecutive to a non-specified exocytosis and does not result from the exocytosis of transport vesicles at postsynaptic sites. The available data imply that cell surface insertion of GlyR is constitutive and independent, in contrast to synaptic stabilization, of the availability of-, or affinity for pre-existing postsynaptic acceptor sites. This also implies that two types of targeting information only are used, respectively specifying sorting into vesicles destined for the somatic plasma membrane, and retention in postsynaptic domains. Therefore, there would be no obligatory segregation of GlyR in the TGN from other somatodendritic proteins into specific transport vesicles.

Altogether, this puts forward the key role of the retention signal and in turn the anchoring role of gephyrin (for a discussion see, ref. 56). It remains to be determined whether this model, which posits that vesicular transport is not vectorial, also holds for the formation of other dendritic receptor- or channel-enriched domains[60]. Indeed, the intriguing fact that AMPA receptors of different compositions accumulate at excitatory synapses with different apparent rates[61] provides an example of the complexity of understanding the kinetics of postsynaptic anchoring process since it may as well rely on the exocytosis site, the molecules diffusion velocities, as on the concentration of synaptic binding sites.

4.3. Diffusion properties of cell surface GlyR

To further investigate these points, the movement of receptors (individual molecules or small clusters) has been recently analyzed by the widely used single-particle tracking (SPT) approach. Receptor mobility has been monitored by videomicroscopy using latex particles (0.5-μm) linked to cell surface glycine, AMPA, or metabotropic type 5 glutamate (mGluR5) receptors via specific antibodies coated, and traces were recorded in cultured neurons expressing or not the receptors scaffolding partners[33, 62-63]. In all cases movements displays interspersed periods of high and low diffusion rates differing by one to two orders of magnitude, indicating that receptors can rapidly explore domains areas in the micrometer range or be relatively immobile. These periods are detectable using an index function (confinement index) inversely related to the surface explored by the particle per time unit. In the absence of any GlyR–gephyrin interaction, GlyR were freely diffusive more than 80% of the time, but in the presence of a GlyR–gephyrin interaction, GlyR spent 40 to 65% of the time in a state of reduced diffusion. Increased confinement indices usually arises both from a decrease in the diffusion coefficient or from a confinement by membrane-associated structures.

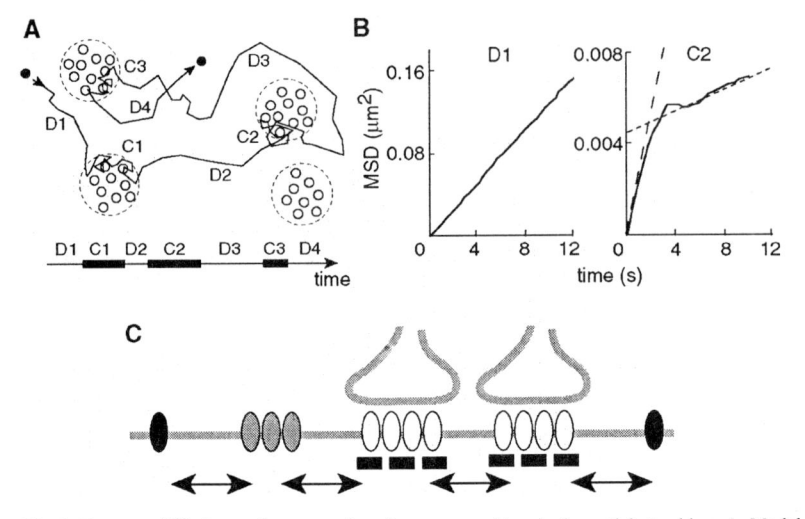

Fig. 3. Receptor diffusion at the neuronal surface measured by single particle tracking. **A**, Model of trajectory showing the exchange of receptors between dispersed (freely diffusing, D1-D4) and clustered (confined movement, C1-C3) states. Receptors can also diffuse within the clusters (encircled by dashed line).The time scale shows the alternance of of the two states. **B**, Analysis of confined and diffusive events in GlyR movement. Mean square displacement (MSD) plotted as function of time for diffusive (D1) and confined (C2) events in the trajectory shown in A. Whereas the confined MSD exhibits a characteristic negative curvature, the diffusive MSD is linear with time. Two key parameters can be extracted from these plots: the instantaneous diffusion coefficient derived from the slope at the origin (dashed line), and the asymptotic diffusion coefficient (dotted line). For the diffusive MSD, the two parameters are identical. **C**, Model for the exchange of receptors between postsynaptic domains through lateral diffusion. Modified from [22; 62; 63].

Interestingly, as shown for GlyR and mGluR5, whereas periods of fast diffusion correspond to Brownian movement those of slow mobility mainly resulted from transient association with clusters of scaffold proteins[33, 63]. It is worth noting that this association, which reduces diffusion coefficients from up to 0.5 $\mu m^2/sec$ to less than 10^{-2} $\mu m^2/sec$, imposes a restricted area of exploration (confinement) similar to that generated for other proteins by insertion into lipid rafts or by contact with protein fences or obstacles[64-66].

Short periods of reduced GlyR diffusion can exist in the absence of gephyrin, characterized a lower level of confinement than in the presence of gephyrin. As mentioned above, GlyR in neurons form small aggregates independently of an interaction with gephyrin[57]. Therefore, short periods of reduced diffusion with low levels of confinement may correspond to transition of GlyR from being freely diffusing to being aggregated but not gephyrin associated[33]. In experiments performed on GlyR-transfected cells (neuronal or not), receptor-enriched domains were obtained that were comparable to postsynaptic clusters found in mature synapses. The fact that the receptors are not irreversibly retained in these domains but can also exit from them is a key finding because it phenomenon sheds light on how receptor clusters are formed or modified during plastic processes[67]. Furthermore, it favors the new notion that any cluster of gephyrin can behave as donor or/and acceptor of GlyR. Several data on AMPA and GABAA receptors[68] had suggested that such property could underlie the dramatic changes of receptor amounts during plastic phenomenons.

Indeed, AMPAR and mGluR5 behave similarly to GlyR with respect to synaptic loci or to their own scaffolding proteins[62,63]. According to this concept, if GlyR can exit the postsynaptic density and recombinant gephyrin clusters at similar rates, the dynamic properties of the GlyR–gephyrin interaction are the basis for the regulation of the number of synaptic receptors. Therefore, a receptor escaping synaptic confinement might either be exchanged between synapses, or diffuse outside postsynaptic densities to enter the constitutive endocytotic pathway, thus contributing to receptor global turnover. An important issue will be to delineate how the stabilizing scaffold molecules through their local density and their affinity(ies) for receptors can regulate these transitions, since these mechanisms are likely to operate both in synaptogenesis and synaptic plasticity[67].

4.4. Local synthesis and diffusion process

Local insertion of locally newly-synthesized receptors is also a mean for changing receptors number. mRNA localization is a general mechanism for targeting proteins to particular subcellular domains, and the translation of localized mRNA is often regulated, allowing the possibility of local control of gene expression (ref in 69). The discovery of the dendritic localization of mRNAs encoding for synaptic proteins, as well as the fact that the synaptic activity can trigger the dendritic translocation of certain mRNAs, supported the view that these mechanisms could play a role in synaptic plasticity. Indeed, it was also shown that mRNAs encoding certain neurotransmitter receptors, including GlyR, can be targeted to dendrites[43,70,71] and detected near synaptic sites[43]. Furthermore, a micro-secretory apparatus that may participate in the translation of mRNAs encoding transmembrane proteins, such as synaptic neurotransmitter receptors is present in the sub-synaptic cytoplasm[72].

In agreement with these data, it was demonstrated that dendrites and isolated synapses could undergo protein translation and glycosylation (ref in 69), and synapse formation in the absence of cell bodies requires protein synthesis[73]. Interestingly, experimental protocols leading to LTP of hippocampal synapses were shown to induce the synaptic targeting and the local translation of some mRNAs, and inhibition of protein synthesis disrupts long term synaptic plasticity[74,75]. Local synthesis and insertion of receptors in the vicinity of synapse, could be at the origin of local change in the concentration of receptors in the dendritic membrane. This dendritic phenomenon is not different in nature from the synthesis at the somatic level, but its main functional characteristic is that it may allow local regulations of receptor synthesis and insertion by afferent innervations.

5. CONCLUDING REMARKS

The regulation of receptor number in the postsynaptic differentiation at inhibitory sypapses relies on a set of molecules which comrise the PSD and form a subsynaptic scaffold.The proteins identified on the basis of their interaction with gephyrin, are components of this scaffold, however they are not systematicaly detected at inhibitory synapses. This indicates a variability in the composition of the PSD from one synapse to the other, and may reflect differences in function. The molecular assembly of regulatory proteins within the PSD, therefore, extends to inhibitory synapses the concept of a structure functionning as a molecular machine, an idea initialy postulated for ribosomes, the proton pump and more recently adapted to excitatory synapses[76]. Gephyrin is central to the assembly and activity of this multimolecular structure. Somehow the biological activity of this nanomachine is to regulate the exchange of receptor beween synapti and non synaptic compartments of the neuronal surface[67]. This exchange function operates through diffusion properties of the receptors and the presence of specific binding sites responsible for confinement. This molecular machine also contain the clockwork for the regulation of receptor synthesis and fate by activity.

Many other components of the inhibitory PSDs are still to be discovered. The powerfull tools offered by proteomic approaches are likely to reveval a mechanistic view of the intricate network of interactions necessary for inhibitory receptor number regulation during development and plastic adaptation. This approach must be combined with the tracking of mobile receptors out, around and in synapses in relation to their associations with partners within the scaffold and with respect to synaptic activity. The ultimate goal will be to see the molecular machine in action and to monitor in real time the molecular interactions in relation with synaptic physiology.

6. REFERENCES

1. Moss, S., and Smart, T., (2001) Constructing inhibitory synapses, *Nat Rev Neurosci.* **2**, 240.
2. Triller, A., Cluzeaud, F., Pfeiffer, F., Betz, H. and Korn, H., (1985) Distribution of glycine receptors at central synapses: an immunoelectron microscopy study, *J Cell Biol.* **101**: 683.
3. Nusser, Z., Sieghart, W., and Somogyi, P., (1998) Segregation of different GABAA receptors to synaptic and extrasynaptic membranes of cerebellar granule cells, *J Neurosci.* **18**:1693
4. Sotelo, C., and Triller A., (2002) The central neuron. In *Grenfield's Neuropathology*, ed., Graham, D., and Lantos, P., Arnold publisher (London, NY, New Delhi), pp1-74.
5. Craig, A. M., and Boudin, H., (2001) Molecular heterogeneity of central synapses: afferent and target regulation. *Nat Neurosci.* **4**, 569.
6. Maxwell, D. J., Todd, A. J., and Kerr, R., (1995) Colocalization of glycine and GABA in synapses on spinomedullary neurons. *Brain Res.* **690**: 127.
7. Levi, S., Chesnoy-Marchais, D., Sieghart, W., and Triller, A., (1999) Synaptic control of glycine and GABA(A) receptors and gephyrin expression in cultured motoneurons, *J Neurosci.* **19**: 7434.
8. Pfeiffer, F., Graham, D., and Betz, H., (1982) Purification by affinity chromatography of the glycine receptor of rat spinal cord, *J. Biol. Chem.* **257**: 9389.
9. Schmitt, B., Knaus, P., Becker, C. M., and Betz, H., (1987) The Mr 93,000 polypeptide of the postsynaptic glycine receptor complex is a peripheral membrane protein, *Biochemistry.* **26**: 805.
10. Kirsch, J., Wolters, I., Triller, A., and Betz, H., (1993) Gephyrin antisense oligonucleotides prevent glycine receptor clustering in spinal neurons, *Nature* **366**:745.
11. Feng, G., Tintrup, H., Kirsch, J., Nichol, M. C., Kuhse, J., Betz, H., and Sanes, J., (1998) Dual requirement for gephyrin in glycine receptor clustering and molybdoenzyme activity, *Science.* **282**: 1321-1324.
12. Essrich, C., Lorez, M., Benson, J. A., Fritschy, J. M. and Luscher, B., (1998) Postsynaptic clustering of major GABAA receptor subtypes requires the gamma 2 subunit and gephyrin. *Nat Neurosci.* **1**, 563.
13. Kneussel, M., Brandstatter, J., Laube, B., Stahl, S., Muller, U., and Betz, H., (1999a) Loss of postsynaptic GABA(A) receptor clustering in gephyrin-deficient mice, *J. Neurosci.* **19**: 9289.
14. Wang, H., Bedford, F. K., Brandon, N. J., Moss, S. J., and Olsen, R. W., (1999) GABA(A)-receptor-associated protein links GABA(A) receptors and the cytoskeleton. *Nature* **397**, 69.
15. Kneussel, M. and, Betz, H., (2000) Clustering of inhibitory neurotransmitter receptors at developing postsynaptic sites: the membrane activation model, *Trends Neurosci.* **23**:429.
16. Kittler, J. T. *et al.*, (2001) The subcellular distribution of GABARAP and its ability to interact with NSF suggest a role for this protein in the intracellular transport of GABA(A) receptors. *Mol Cell Neurosci.* **18**: 13.
17. Kirsch, J., Langosch, D., Prior, P., Littauer, U., Schmitt, B., and Betz, H., (1991) The 93-kDa glycine receptor-associated protein binds to tubulin, *J. Biol. Chem.* **266**: 22242.
18. Meyer, G., Kirsch, J., Betz, H. and Langosch, D., (1995) Identification of a gephyrin binding motif on the glycine receptor beta subunit, *Neuron.* **15**: 563.

19. Kneussel, M., Hermann, A., Kirsch, J., and Betz, H., (1999b) Hydrophobic interactions mediate binding of the glycine receptor beta- subunit to gephyrin, *J. Neurochem.* **72:** 1323.
20. Prior, P., Schmitt, B., Grenningloh, G., Pribilla, I., Multhaup, G., Beyreuther, K., Maulet, Y., Werner, P., Langosch, D., Kirsch, J., and Betz, H., (1992) Primary structure and alternative splice variants of gephyrin, a putative glycine receptor-tubulin linker protein, *Neuron.* **8:** 1161.
21. Ramming, M., Kins, S., Werner, N., Hermann, A., Betz, H., and Kirsch, J., (2000) Diversity and phylogeny of gephyrin: tissue-specific splice variants, gene structure, and sequence similarities to molybdenum cofactor-synthesizing and cytoskeleton-associated proteins, *Proc Natl Acad Sci U S A.* **97**:10266.
22. Meier, J., De Chaldee, M., Triller, A., and Vannier, C., (2000a) Functional heterogeneity of gephyrins, *Mol Cell Neurosci.* **16**:566.
23. Kamdar, K. P., Shelton, M. E., and Finnerty, V., (1994) The Drosophila molybdenum cofactor gene cinnamon is homologous to three Escherichia coli cofactor proteins and to the rat protein gephyrin, *Genetics.* **137:** 791.
24. Stallmeyer, B., Nerlich, A., Schiemann, J., Brinkmann, H., and Mendel, R., (1995) Molybdenum co-factor biosynthesis: the Arabidopsis thaliana cDNA cnx1 encodes a multifunctional two-domain protein homologous to a mammalian neuroprotein, the insect protein Cinnamon and three Escherichia coli proteins, *Plant J.* **8:** 751.
25. Stallmeyer, B., Schwarz, G., Schulze, J., Nerlich, A., Reiss, J., Kirsch, J., and Mendel, R., (1999) The neurotransmitter receptor-anchoring protein gephyrin reconstitutes molybdenum cofactor biosynthesis in bacteria, plants, and mammalian cells, *Proc. Natl. Acad. Sci. U S A.* **96:** 1333.
26. Reiss J., Gross-Hardt S., Christensen E., Schmidt P., Mendel R., and Schwarz G., (2001) A mutation in the gene for the neurotransmitter receptor-clustering protein gephyrin causes a novel form of molybdenum cofactor deficiency, *Am J Hum Genet.* **68**:208.
27. Lee, HJ., Adham, IM., Schwarz, G., Kneussel, M., Sass, JO., Engel, W., and Reiss, J., (2002) Molybdenum cofactor-deficient mice resemble the phenotype of human patients, *Hum Mol Genet.* **11**:3309
28. Liu, M., Wuebbens, M., Rajagopalan, K., and Schindelin, H., (2000) Crystal structure of the gephyrin-related molybdenum cofactor biosynthesis protein MogA from Escherichia coli, *J Biol Chem.* **275**:1814.
29. Sola, M., Kneussel, M., Heck, I., Betz, H., and Weissenhorn, W., (2001) X-ray crystal structure of the trimeric N-terminal domain of gephyrin, *J Biol Chem.* **276**:25294.
30. Xiang, S., Nichols, J., Rajagopalan, K., and Schindelin, H., (2001) The crystal structure of Escherichia coli MoeA and its relationship to the multifunctional protein gephyrin, *Structure.* **9**:299.
31. Schwarz, G., Schrader, N., Mendel, R., Hecht, H., and Schindelin, H., (2001) Crystal structures of human gephyrin and plant Cnx1 G domains: comparative analysis and functional implications, *J Mol Biol.* **312**:405.
32. Kneussel, M. and Betz, H., (2000) Clustering of inhibitory neurotransmitter receptors at the developing postsynaptic sites : the membrane activation model. *Trends Neurosci.* **23** (9) : 429-435.
33. Meier, J., Vannier, C., Serge, A., Triller, A., and Choquet, D., (2001) Fast and reversible trapping of surface glycine receptors by gephyrin, *Nat Neurosci.* **4:** 253.
34. Lee, G., Cowan, N., and Kirschner, M., (1988) The primary structure and heterogeneity of tau protein from mouse brain, *Science.* **239**:285.
35. Lewis S., Wang D., and Cowan N., (1988) Microtubule-associated protein MAP2 shares a microtubule binding motif with tau protein, *Science* **242**:936.
36. Kins, S., Betz, H., and Kirsch, J., (2000) Collybistin, a newly identified brain-specific GEF, induces submembrane clustering of gephyrin, *Nat Neurosci.* **3**:22.
37. Sabatini, D., Barrow, R., Blackshaw, S., Burnett, P., Lai, M., Field, M., Bahr, B., Kirsch, J., Betz, H., and Snyder, S., (1999) Interaction of RAFT1 with gephyrin required for rapamycin-sensitive signaling, *Science.* **284**:1161.
38. Fuhrmann, J., Kins, S., Rostaing, P., El Far, O., Kirsch, J., Sheng, M., Triller, A., Betz, H., and Kneussel, M., (2002) Gephyrin interacts with Dynein light chains 1 and 2, components of motor protein complexes, *J Neurosci.* **22**:5393.
39. Reid, T., Bathoorn, A., Ahmadian, M., and Collard, J., (1999) Identification and characterization of hPEM-2, a guanine nucleotide exchange factor specific for Cdc42, *J Biol Chem.* **274**:33587.
40. Bishop, A., and Hall, A., (2000) Rho GTPases and their effector proteins, *Biochem J.* **348**:241.
41. Cullen, P., and Chardin, P., (2000) Membrane targeting: what a difference a G makes, *Curr Biol.* **10**:876.
42. Grosskreutz, Y., Hermann, A., Kins, S., Fuhrmann, J., Betz, H., and Kneussel, M., (2001) Identification of a gephyrin-binding motif in the GDP/GTP exchange factor collybistin, *Biol Chem.* **382**:1455.
43. Racca, C., Gardiol, A., and Triller, A., (1977) Dendritic and postsynaptic localizations of glycine receptor alpha subunit mRNAs, *J Neurosci.* **17**:1691.
44. Setou, M., Nakagawa, T., Seog, DH., and Hirokawa, N., (2000) Kinesin superfamily motor protein KIF17 and mLin-10 in NMDA receptor-containing vesicle transport, *Science.* **288**:1796.
45. Naisbitt, S., Valtschanoff, J., Allison, D., Sala, C., Kim, E., Craig, AM., Weinberg, R., and Sheng, M., (2000) Interaction of the postsynaptic density-95/guanylate kinase domain-associated protein complex with a light chain of myosin-V and dynein, *J Neurosci.* **20**:4524.
46. Mammoto, A., Sasaki, T., Asakura, T., Hotta, I., Imamura, H., Takahashi, K., Matsuura, Y., Shirao, T., and Takai, Y., (1998) Interactions of drebrin and gephyrin with profilin, *Biochem. Biophys. Res. Commun.* **243:** 86.

47. Schluter, K., Jockusch, B., and Rothkegel, M., (1997) Profilins as regulators of actin dynamics, *Biochim Biophys Acta.* **1359**:97.
48. Witke, W., Podtelejnikov, A., Di Nardo, A., Sutherland, J., Gurniak, C., Dotti C., and Mann M., (1998) In mouse brain profilin I and profilin II associate with regulators of the endocytic pathway and actin assembly, *EMBO J.* **17**:967.
49. Jeffery C., (1999) Moonlighting proteins, *Trends Biochem Sci.* **24**:8.
50. Allison, D., Chervin, A., Gelfand, V., and Craig, AM., (2000) Postsynaptic scaffolds of excitatory and inhibitory synapses in hippocampal neurons: maintenance of core components independent of actin filaments and microtubules, *J Neurosci.* **20**:4545.
51. Kirsch, J., and Betz, H., (1995) The postsynaptic localization of the glycine receptor-associated protein gephyrin is regulated by the cytoskeleton, *J. Neurosci.* **15**: 4148.
52. Kirsch, J. and Betz, H., (1998) Glycine-receptor activation is required for receptor clustering in spinal neurons. *Nature* **392**: 717-720.
53. Levi, S., Vannier, C., and Triller, A., (1998). Strychnine-sensitive stabilization of postsynaptic glycine receptor clusters, *J Cell Sci.* **111**: 335-345.
54. Rasmussen, H., Rasmussen, T., Triller, A., and Vannier, C., (2002) Strychnine-blocked glycine receptor is removed from synapses by a shift in insertion/degradation equilibrium, *Mol Cell Neurosci.* **19**:201.
55. Ehlers, M., 2000, Reinsertion or degradation of AMPA receptors determined by activity-dependent endocytic sorting, *Neuron.* **28**: 511.
56. Rosenberg, M., Meier, J., Triller, A., and Vannier, C., (2001) Dynamics of glycine receptor insertion in the neuronal plasma membrane, *J Neurosci.* **21**: 5036.
57. Meier, J., Meunier-Durmort, C., Forest, C., Triller, A., and Vannier, C., (2000b) Formation of glycine receptor clusters and their accumulation at synapses, *J Cell Sci.* **113**: 2783.
58. Buttner, C., Sadtler, S., Leyendecker, A., Laube, B., Griffon, N., Betz, H., and Schmalzing, G., (2001) Ubiquitination precedes internalization and proteolytic cleavage of plasma membrane-bound glycine receptors, *J Biol Chem.* **276**:42978.
59. Dumoulin, A., Rostaing, P., Bedet, C., Levi, S., Isambert, M., Henry, JP., and Triller, A., Gasnier, B., (1999) Presence of the vesicular inhibitory amino acid transporter in GABAergic and glycinergic synaptic terminal boutons, *J Cell Sci.* **112**: 811.
60. Fanning, A., and Anderson, J., (1999) Protein modules as organizers of membrane structure, *Curr. Opin. Cell Biol.* **11**: 432.
61. Passafaro, M., Piech, V., and Sheng, M., (2001) Subunit-specific temporal and spatial patterns of AMPA receptor exocytosis in hippocampal neurons, *Nat Neurosci* **4**, 917.
62. Borgdorff, A., and Choquet, D., (2002) Regulation of AMPA receptor lateral movements, *Nature* **417**:649.
63. Serge, A., Fourgeaud, L., Hemar, A., and Choquet, D., (2002) Receptor activation and homer differentially control the lateral mobility of metabotropic glutamate receptor 5 in the neuronal membrane, *J. Neurosci.* **22**:3910.
64. Dietrich, C., Yang, B., Fujiwara, T., Kusumi, A. and Jacobson, K. (2002) Relationship of lipid rafts to transient confinement zones detected by single particle tracking. *Biophys J* **82**: 1, 274-284.
65. Kusumi, A., and Sako, Y., (1996) Cell surface organization by the membrane skeleton, *Curr Opin Cell Biol* **8**: 566.
66. Daumas, F., Destainville, N., Millot, C., Lopez, A., Dean, D., and Salome, L., (2003) Confined diffusion without fences of a G-protein-coupled receptor as revealed by single particle tracking, *Biophys J.* **84**:356.
67. Choquet D., and Triller, A., (2003) The role of receptor diffusion in the organization of the postsynaptic membrane. Nat Rev Neurosci. **4**: *in press.*
68. Kittler, J., and Moss, S., (2001) Neurotransmitter receptor trafficking and the regulation of synaptic strength, *Traffic* **2**:437.
69. Steward, O. and Schuman, E.M., (2001) Protein synthesis at synaptic sites on dendrites, *Annu Rev Neurosci.* **24**: 299.
70. Gazzaley, A.H., Benson, D.L., Huntley, G.W. and Morrison J.H., (1997) Differential subcellular regulation of NMDAR1 protein and mRNA in dendrites of dentate gyrus granule cells after perforant path transection, *J Neurosci.* **17**: 2006.
71. Steward, O. (2002) mRNA at synapses, synaptic plasticity, and memory consolidation. *Neuron* **36**, 338.
72. Gardiol, A., Racca, C., and Triller, A.,(1999) Dendritic and postsynaptic protein synthetic machinery, *J Neurosci.* **19**:168.
73. Schacher, S., Wu, F., (2002) Synapse formation in the absence of cell bodies requires protein synthesis, *J Neurosci.* **22**:1831.
74. Kang, H., and Schuman, E.M., (1996) A requirement for local protein synthesis in neurotrophin-induced hippocampal synaptic plasticity, *Science* **273**: 1402.
75. Martin, K.C., Barad, M., and Kandel, E.R., (2000) Local protein synthesis and its role in synapse-specific plasticity, *Curr Opin Neurobiol.* **10**: 587.
76. Kennedy, M.B., (2000) Signal-processing machines at the postsynaptic density, *Science* **290**: 750.

4

LONG-TERM MODIFICATION AT VISUAL CORTICAL INHIBITORY SYNAPSES

Yukio Komatsu and Yumiko Yoshimura[*]

1. INTRODUCTION

Selective responsiveness of visual cortical cells is shaped by visual experience during the postnatal critical period[1-2]. It is thought that experience modifies the responsiveness by producing enduring changes in the strength of cortical synapses activity-dependently. At first, only excitatory synapses were considered to be modifiable and they were examined in slice preparations. In fact, multiple types of long-term plasticity were demonstrated at excitatory synapses of visual cortical cells[3-4]. The postsynaptic membrane potential level is a crucial factor in the determination of the initiation and direction of plastic changes similarly in excitatory synaptic transmission[5-7] and visual responses[8-9]. On the other hand, it was almost implicitly considered that inhibitory synapses themselves did not undergo long-term modification physiologically, and that their role in synaptic plasticity was to regulate the modification of excitatory synaptic strength by modulating postsynaptic depolarization associated with excitatory inputs.

Visual responsiveness of cortical cells emerges from the spatiotemporal integration of excitatory and inhibitory synaptic inputs. Unless these two types of connections are well balanced, visual stimulation may produce excessive or poor responses in cortical cells. If only excitatory synapses on pyramidal cells are plastic, it may be not so easy to establish balanced excitatory and inhibitory connections producing an adequate level of visual responses, which are selective for various stimulus features. This suggests that plastic changes could occur at inhibitory synapses on pyramidal cells and/or excitatory/inhibitory synapses on inhibitory interneurons. Because neocortical neural connections are very complicated, it is technically difficult to study these synapses in isolation. Although little is known about synaptic plasticity in inhibitory interneurons, we have shown that inhibitory synapses on pyramidal cells are highly modifiable.

---

[*] Y. Komatsu and Y. Yoshimura, Department of Visual Neuroscience, Research Institute of Environmental Medicine, Nagoya University, Nagoya 464-8601, Japan

To study plasticity at γ-aminobutyric acid (GABA)-mediated inhibitory synapses, inhibitory postsynaptic potentials (IPSPs) evoked by layer 4 stimulation were recorded from layer 5, presumably pyramidal, cells in rat visual cortical slices under a blockade of excitatory synaptic transmission using a non-NMDA receptor antagonist DNQX and an NMDA receptor antagonist APV (Fig. 1). The description below deals mainly with long-term potentiation (LTP) and partly long-term depression (LTD) at these inhibitory synapses in comparison with NMDA receptor-dependent LTP and LTD at excitatory synapses of hippocampal CA1 pyramidal cells, which have been studied most thoroughly.

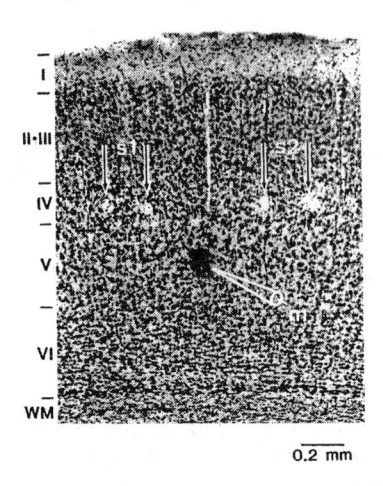

Fig. 1. Analysis of inhibitory synaptic plasticity in rat visual cortical slices. The experimental arrangement of stimulating (s1 and s2) and recording electrodes (m) illustrated on a histological section stained with cresyl violet. Laminar boundaries are indicated to the left of the section. Layers 2-4 between two stimulating electrodes were surgically cut to ensure that they activated separate groups of fibers. Test stimulation was applied alternately to s1 and s2 at intervals of 5 s. As a conditioning stimulation, HFS (50 Hz) was applied for 1 sec ten times at intervals of 10 sec to one of the stimulating electrodes. The other stimulating electrode served as a control. The intensity of test stimulation was adjusted to a weak value, 1.2-1.5 times the threshold intensity evoking IPSPs (1.2-1.5T). The intensity of HFS was usually adjusted to a strong intensity (5T), which consistently produced LTP at inhibitory synapses in young rats. To record IPSPs isolated from excitatory postsynaptic potentials, 40 μM 6,7-dinitroquinoxaline-2,3-dione (DNQX), a non-NMDA receptor antagonist, and 100 μM DL-2-amino-5-phosphonovaleric acid (APV) were added to the extracellular solution. Experiments were conducted using developing rats at postnatal 20-30 days, except for the age dependence study. Taken from Komatsu[10].

2. BIDIRECTIONAL MODIFICATION AT INHIBITORY SYNAPSES

Inhibitory synaptic transmission could undergo LTP and LTD depending on the pharmacological manipulation when high-frequency stimulation (HFS) was applied to presynaptic fibers, as shown in Fig. 2[11]. Under the blockade of non-NMDA and NMDA receptors, HFS, eliciting hyperpolarizing responses in the postsynaptic cell, induced LTP of IPSPs (Fig. 2A). When the same HFS was applied while the NMDA receptor antagonist APV was removed, it elicited hyperpolarizing or small depolarizing responses (Fig. 2B). After this stimulation, LTP also occurred at inhibitory synapses. In this situation, HFS likely failed to activate NMDA receptors because the $GABA_A$ receptor activation prevents depolarization. However, when $GABA_A$ receptors were blocked by bicuculline methiodide (BMI) together with the removal of APV, HFS elicited large depolarizing responses (Fig. 2C). This induced LTD of IPSPs instead. Because of the difficulty involved in washing out the non-NMDA receptor antagonist, we have not yet tested whether HFS produces LTP or LTD at inhibitory synapses, when it is applied without any blockers. However, it is likely that the stimulation produces LTD, since such stimulation usually elicits large depolarizing responses in pyramidal cells in a normal solution. To test this possibility, we recorded unitary IPSPs from pyramidal cells in response to stimulation of single presynaptic inhibitory cells in a solution containing no glutamate receptor antagonists and found that HFS induced LTD of IPSPs (Yoshimura et al., unpublished observation).

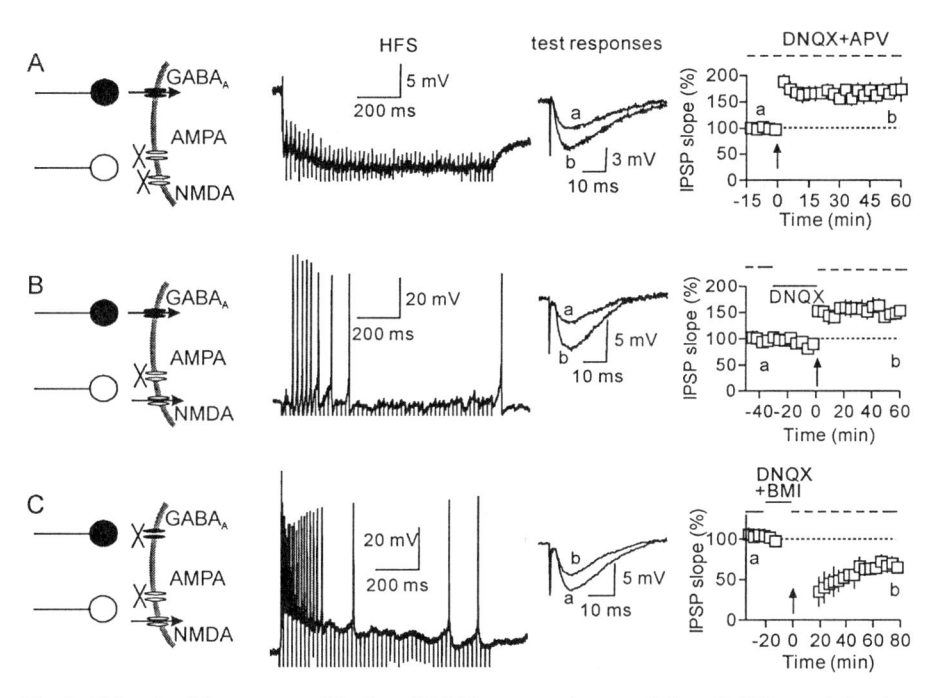

Fig. 2. Bidirectional long-term modification of inhibitory synaptic transmission. A) LTP was induced at inhibitory synapses when HFS was applied under the blockade of non-NMDA and NMDA receptors. B) LTP was induced at inhibitory synapses when HFS was applied under the blockade of non-NMDA receptors. C) LTD was induced at inhibitory synapses when HFS was applied under the blockade of non-NMDA and $GABA_A$ receptors (BMI, 20 µM). In A-C, the left and middle traces show responses evoked by HFS and superimposed test responses before (a) and after HFS (b), respectively. The time course of the LTP or LTD is plotted in the right graph. HFS was applied at time 0 (arrows). HFS elicited hyperpolarizing responses in A, small depolarizing or hyperpolarizing responses in B and large depolarizing responses in C.

These results suggest that the direction of changes in the strength of inhibitory synapses is determined by the level of postsynaptic depolarization during HFS. When HFS produces postsynaptic depolarization sufficient to activate NMDA receptors, it induces inhibitory LTD. On the other hand, when it fails to produce depolarization necessary to activate NMDA receptors, it produces inhibitory LTP instead. Thus, the NMDA receptor seems to determine the direction of changes in inhibitory synaptic strength. This is similar to the case in excitatory synapses of CA1 pyramidal cells[12-13]. However, a clear difference is present between these two synapses. At the excitatory synapses, LTD requires NMDA receptor activation weaker than that required for LTP induction. LTP at inhibitory synapses requires no NMDA receptor activation. Similar bidirectional modification was demonstrated at GABAergic synapses on neonate CA3 pyramidal cells[14].

3. PROPERTIES OF LTP AT INHIBITORY SYNAPSES

Visual cortical inhibitory LTP has properties similar to those for LTP at excitatory synapses of CA1 pyramidal cells[10]. The common properties are input specificity, associativity, quick induction and long maintenance, which may be important properties for information storage. The intensity of HFS has to exceed a threshold value to induce LTP, indicating that LTP is induced only when more than a threshold number of presynaptic fibers are activated simultaneously. This may provide associativity of input fibers to induce LTP as shown for NMDA receptor-dependent excitatory LTP[15-16]. Indeed, weak HFS, which alone failed to induce LTP, induced LTP when the same HFS was applied together with strong HFS to separate presynaptic fibers converging to the same cell.

Although these basic properties are common, the stimulation frequency capable of inducing LTP is different[10]. At CA1 excitatory synapses, brief HFS (10-100 Hz) induced LTP while low-frequency stimulation (a few Hz) continued for a longer period induced LTD[13]. At visual cortical inhibitory synapses, both types of stimulation could induce LTP. Another important difference is age-dependence. CA1 excitatory LTP is robustly induced

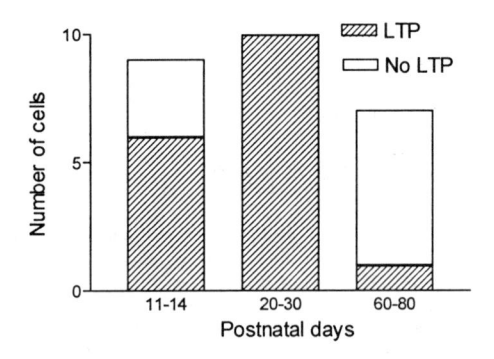

Fig. 3. Age-dependence of LTP at inhibitory synapses. Incidence of LTP was determined with strong (5 T) HFS.

in matured animals[16]. In contrast, visual cortical inhibitory LTP is induced more easily during development than in adulthood (Fig. 3). $GABA_A$ receptor-mediated postsynaptic potentials first appear around the time of eye opening. The incidence of LTP was considerably high at this time, highest during the critical period of visual response modification and low in adulthood. This age dependence suggests that this LTP is involved in experience-dependent development of visual cortical functions. However, when very strong HFS was applied, LTP occurred in a part of the cells tested, indicating that modifiability remains in a restricted degree even in mature cortex. In addition to these differences, the maintenance of LTP at inhibitory synapses is very different from other types of LTP, as described below[18].

4. INDUCTION MECHANISM OF LTP

The induction mechanism of LTP at visual cortical inhibitory synapses examined so far[19] is schematically illustrated in Fig 4. LTP induction requires a postsynaptic elevation of

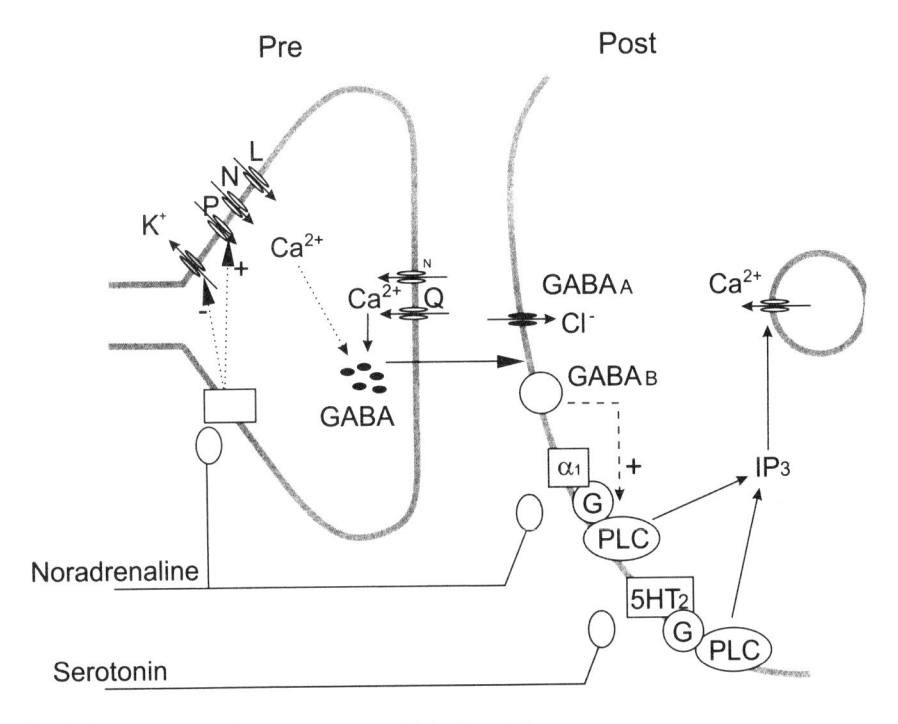

Fig. 4. Induction and maintenance mechanism of inhibitory LTP.

Ca^{2+}, as is the case in most long-term synaptic plasticity, because postsynaptic loading of Ca^{2+} chelator BAPTA completely abolished LTP production. In our experimental conditions, where non-NMDA and NMDA receptors were completely blocked, LTP was induced by HFS applied under voltage clamp at either hyperpolarized (-90 mV) or depolarized membrane potentials (+20 to +40 mV). Furthermore, Ca^{2+} entering through NMDA receptors is likely to contribute to initiating LTD[11]. This suggests that the source of Ca^{2+} responsible for the induction originates from the internal Ca^{2+} store, but not outside of the cells, in contrast with most synaptic plasticity, which requires Ca^{2+} entry[20, 12]. Indeed, postsynaptic loading of any of the following agents, phospholipase C (PLC) inhibitor U73122, G-protein inhibitor GDPβS and IP_3 receptor blocker heparin, abolished LTP production. This suggests that the following postsynaptic biochemical cascade is involved in LTP induction: activation of G protein coupled receptors, activation of PLC, production of IP_3 and IP_3 receptor-mediated Ca^{2+} release from the internal Ca^{2+} store (Fig. 4).

In addition, GABAergic synaptic transmission is necessary to induce LTP (Fig. 5). Although IPSPs evoked by a single test stimulation applied at a low frequency is mediated only by $GABA_A$ receptors, LTP induction requires the activation of $GABA_B$, but not $GABA_A$, receptors. $GABA_B$ receptors are known to couple to adenylate cyclase,

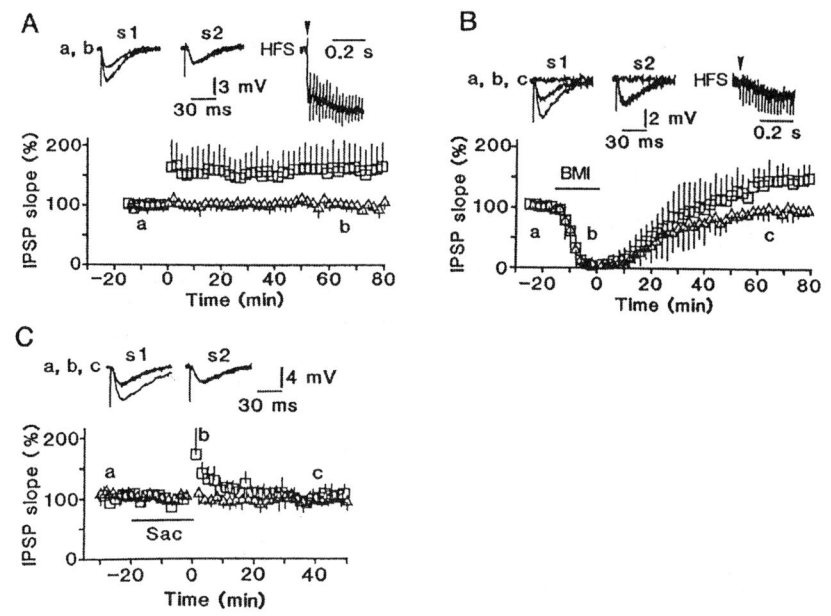

Fig. 5. LTP induction requires activation of $GABA_B$ but not $GABA_A$ receptors. A) Control LTP. B) HFS was applied in the presence of a high-dose of $GABA_A$ receptor antagonist BMI (20 μM). C) HFS was applied in the presence of $GABA_B$ receptor antagonist 2-Hydroxysaclofen (Sac). Squares and triangles represent test and control pathway responses, respectively. HFS was applied at time 0. Taken from Komatsu[19].

but not PLC, via G proteins in cerebral cortex[21]. A blockade of metabotropic glutamate receptors coupled with IP_3 formation did not affect LTP induction. However, it is known that $GABA_B$ receptor activation facilitates α_1 adrenoceptor-mediated IP_3 formation[21-22]. Therefore, we examined the involvement of aminergic receptors coupled with IP_3 formation and found that activation of α_1 adrenoceptors and serotonin $5-HT_2$ receptors was required for LTP induction. Thus, we consider, tentatively, that HFS strongly activates $GABA_B$ receptors, which facilitates α_1 adrenoceptor- and/or $5-HT_2$ receptor-mediated IP_3 formation and produces IP_3 sufficient for IP_3 receptor-mediated

Ca^{2+} release (Fig. 4). If there is a threshold in some steps of this cascade, it may explain the associativity of LTP induction. To substantiate this supposition, further studies may be necessary to investigate the relationship between $GABA_B$ receptor activation and IP_3 formation.

5. MAINTENANCE MECHANISM OF LTP

Neural activity producing a transient increase in intracellular Ca^{2+} concentration can initiate long-term modification of synaptic strength. It has been thought that modified synaptic strength is maintained independent of neural activity. In regard to NMDA receptor-dependent LTP, it has been proposed that a temporary Ca^{2+} elevation mediated by NMDA receptor activation produces autophosphorylation of CaM kinase II, converting the enzyme into a Ca^{2+}-independnt active form, which phosphorylates its substrates persistently[23]. This biochemical process may maintain potentiation long-lastingly without any following neural activity. However, we have found that inhibitory LTP requires neural activity for maintenance, although the necessary frequency is low[18].

In our studies of inhibitory LTP, the extracellular solution contained a concentration of Ca^{2+} (4 mM) higher than those we usually used (2.4 mM), because LTP was consistently produced in solutions with a high concentration of Ca^{2+}. The incidence of LTP was highly dependent on that concentration (Fig. 6). LTP occurred rarely at a normal

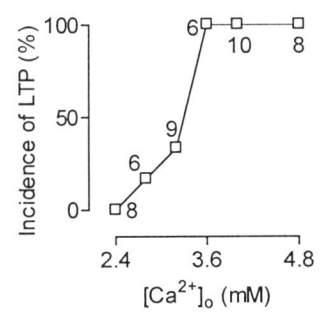

Fig. 6. LTP incidence is highly dependent on extracellular Ca^{2+} concentration. The figures attached to the symbols indicate the number of tested cells. Taken from Komatsu and Yoshimura[18].

value (2.4 mM), more frequently with the increase in the concentration and always at 3.6-4.8 mM. A high Ca^{2+} concentration was required for maintenance, but not induction, because HFS, applied under normal Ca^{2+} concentration, produced LTP if high Ca^{2+} concentration was resumed immediately after HFS (Fig. 7A). On the other hand, LTP was abolished when the concentration was temporary reduced to the normal value (Fig. 7B), suggesting that the Ca^{2+} entry associated with the test stimulation is required to maintain LTP.

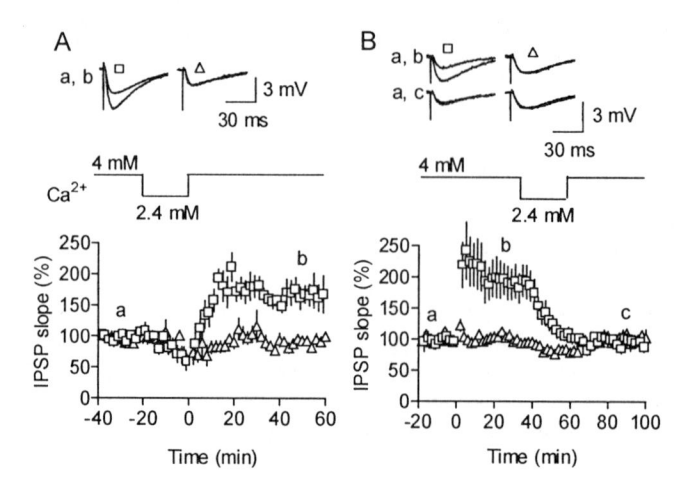

Fig. 7. A high Ca^{2+} concentration is required for maintenance, but not induction of inhibitory LTP. Taken from Komatsu and Yoshimura[18].

This supposition was tested by stopping test stimulation for 30 minutes after LTP induction. LTP was maintained more than 3 hours, as long as the test stimulation was continued at 0.1 Hz. LTP was abolished after temporal cessation of test stimulation in about two thirds of the cells when the cessation was started immediately, 30 or 120 minutes after HFS. LTP persisted even after the test stimulation cessation was abolished with a temporary application of Na^+ channel blocker saxitoxin. Therefore, we consider that test stimulus-evoked and spontaneous action potentials both contribute to the maintenance of LTP.

To test whether pre or postsynaptic activity is required, postsynaptic responses were completely blocked by GABA receptor antagonists after LTP induction. LTP persisted after these blockers were washed out, indicating that LTP maintenance requires presynaptic, but not postsynaptic, neural activity. We have confirmed that presynaptic inhibitory cells fire spontaneously even under our experimental conditions, where excitatory synaptic transmission was completely blocked pharmacologically. Bath application of the Na^+ channel blocker tetrodotoxin reduced the frequency of spontaneous inhibitory postsynaptic currents (IPSCs) recorded from layer 5 pyramidal cells by about 7 Hz on average, which may correspond to the frequency of IPSCs resulted from spontaneous firing of presynaptic inhibitory cells.

It is likely that the effect of the presynaptic activity necessary for LTP maintenance is mediated by multiple types of high-threshold voltage-gated Ca^{2+} channels (Fig. 4). Transmitter release at these inhibitory synapses was mediated mostly by Q type Ca^{2+} channels, and partially by N type Ca^{2+} channels (Fig. 8C-D). LTP maintenance was completely blocked by a temporary pharmacological blockade of either P, L or N type Ca^{2+} channels, while it was not affected by a blockade of T or R type Ca^{2+} channels (Fig. 8A-E). We could not draw any conclusions about Q type Ca^{2+} channels, because the blockade of those channels completely abolished IPSPs. In addition, a brief application of

PLASTICITY AT VISUAL CORTICAL INHIBITORY SYNAPSES

membrane permeable slow Ca^{2+} chelator EGTA-AM, which would prevent Ca^{2+} elevation at both pre and postsynaptic sites, abolished LTP without affecting basal synaptic transmission (Fig. 8F). However, LTP was maintained when EGTA was loaded into postsynaptic cells from patch electrodes. These results suggest that LTP maintenance is mediated by a slow presynaptic Ca^{2+} elevation, which activates Ca^{2+}-dependent reactions different from those triggering transmitter release.

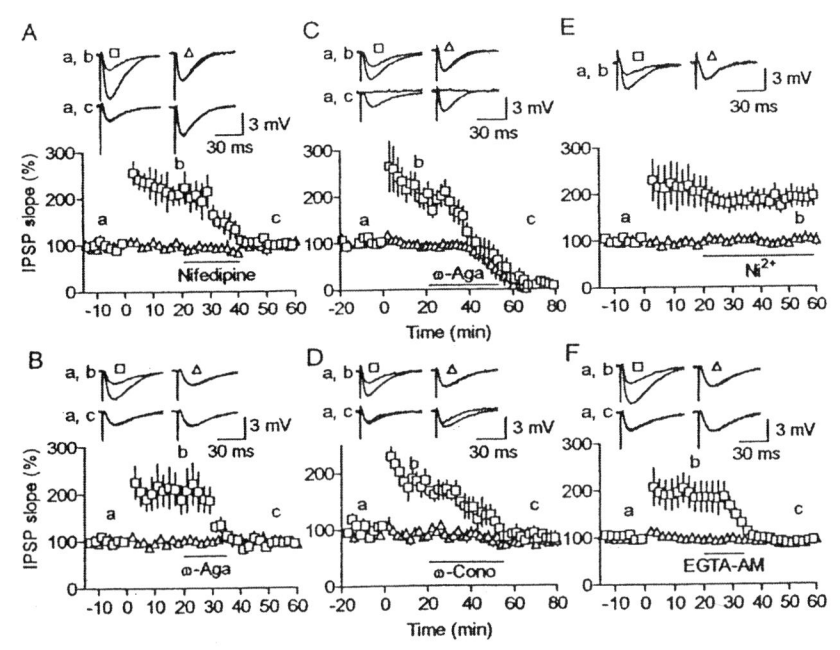

Fig. 8. Effects of Ca^{2+} channel blockers and chelators on LTP maintenance. A) Nifedipine (20 μM) was applied for the period indicated by the horizontal bar. B-F) Similar to A, but 30-50 nM (B) or 1 μM ω-Agatoxin IVA (C), 1 μM ω-Conotoxin GVIA (D), 50 μM Ni^{2+}(E), or 2 μM EGTA-AM (F) was applied instead. L (A), P (B), Q (C), N (D), and R and T type Ca^{2+} channels (E) were blocked, respectively. Taken from Komatsu and Yoshimura[18].

Even under a normal Ca^{2+} concentration, LTP was maintained in the presence of iberiotoxin, charybdotoxin or Cs^+, which are K^+ channel blockers known to block high conductance Ca^{2+}-activated K^+ channels (BK channels). Bath application of 1 mM Cs^+ did not affect baseline responses, but did facilitate LTP maintenance, whereas application of 4-aminopyridine facilitated baseline responses, but not LTP maintenance. Transmitter release is triggered by a rapid Ca^{2+} elevation in association with action potentials, while LTP seems to be maintained by a slow Ca^{2+} elevation. Thus, it is likely that BK or related K^+ channels regulate the later slow phase of Ca^{2+} elevation associated with spikes, whereas 4-amiropyridine sensitive K^+ channels modulate the early quick phase. Another possibility is that the BK or related K^+ channels and Ca^{2+} channels involved in LTP maintenance are located at nearby sites, which are yet apart from the transmitter release site (Fig. 4).

In the presence of noradrenaline (5 μM), LTP was produced under a normal Ca^{2+} concentration (Komatsu and Yoshimura, unpublished observation). When noradrenaline was washed out after HFS, LTP was still maintained under a normal Ca^{2+} concentration (Komatsu and Yoshimura, unpublished observation). This suggests that simultaneous activation of inhibitory synapses and adrenergic receptors facilitates the maintenance mechanism. Possible changes taking place in the maintenance mechanism may be up-regulation of Ca^{2+} channels, down-regulation of K^+ channels, and a reduction in sensitivity to Ca^{2+} (Fig. 4). Although it has not been investigated, some modulatory substances other than noradrenaline might terminate LTP, for example, by reducing presynaptic Ca^{2+} entry via modulation of the Ca^{2+} or K^+ channels involved in LTP maintenance. Since LTP maintenance requires presynaptic cells to continue to fire, at least at some low frequency, it is difficult to consider that this plasticity acts to store information for a very long time. Thus, it may have a rather temporary nature.

The reversibility of modified synaptic strength may provide flexibility to a cellular process functioning for information storage. At CA1 excitatory synapses, HFS (10-100 Hz)-induced LTP is reversed by a low-frequency (1-5 Hz) stimulation continued for 10-15 minutes[24-25]. This type of LTP reversal has been reported for other excitatory synapses[26-28]. In contrast, at visual cortical inhibitory synapses, LTP is induced by stimulation of a wide range (2-50 Hz) of frequencies and reversal is attained by reducing the frequency of stimulation to a level lower than that for the test stimulation. This difference suggests that the two forms of LTP operate very differently.

6. POSSIBLE FUNCTIONAL ROLES OF PLASTICITY AT INHIBITORY SYNAPSES

The production of LTP at visual cortical inhibitory synapses is strongly influenced by noradrenergic and serotonergic receptors. Since locus ceruleus and raphe cells maintain high-frequency spike activities during awake, but not sleep, states[29-31], it is likely that inhibitory LTP effectively occurs when animals are attentively looking at their visual environment. In addition, this LTP occurred mostly during development, suggesting that it could contribute to experience-dependent maturation of selective responsiveness of visual cortical cells. This idea is consistent with findings that noradrenaline and serotonin contribute to visual response plasticity of cortical cells[32-33].

Inhibitory synaptic connections could enhance selective responsiveness established by excitatory connections[34-37]. Repetitive visual stimulation, evoking depolarizing responses large enough to activate NMDA receptors of a postsynaptic cell, could produce LTD at inhibitory synapses on the cell. On the contrary, repetitive visual stimulation, evoking postsynaptic responses smaller than those necessary to activate NMDA receptors, could produce LTP at inhibitory synapses. If visual stimulation induces LTP at inhibitory synapses in a cortical cell, the stimulation evokes smaller visual responses in the cell thereafter. However, if visual stimulation induces LTD instead, the stimulation evokes larger visual responses thereafter. Thus, small visual responses get smaller, whereas large visual responses get larger, leading to the enhancement of selectivity.

Even though visual response selectivity is improved by inhibitory synaptic plasticity as postulated above, increased selectivity may be lost unless the same visual stimulation, which induced the changes, is repeated thereafter, at least at a low frequency. Indeed, orientation selectivity is improved to a level almost the same as that for adults in the middle of the critical period[2] and a following deprivation of light, even for a few days, degrades the selectivity[38]. This observation supports the idea that neural activity plays a crucial role in the maintenance of visual responsiveness, which is refined depending on experience. The modification in the strength of inhibitory synapses might be converted from the temporary to an enduring form, which is independent of activity, when those synapses are activated repetitively during the critical period. This process could require protein synthesis, as is the case in the late phase of hippocampal NMDA receptor-dependent LTP[39-40].

However, it is not so plausible that this type of conversion takes place, because it has been reported that visual response selectivity assessed from action potentials is often very similar to that assessed from excitatory postsynaptic potentials recorded under a blockade of inhibition in mature visual cortical cells[41]. To consider this issue, it is notable that the time course of the two forms of LTP at excitatory and inhibitory synapses, which both occur mostly during development[10, 42-43], is very different. Inhibitory LTP can be induced by brief HFS and develops immediately after that stimulation[10], whereas voltage-gated Ca^{2+} channel-dependent LTP at excitatory synapses is induced by low-frequency stimulation continued for a long period and develops slowly after that stimulation[44, 43]. Thus, visual stimulation could quickly modify visual response selectivity via the modification of inhibitory synaptic strength. This alters the relationship between excitatory inputs and postsynaptic responses in association with visual stimulation. The altered relationship may gradually adjust the strength of excitatory synapses according to the modification rule depending on postsynaptic depolarization level, if modified inhibitory synaptic strength is maintained for a while by spontaneous or visually-evoked activity of inhibitory presynaptic cells. The induction of voltage-gated Ca^{2+} channel-dependent excitatory LTP also requires postsynaptic depolarization[45]. Therefore, visual response properties shaped initially by inhibitory synaptic plasticity could be maintained eventually by modified excitatory synapses. However, we have not yet tested whether modification at excitatory synapses is far more persistent than that at inhibitory synapses.

Although it is uncertain whether the speculation noted above is the case or not, it is likely that long-term modifications at excitatory and inhibitory synapses are not independent processes, although each may have distinct roles. The direction of changes in both excitatory and inhibitory synapses is regulated by the postsynaptic depolarization level. This regulation in postsynaptic cells may function to maintain the balance between excitatory and inhibitory synaptic transmissions and lead to the establishment of coordinated neural circuits.

7. REFERENCES

1. Wiesel, T. N., 1982, Postnatal development of the visual cortex and the influence of environment, *Nature* **299**: 583.
2. Fregnac, Y., and Imbert, M., 1984, Development of neuronal selectivity in primary visual cortex of cat, *Physiol. Rev.* **64**: 325.
3. Tsumoto, T., 1992, Long-term potentiation and long-term depression in the neocortex, *Prog. Neurobiol.*, **39**: 209.
4. Bear, M. F., and Kirkwood, A., 1993, Neocortical long-term potentiation, *Curr. Opin. Neurobiol.* **3**: 197.
5. Malinow, R., and Miller, J. P., 1986, Postsynaptic hyperpolarization during conditioning reversibly blocks induction of long-term potentiation, *Nature* **320**: 529.
6. Gustafsson, B., Wigstrom, H., Abraham, W. C., and Huang, Y.-Y., 1987, Long-term potentiation in the

hippocampus using depolarizing current pulses as the conditioning stimulus to single volley synaptic potentials, *J. Neurosci.* **7**: 774.
7. Artola, A., Brocher, S., and Singer, W., 1990, Different voltage-dependent thresholds for inducing long-term depression and long-term potentiation in slices of rat visual cortex, *Nature* **347**: 69.
8. Fregnac, Y., Shulz, D., Thorpe, S., and Bienenstock, E., 1988, A cellular analogue of visual cortical plasticity, *Nature* **333**: 367.
9. Reiter, H. O., and Stryker, M. P., 1988, Neural plasticity without postsynaptic action potentials: less-active inputs become dominant when kitten visual cortical cells are pharmacologically inhibited, *Proc. Natl. Acd. Sci. USA* **85**: 3623.
10. Komatsu, Y., 1994, Age-dependent long-term potentiation of inhibitory synaptic transmission in rat visual cortex, *J. Neurosci.* **14**: 6488.
11. Komatsu, Y., and Iwakiri, M., 1993, Long-term modification of inhibitory synaptic transmission in developing visual cortex, *Neuroreport* **4**: 907.
12. Malenka, R. C., and Nicoll, R. A., 1993, NMDA-receptor-dependent synaptic plasticity: multiple forms and mechanisms, *Trends Neurosci.* **16**: 521.
13. Bear, F. M., and Abraham, W. C., 1996, Long-term depression in hippocampus, *Annu. Rev. Neurosci.* **19**: 437.
14. McLean, H. A., Caillard, O., Ben-Ari, Y., and Gaiarsa, J.-L., 1996, Bidirectional plasticity expressed GABAergic synapses in the neonatal rat hippocampus, *J. Physiol. Lond.* **496**: 471.
15. McNaughton, B. L., Douglas, R. M., Goddard, G. V., 1978, Synaptic enhancement in fascia dentata: cooperativity among coactive afferents, *Brain Res.* **157**: 277.
16. Levy, W. B., and Steward, O., 1979, Synapses as associative memory elements in the hippocampal formation, *Brain Res.* **175**: 233.
17. Teyler, T.J., and DiScenna, p., 1987, Long-term potentiation, *Annu. Rev. Neurosci.* **10**: 131.
18. Komatsu, Y., and Yoshimura, Y., 2000, Activity-dependent maintenance of long-term potentiation at visual cortical inhibitory synapses, *J. Neurosci.* **20**: 7539.
19. Komatsu, Y., 1996, $GABA_B$ receptors, monoamine receptors, and postsynaptic inositol trisphosphate-induced Ca^{2+} release are involved in the induction of long-term potentiation at visual cortical inhibitory synapses, *J. Neurosci.* **16**: 6342.
20. Johnston, D., Williams, S., Jaffe, D., and Gray, R., 1992, NMDA-receptor-independent long-term potentiation, *Annu. Rev. Physiol.* **54**: 489.
21. Crawford, M. L. A., and Young, J. M., 1988, $GABA_B$ receptor-mediated inhibition of histamine H_1-receptor-induced inositol phosphate formation in slices of rat cerebral cortex, *J. Neurochem.* **51**: 1441.
22. Crawford, M. L. A., and Young, J. M., 1990, Potentiation by γ-aminobutyric acid of α_1-agonist-induced accumulation of inositol phosphates in slices of rat cerebral cortex, *J. Neurochem.* **54**: 2100.
23. Lisman, J., 1994, The CaM kinase II hypothesis for the storage of synaptic memory, *Trends Neurosci.* **17**: 406.
24. Staubli, U., and Lynch, G., 1990, Stable depression of potentiated synaptic responses in the hippocampus with 1-5 Hz stimulation, Brain Res. **513**: 113.
25. Fujii, S., Saito, K., Miyakawa, H., Ito, K., and Kato, H., 1991, Reversal of long-term potentiation (depotentiation) induced by tetanus stimulation of the input to CA1 neurons of guniea pig hippocampal slices, *Brain Res.* **555**: 112.
26. Kirkwood, A., Dudek, S. M., Gold, J. T., Aizenman, C. D., and Bear, M. F., 1993, Common forms of synaptic plasticity in the hippocampus and neocortex in vitro, *Science* **260**: 1518.
27. Chen, W. R., Lee, S., Kato, K., Spencer, D. D., Shepherd, G. M., and Williamson, A., 1996, Long-term modifications of synaptic efficacy in the human inferior and middle temporal cortex, *Proc. Natl. Acad. Sci. USA* **93**: 8011.
28. Tzounopoulos, T., Janz, R., Sudhof, T. C., Nicoll, R. A., and Malenka, R. C., 1998, A role for cAMP in long-term depression at hippocampal mossy fiber synapses, *Neuron* **21**: 837.
29. McGinty, D. J., and Harper, R. M., 1976, Dorsal raphe neurons: depression of firing during sleep in cats, *Brain Res.* **101**: 569.
30. Sakai, K., and Jouvet, M., 1980, Brain stem PGO-on cells projecting directly to the cat dorsal lateral geniculate nucleus, *Brain Res.* **194**: 500.
31. Cespuglio, R., Faradji, H., Gomez, M. E., and Juvet, M., 1981, Single unit recordings in the nuclei raphe dorsalis and magnus during the sleep waking cycle of semi-chronic prepared cats, *Neurosci. Lett.* **24**: 133.
32. Kasamatsu, T., and Pettigrew, J. D., 1976, Depletion of brain catecholamine: failure of ocular dominance shift after monocular occlusion in kitten, *Science* **194**: 206.
33. Gu, Q., and Singer, W., 1995, Involvement of serotonin in developmental plasticity of kitten visual cortex, *Eur. J. Neurosci.* **7**: 1146.

34. Sillito, A. M., 1975, The contribution of inhibitory mechanisms to the receptive field properties of neurones in the striate cortex of the cat, *J. Physiol. Lond.* **250**: 305.
35. Tsumoto, T., Eckart, W., and Creutzfeldt, O. D., 1979, Modification of orientation sensitivity of cat visual cortex neurons by removal of GABA-mediated inhibition, *Exp. Brain Res.* **34**: 351.
36. Crook, J. M., and Eysel, U. T., 1992, GABA-induced inactivation of functionally characterized sites in cat visual cortex (area 18): effects on orientation tuning, *J. Neurosci.* **12**: 1816.
37. Sato, H., Katsuyama, N., Tamura, H., Hata, Y., and Tsumoto, T., 1994, Broad-tuned chromatic inputs to color-selective neurons in the monkey visual cortex, *J. Neurophysiol.* **72**: 163
38. Freeman, R. D., Mallach, R., and Hartley, S., 1981, Responsivity of normal kitten striate cortex deteriorates after brief binocular deprivation, *J. Neurophysiol.* **45**: 1074.
39. Frey, U., Krug, M., Reymann, K. G., and Matthies, H., 1988, Anisomycin, an inhibitor of protein synthesis, blocks late phases of LTP phenomena in the hippocampal CA1 region in vitro, *Brain Res.* **452**: 57.
40. Nguyen, P. V., Abel, T., and Kandel, E. R., 1994, Requirement of a critical period of transcription for induction of a late phase of LTP, *Science* **265**: 1104.
41. Ferster, D., and Miller, K. D., 2000, Neural mechanisms of orientation selectivity in the visual cortex, *Annu. Rev. Neurosci.* **23**: 441.
42. Komatsu, Y., Fujii, K., Maeda, J., Sakaguchi, H., and Toyama, K., 1988, Long-term potentiation of synaptic transmission in kitten visual cortex, *J. Neurophysiol.* **59**: 124.
43. Ohmura, T., Ming, R., Yoshimura, Y., and Komatsu, Y., 2003, Age and experience dependence of N-methyl-D-aspartate receptor-independent long-term potentiation in rat visual cortex, *Neurosci. Lett.* in press
44. Komatsu, Y., Nakajima, S., and Toyama, K., 1991, Induction of long-term potentiation without participation of N-methyl-D-aspartate receptors in kitten visual cortex, *J. Neurophysiol.* **65**: 20.
45. Komatsu, Y., and Iwakiri, M., 1992, Low-threshold Ca^{2+} channels mediate induction of long-term potentiation in kitten visual cortex, *J. Neurophysiol.* **67**: 401.

5

ACTIVITY-DEPENDENT MODIFICATION OF CATION-CHLORIDE COTRANSPORTERS UNDERLYING PLASTICITY OF GABAERGIC SYNAPTIC TRANSMISSION

Melanie A. Woodin and Mu-ming Poo [*]

1. INTRODUCTION

In different brain regions and at different developmental stages, γ-aminobutyric acid (GABA)-mediated synaptic transmission may undergo long-term changes in its efficacy[1,2]. Long-term potentiation and depression (LTP and LTD) of GABAergic synapses are frequently associated with changes in total synaptic conductance, caused by alterations in either presynaptic transmitter release or postsynaptic GABA receptors. However, the amplitude of GABAergic postsynaptic currents may also be modified by changing the driving force for Cl^- through activated GABA receptor-channels, as a result of alterations in the electrochemical gradient of Cl^- and consequent changes in the reversal potential of postsynaptic currents[3-6]. This modulation can be bi-directional, resulting in either an increase or decrease in neuronal inhibition by GABA. This review describes recent evidence suggesting that changes in the activity of cation-chloride cotransporters (CCCs) may serve as a mechanism for synaptic plasticity in both the developing and mature nervous systems.

2. NEURONAL CHLORIDE TRANSPORTERS

Neuronal $[Cl^-]_i$ is regulated by both passive and active transport mechanisms, including several families of CCCs as well as Cl^--HCO_3^- and Cl^--ATPase exchangers[3,7-9]. To date five CCCs have been identified in the central nervous system (CNS), one loop

---

[*] M.A. Woodin and M-m. Poo, Department of Molecular and Cell Biology, University of California, Berkeley, CA 94720-3200.USA

diuretic-sensitive Na^+-K^+-$2Cl^-$ cotransporter (NKCC1) and four K^+-Cl^- cotransporters (KCC1-4)[10]. NKCC1 is localized to choroids plexus, oligodendrocytes and neurons[11], while KCC1 has a wide expression in many neural and non-neural tissues. Both NKCC1 and KCC1 are known to be important for cell volume maintenance and regulation[12]. With the exception of dorsal root ganglion (DRG) neurons[13-16], KCC2 expression has been exclusively localized to CNS neurons[16-18]. KCC3 has been found in multiple tissues including brain[7,19-22], and the expression pattern of KCC4 has yet to be determined. Examination of these various CCCs suggests that NKCC1 and KCC2 play most active roles in the CNS[8,23], and thus will be the primary focus of this review.

Using energy stored in the Na^+ and K^+ gradients across the plasma membrane, NKCC1 and KCC2 normally work in an opposing manner to regulate $[Cl^-]_i$: NKCC1 actively pumps 2 Cl^- into the cell along with 1 Na^+ and 1 K^{+11}, while KCC2 couples extrusion of 1 Cl^- with 1 $K^{+8,15}$. GABAergic synaptic transmission is mediated by Cl^- currents through GABA – activated channels. Because the magnitude and direction of these Cl^- currents are set by their reversal potential, which in turn depends on the level of $[Cl^-]_i$, regulation of NKCC1 and KCC2 may provide a mechanism for modulating the efficacy of these synapses.

3. DEPOLARIZING GABAERGIC SYNAPTIC TRANSMISSION EARLY DURING DEVELOPMENT

During embryonic and neonatal development, CNS neurons normally express high levels of NKCC1, resulting in an elevated $[Cl^-]_i^{8,24}$. GABAergic synaptic transmission results in a net efflux of Cl^- and accompanying depolarization of the membrane that may be significant enough to fire action potentials[25-32]. Thus, GABAergic synaptic transmission is depolarizing in a wide range of developing brain structures including the hippocampus[33,34], neocortex[35,36] hypothalamus[28,29], spinal cord[27,37], ventral tegmental area and cerebellum[38,39]. Despite its membrane depolarizing effect, GABA may exert inhibitory actions by shunting excitatory inputs and by clamping the membrane potential near the equilibrium potential for the Cl^- currents, which normally sits below the threshold for firing action potentials.

During development, GABAergic transmission appears prior to the emergence of glutamatergic transmission. It has been proposed that the early appearance of GABAergic synapses found in a wide variety of species provides the excitatory drive necessary for proper development of neurons and neural circuits[23]. The excitatory actions of GABA take part in generating giant depolarizing potentials (GDPs), a network-driven pattern of electrical activity observed in many developing circuits[23,33]. Such GDPs are responsible for producing large oscillations in intracellular Ca^{2+} by activating voltage-dependent Ca^{2+} channels (VDCCs)[40]. Through Ca^{2+}-dependent gene regulation[41], GDPs may be responsible for regulating cell proliferation, neuronal migration, morphological development, synapse formation and neurotrophic factor expression[23]. As the nervous system matures and excitatory glutamatergic transmission is established, GABAergic transmission is switched from excitation to inhibition. Proper balance between excitation and inhibition becomes important for normal functioning of neural circuits.

4. INTRACELLULAR CHLORIDE DECREASES DURING POSTNATAL DEVELOPMENT

During the first three weeks of postnatal development, the expression of NKCC1 decreases[11], while that of KCC2 increases[15-17,42-44]. The predominant action of KCC2 in driving Cl$^-$ efflux results in a drop of [Cl$^-$]$_i$ by approximately 20-40 mM and a shift in the reversal potential for Cl$^-$ currents towards more negative values[23]. When the reversal potential becomes more negative than the resting membrane potential, GABAergic synaptic transmission is accompanied by a hyperpolarization of the postsynaptic membrane[23,33,36,45-47]. In cultures of developing hippocampal neurons, blockade of GABA$_A$ receptors prevented the switch in GABAergic transmission from being depolarizing to hyperpolarizing, suggesting that GABA itself promotes the developmental change in [Cl$^-$]$_i$[42]. Furthermore, GABAergic activity appeared to be responsible for an up-regulation of KCC2, through a Ca^{2+}-dependent mechanism.

5. ACTIVITY-DEPENDENT MODIFICATION OF CATION-CHLORIDE COTRANSPORTERS

Given the role of KCC2 in maintaining low [Cl$^-$]$_i$ in mature neurons, small changes in the level of expression or activity of this transporter could have a significant impact on the efficacy of inhibitory synaptic transmission. A recent study has shown that electrical activity may acutely modulate the function of CCCs[5]. Repetitive postsynaptic spiking (5 Hz, 30 s) of cultured hippocampal neurons within 20 ms of GABAergic synaptic transmission led to a persistent reduction in the strength of inhibition by a local decrease in CCC activity. This appeared to be a rapid modulation of CCCs and reversal potential for GABA-induced Cl$^-$ currents (E$_{GABA}$) requiring activation of L-type Ca^{2+} channels[5]. When the time interval between synaptic activation and postsynaptic spiking was increased to beyond 50 ms there was no significant change in E$_{GABA}$, suggesting that alteration of the CCC function requires specific temporal patterns of activity. This activity-dependent modification of CCCs may provide a mechanism for GABAergic synaptic plasticity that allows the level of inhibition to be modulated in accordance with the temporal pattern of postsynaptic excitation.

Whereas neuronal activity may alter CCC functions, changes in CCC expression in the mature nervous system are also known to alter the activity of neural circuits, as shown by neuronal responses to injury. For example, axotomy of vagus nerve reduces KCC2 expression in dorsal motor neurons, leading to an increase in [Cl$^-$]$_i$ and a GABA$_A$ receptor-mediated excitation[48]. Axonal injury specifically altered KCC2 expression, without changing NKCC1[48]. The depolarizing action of GABA and its ability to increase [Ca^{2+}]$_i$ after axonal injury may participate in neuronal recovery through changes in gene expression and resumption of neuronal growth and regeneration, characteristics associated with developing neurons[49,50].

The suprachiasmatic nucleus (SCN), a brain region that contains the circadian clock critical for controlling diurnal rhythmicity, contains a high density of GABAergic synapses[51]. Adult SCN neurons in brain slice respond to GABA with an increase in firing frequency during the day and a decrease at night, an oscillation that has been attributed to an oscillation in [Cl$^-$]$_i$[52]. Significant fluctuations in [Cl$^-$]$_i$ over a 24 hr period may be controlled by changes in CCC activity. Taken together, these findings from different systems provide strong evidence that regulation of CCC activity and expression are potential mechanisms underlying activity-dependent synaptic plasticity.

6. BDNF MODULATES CATION-CHLORIDE COTRANSPORTER ACTIVITY

In addition to their roles in neuronal survival and differentiation, neurotrophins participate in activity-dependent synaptic plasticity[53]. Recently, brain-derived neurotrophic factor (BDNF), acting via TrkB receptors, has been shown to regulate KCC2 activity and expression in both the developing and mature nervous systems. Overexpression of BDNF in embryonic hippocampal neurons increased KCC2 expression, enhanced the rate at which GABA responses are converted from depolarizing to hyperpolarizing, and regulated spontaneous synaptic activity[54]. The effect of BDNF on KCC2 appears to be reversed at later developmental stages when BDNF inhibits GABA$_A$ receptor - mediated synaptic transmission[55-58]. Recently, acute and chronic application of either BDNF or a related protein neurotrophin-4 (NT-4) to rat organotypic hippocampal cultures were shown to reduce KCC2 mRNA and protein, leading to decreased neuronal Cl$^-$ extrusion and suppressed GABAergic inhibition[59]. Furthermore, in vivo kindling-induced seizures, which lead to a massive up-regulation and release of BDNF[60,61], also decreased KCC2 mRNA throughout the hippocampus[59]. In hippocampal cell cultures, the ability of BDNF to modulate GABAergic synaptic transmission correlates directly with the level of CCC activity, as determined by the shift in E_{GABA}[62].

7. SYNAPSE SPECIFICITY: LOCALIZED EXPRESSION AND REGULATION OF CATION-CHLORIDE COTRANSPORTERS

One of the important characteristics of activity-dependent synaptic plasticity is synapse specificity – only activated synapses are modified. For CCCs to play an active role in synapse-specific plasticity, modulation of CCC activity must be relatively local. Immunostaining with specific antibodies showed that KCC2 molecules are concentrated in the dendrites of hippocampal pyramidal cells[16]. In the mouse spinal cord KCC2 was found to flank synaptic clefts and colocalize with gephyrin, a protein responsible for maintaining postsynaptic clusters of glycine receptors[63]. This localized pattern of KCC2 expression may result in a lower [Cl$^-$]$_i$ at the dendrites than the soma.
Evidence for a [Cl$^-$]$_i$ gradient in the neuron was obtained by comparing the reversal potentials for GABA-induced Cl$^-$ currents at the soma and at the synapse, using gramicidin perforated patch-clamp recordings from cultured hippocampal neurons[5]. This reversal potential was used to estimate [Cl$^-$]$_i$ at the synapse, whereas the reversal potential for Cl$^-$ currents at the soma was determined by puffing GABA onto the cell body. These studies have shown that, at least in a subpopulation of hippocampal neurons, the reversal potential at the synapse was 5 mV more hyperpolarized than that at the soma, reflecting existence of a steady state gradient of dendritic [Cl$^-$]$_i$ within the same neuron.

Inhibitory synaptic inputs have a well-defined function in visual information processing in the retina. Recently the findings that KCC2 is localized to specific neuronal types as well as to subcellular regions of a neuron have revealed in part how

GABAergic synaptic inputs produce either depolarizations or hyperpolarizations in different classes of bipolar neurons and in different regions of the same neuron[13,14]. When surround illumination of receptive field increases GABA release from horizontal cells onto bipolar dendrites, OFF cells hyperpolarize and ON cells depolarize. This differential bipolar neuron response to GABA is attributed to the fact that OFF cells express KCC2 and ON cells express NKCC1[13]. In addition to the differential cellular localization of KCC2 and NKCC1 in the retina, these CCCs are also differentially distributed within the same neuron. In ON rod bipolar cells NKCC1 is targeted to the dendrite while KCC2 is targeted to the axon, consistent with the finding that GABA depolarizes the dendrites and hyperpolarizes the axon[13]. Specific targeting and localization of KCC2 to postsynaptic sites suggests that it is strategically positioned in the neuron to maintain local postsynaptic $[Cl^-]_i$ and to play a role in synaptic plasticity.

Can coincident activity induce synapse-specific modification through local regulation of CCCs? We have examined this issue by simultaneous triple whole-cell recordings from neurons making divergent or converging GABAergic connections[5]. When a GABAergic presynaptic neuron made divergent connections onto two postsynaptic cells, only the pair of neurons that underwent repetitive coincident pre- and postsynaptic activity showed significant synaptic modification, as a result of depolarizing shift of E_{GABA} at stimulated synapses. For a postsynaptic cell receiving two converging GABAergic inputs, the coincident pre- and postsynaptic activity resulted in a significant synaptic modification at the stimulated input, often leaving the input from the unstimulated neuron unchanged or depressed. These results, together with the existence of steady $[Cl^-]_i$ gradient in neurons[5,44], suggest that activity-induced regulation of KCC2 and the consequent changes in dendritic $[Cl^-]_i$ can be synapse specific.

8. MECHANISMS FOR CATION-CHLORIDE COTRANSPORTER MODIFICATION

How activity and neurotrophins modulate CCC levels and functions is still largely unclear. In hippocampal cultures, repetitive coincident pre- and postsynaptic activity led to a decrease in KCC2 activity almost immediately[5] and thus are likely to involve posttranslational modification or membrane trafficking of KCC2, rather than changes in gene transcription or protein synthesis. Induction of LTP[64] and LTD[65] at hippocampal CA1 excitatory synapses results from the insertion or endocytotic removal of synaptic AMPA receptors, respectively. Furthermore, regulated trafficking of transmitter transporters can have a significant impact on synaptic transmission[66]. The decrease in Cl⁻ transporter activity may result from a removal of KCC2 from the plasma membrane. Regulation of KCC2 transport function may also be achieved by a phosphorylation or dephosphorylation at one of its consensus sites for protein kinase C and tyrosine kinase[17]. Serine/threonine phosphorylation has been shown to regulate KCC2 activity in oocytes[67], while cytosolic tyrosine kinase activity regulates CCC activity during the maturation of hippocampal neurons[44]. Activity-induced modification of KCC2 may result from interaction between $GABA_A$ receptor activation and increased Ca^{2+} levels associated with spike-induced L-type Ca^{2+} channel activation, leading to postsynaptic signaling that down-regulate KCC2 activity by dephosphorylation, in a manner similar to the dephosphorylation of AMPA receptors following the induction of LTD[68].

Sustained activity can also alter ionic gradients that may directly alter the function of CCCs. For example, on the basis of the high affinity of KCC2 for external K^+ and the thermodynamic characteristics of KCC2 when it is expressed in human embryonic kidney cells, it has been hypothesized that high frequency spiking can elevate $[K^+]_o$ sufficiently to reverse KCC2 cotransport, resulting in the accumulation of $[Cl^-]_i$[69]. However it remains to be shown that physiological synaptic activity can alter $[K^+]_o$ sufficiently to reverse the transport function of KCC2.

9. A ROLE FOR CATION-CHLORIDE COTRANSPORTERS IN EPILEPSY

Disruption of the balance between neuronal excitation and inhibition in neural circuits can result in pathological consequences. In particular, abnormal inhibitory signals can produce hyperexcitability and hypersynchronized electrical activity in the brain that leads to epileptic seizures[70]. Strong evidence for a role of KCC2 activity in epilepsy has been obtained from studies using a targeted deletion of the KCC2 gene in mouse[71]. A complete deletion of the KCC2 gene produced mice that exhibited severe motor deficits that lead to respiratory failure at birth[63]. However, in mice possessing only 5% of the normal level of KCC2, frequent seizure activity was observed and severe brain injury occurred[71]. *In vivo* hippocampal kindling-induced seizures, which lead to a marked up-regulation and release of BDNF, also decreased KCC2 mRNA[59]. However, whether normal pathogenesis of epilepsy involves regulation of KCC2 expression in neurons remains to be established.

10. CONCLUDING REMARKS

Modification of CCC function in the mature nervous system may represent a mechanism underlying GABAergic synaptic plasticity. Coincident pre- and postsynaptic activity, BDNF and neuronal injury all decrease the Cl^- extrusion capacity of CCCs and increase $[Cl^-]_i$, which in turn shifts the reversal potential of Cl^- currents to a more depolarized value, reducing synaptic inhibition. While the underlying mechanisms remain largely unclear, posttranslational modification or membrane trafficking of CCCs most likely account for the synaptic modification. The decrease in CCC function observed in the mature nervous system represents a reversion to a more immature state, suggesting that the developmental remodeling of synapses and activity-dependent synaptic plasticity share common cellular mechanisms (Fig. 1). A further understanding of the specific role CCCs play in synaptic plasticity will not only provide insights into proper functions of neural circuits, but also further our understanding of how CCCs contribute to brain disorders such as epilepsy.

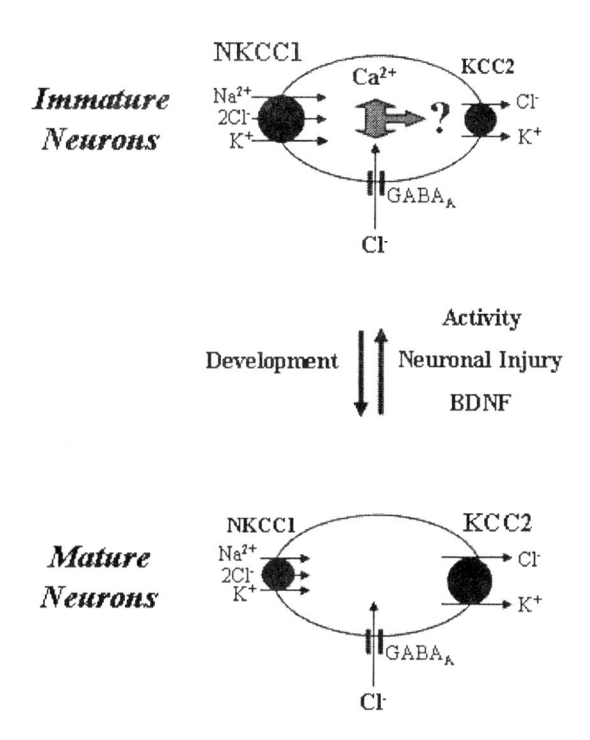

Fig. 1. Activity-dependent modification of CCCs in the mature nervous system reverts neurons to a more immature state. In immature neurons NKCC1 is more highly expressed than KCC2 and is responsible for maintaining a relatively high $[Cl^-]_i$. In more mature neurons the higher expression level of KCC2 maintains a lower $[Cl^-]_i$ responsible for producing inhibitory GABAergic synaptic transmission. In the mature neuron, coincident pre- and postsynaptic activity, BDNF and neuronal injury are all capable of decreasing KCC2 function, leading to an increase in $[Cl^-]_i$ and subsequent decrease in synaptic inhibition. In immature neurons and in mature neurons with decreased KCC2 function, coincidence between $GABA_A$ receptor activation and an increase in $[Ca^{2+}]_i$ is thought to modulate CCC activity.

11. REFERENCES

1. Gaiarsa, J. L., Caillard, O. and Ben-Ari, Y. (2002) *Trends Neurosci.* **25**, 564.
2. Marty, A. and Llano, I. (1995) *Curr. Opin. Neurobiol.* **5**, 335.
3. Kaila, K. (1994) *Prog. Neurobiol.* **42**, 489.
4. Ling, D. S. and Benardo, L. S. (1995) *Brain Res.* **670**, 142.
5. Woodin, M. A. and Poo, M. M. (2003), *Manuscript submitted for publication.*
6. Thompson, S. M. and Gahwiler, B. H. (1989) *J. Neurophysiol.* **61**, 501.

7. Voipio, J. and Kaila, K. (2000) *Prog. Brain. Res.* **125**, 329.
8. Delpire, E. (2000) *News Physiol. Sci.* **15**, 309.
9. Inagaki, C., Hara, M. and Zeng, X. T. (1996) *J. Exp. Zool.* **275**, 262.
10. Delpire, E. and Mount, D. B. (2002) *Annu. Rev. Physiol.* **64**, 803.
11. Plotkin, M. D., Snyder, E.Y., Hebert, S.C. and Delpire, E. (1997) *J. Neurobiol.* **33**, 781.
12. Gillen, C. M., Brill, S., Payne, J. A. and Forbush, B. (1996) *J. Biol. Chem.* **271**, 16237.
13. Vardi, N., Zhang, L. L., Payne, J. A. and Sterling, P. (2000) *J. Neurosci.* **20**, 7657.
14. Vu, T. Q., Payne, J. A. and Copenhagen, D. R. (2000) *J. Neurosci.* **20**, 1414.
15. Rivera, C., Voipio, J., Payne, J. A., Ruusuvuori, E., Lahtinen, H., Lamsa, K., Pirvola, U., Saarma, M. and Kaila, K. (1999) *Nature* **397**, 251.
16. Lu, J., Karadsheh, M. and Delpire, E. (1999) *J.Neurobiol.* **39**, 558.
17. Payne, J. A., Stevenson, T. J. and Donaldson, L. F. (1996), *J. Biol.Chem.* **271**, 16245.
18. Williams, J. R., Sharp, J. W., Kumari, V.G., Wilson, M. and Payne, J. A. (1999) *J. Biol. Chem.* **274**, 12656.
19. Hiki, K., D'Andrea, R. J., Furze, J., Crawford, J., Woollatt, E., Sutherland, G. R., Vadas, M. A. and Gamble, J. R. (1999) *J. Biol. Chem.* **274**, 10661.
20. Race, J. E., Makhlouf, F. N., Logue, P. J., Wilson, F. H., Dunham, P. B. and Holtzman, E. J. (1999) *Am. J.Physiol.* **277**, C1210.
21. Mount, D. B., Mercado, A., Song, L., Xu, J., George Jr., A. L., Delpire, E. and Gamba, G. (1999) *J. Biol. Chem.* **274**, 16355.
22. Delpire, E., Mount, D. B., Lu, J., England, R., Kirby, M. and McDonald, M. P. (2001) *FASEB J.* **15**, A440.
23. Ben-Ari, Y. (2002) *Nat. Rev. Neurosci.* **3**, 728.
24. Clayton, G. H., Owens, G. C., Wolff, J. F. and Smith, R. L. (1998) *Brain Res. Dev. Brain Res.* **109**, 281.
25. Khazipov, R., Leinekugel, X., Khalilov, I., Gairasa, J. L. and Ben-Ari, Y. (1997) *J. Physiol.* **498** (Pt 3), 763.
26. Gao, X. B. and van den Pol, A. N. (2001) *J. Neurophysiol.* **85**, 425.
27. Serafini, R. et al., (1995) *J Physiol* **488** (Pt 2) 371.
28. Chen, G., Trombley, P. Q. and van den Pol, A. N. (1996) *J. Physiol.* **494** (Pt 2), 451.
29. Wang, Y. F., Gao, X. B. and van den Pol, A. N. (2001) *J. Neurophysiol.* **86**, 1252.
30. Obrietan, K. and van den Pol, A. (1999) *J. Neurophysiol.* **82**, 94.
31. Cherubini, E., Martina, M., Sciancalepore, M. and Strata, F. (1998) *Perspect. Dev. Neurobiol.* **5**, 289.
32. Khalilov, I., Dzhala, V., Ben-Ari, Y. and Khazipov, R. (1999) *Dev. Neurosci.* **21**, 310.
33. Ben-Ari, Y., Cherubini, E., Corradetti, R. and Gairasa, J. L. (1989) *J. Physiol.* **416**, 303.
34. Leinekugel, X., Khalilov, I., McLean, H., Caillard, O., Gairasa, J. L., Ben-Ari, Y. and Khazipov, R. (1999) *Adv. Neurol.* **79**, 189.
35. Maric, D., Liu, Q. Y., Maric, I., Chaudry, S., Chang, Y. H., Smith, S. V., Seighart, W., Fritschy, J. M. and Barker, J. L. (2001) *J. Neurosci.* **21**, 2343.
36. Luhmann, H. J. and Prince, D. A. (1991) *J. Neurophysiol.* **65**, 247.
37. Obata, K., Oide, M. and Tanaka, H. (1978) *Brain. Res.* **144**, 179.
38. Eilers, J., Plant, T. D., Morandi, N. and Konnerth, A. (2001) *J. Physiol.* **536**, 429.
39. Yuste, R. and Katz, L. C. (1991) *Neuron* **6**, 333.
40. Leinekugel, X., Tseeb, V., Ben-Ari, Y. and Bregestovski, P. (1995) *J. Physiol.* **487** (Pt 2), 319.
41. Cherubini, E., Gaiarsa, J. L. and Ben-Ari, Y. (1991) *Trends Neurosci.* **14**, 515.
42. Ganguly, K., Schinder, A. F., Wong, S. T. and Poo, M. (2001) *Cell* **105**, 521.
43. Kakazu, Y., Akaike, N., Komiyama, S. and Nabekura, J. (1999) *J. Neurosci.* **19**, 2843.
44. Kelsch, W., Hormuzdi, S., Straube, E., Lewen, A., Monyer, H. and Misgeld, U. (2001) *J. Neurosci.* **21**, 8339.
45. Gaiarsa, J. L., Tseeb, V. and Ben-Ari, Y. (1995), *J. Neurophysiol.* **73**, 246.
46. Owens, D. F., Boyce, L. H., Davis, M. B. and Kriegstein, A. R. (1996) *J. Neurosci.* **16**, 6414.
47. DeFazio, R. A., Keros, S., Quick, M. W. and Hablitz, J. J. (2000) *J. Neurosci.* **20**, 8069.
48. Nabekura, J., Ueno, T., Okabe, A., Furuta, A., Iwaki, T., Shimizu-Okabe, C., Fukuda, A. and Akaike, N. (2002) *J. Neurosci.* **22**, 4412.
49. van den Pol, A. N., Obrietan, K. and Chen, G. (1996) *J. Neurosci.* **16**, 4283.
50. Toyoda, H., Ohno, K., Yamada, J., Ikeda, M., Okabe, A., Sato, K., Hashimoto, K. and Fukuda, A. (2003) *J. Neurophysiol.* **89**, 1353.
51. Klein, D. C., Moore, R. Y. and Reppert, S. M. (1991) *Suprachiasmatic Nucleus: The Mind's Clock.* (Oxford University Press, New York).
52. Wagner, S., Castel, M., Gainer, H. and Yarom,Y. (1997) *Nature* **387**, 598.
53. Schinder, A. F. and Poo, M. (2000) *Trends Neurosci.* **23**, 639.

54. Aguado, F., Carmona, M. A., Pozas, E., Aguilo, A., Martinez-Guijarro, F. J., Alcantara, S., Borrell, V., Yuste, R., Ibanez, C. F. and Soriano, E. (2003) *Development* **130**, 1267.
55. Kim, H. G., Wang, T., Olafsson, P. and Lu, B. (1994) *Proc. Natl. Acad. Sci. USA* **91**, 12341.
56. Tanaka, T., Saito, H. and Matsuki, N. (1997) *J. Neurosci.* **17**, 2959.
57. Frerking, M., Malenka, R. C. and Nicoll, R. A. (1998) *J. Neurophysiol.* **80**, 3383.
58. Brunig, I., Penschuck, S., Berninger, B., Benson, J. and Fritschy, J. R. (2001) *Eur. J. Neurosci.* **13**, 1320.
59. Rivera, C., Li, H., Thomas-Crusells, J., Lahtinen, H., Viitanen, T., Nanobashvili, A., Kokaia, Z., Airaksinen, M. S., Voipo, J., Kaila, K. and Saarma, M. (2002) *J. Cell. Biol.* **159**, 747.
60. Binder, D. K., Croll, S. D., Gall, C. M. and Scharfman, H. E. (2001) *Trends Neurosci.* **24**, 47.
61. Huang, E. J. and Reichardt, L. F. (2001) *Annu. Rev. Neurosci.* **24**, 677.
62. Wardle, R. and Poo, M. M. (2003) *Manuscript submitted for publication.*
63. Hubner, C. A., Stein, V., Hermans-Borgmeyer, I., Meyer, T., Ballanyi, K. and Jentsch, T. J. (2001) *Neuron* **30**, 515.
64. Hayashi, Y., Shi, S. H., Esteban, J. A., Piccini, A., J. C. Poncer and Malinow R. (2000) *Science* **287**, 2262.
65. Carroll, R. C., Lissin, D. V., von Zastrow, M., Nicoll, R. A. and Malenka, R. C. (1999) *Nat. Neurosci.* **2**, 454.
66. M. B. Robinson (2002) *J. Neurochem.* **80**, 1.
67. Strange, K., Singer, T. D., Morrison, R. and Delpire, E. (2000) *Am. J. Physiol. Cell Physiol.* **279**, C860.
68. Lee, H. K., Kameyama, K., Huganir, R. L. and Bear, M. F. (1998) *Neuron* **21**, 1151.
69. Payne, J. A. (1997) *Am. J. Physiol.* **273**, C1516.
70. Hochman, D. W., Baraban, S. C., Owens, J. W. and Schwartzkroin, P. A. (1995) *Science* **270**, 99.
71. Woo, N. S., Lu, J., England, R., McClellan, R., Dufour, R., Mount, D. B., Deutsh, A. Y., Lovinger, D. M. and Delpire, E. (2002) *Hippocampus* **12**, 258.

6

ENDOCANNABINOID-MEDIATED MODULATION OF EXCITATORY AND INHIBITORY SYNAPTIC TRANSMISSION

Masanobu Kano[*], Takako Ohno-Shosaku, Takashi Maejima and Takayuki Yoshida

1. INTRODUCTION

Marijuana exerts variable behavioral effects through the interaction of its active component Δ^9-tetrahydrocannabinol with specific cannabinoid receptors. Accumulated evidence suggests that endogenous cannabinoids function as diffusible and short-lived intercellular messengers that modulate synaptic transmission. Recent studies have provided strong experimental evidence that endogenous cannabinoids (endocannabinoids) are released from postsynaptic neurons following depolarization-induced elevation of intracellular Ca^{2+} concentration ($[Ca^{2+}]_i$), activation of group I metabotropic glutamate receptors (mGluRs) or activation of muscarinic acetylcholine receptors (mAChRs). The released endocannabinoids mediate signals retrogradely from postsynaptic neurons to presynaptic terminals to suppress subsequent neurotransmitter release, driving the synapse into an altered state. In this chapter, we review recent studies including ours and propose a possible scheme for the mechanisms of the endocannabinoid-mediated retrograde signaling. We will also discuss about possible physiological significance of this signaling on excitatory-inhibitory balance of the neural circuitry.

2. CANNABINOID RECEPTORS AND ENDOCANNABINOIDS

Several lines of evidence support a possible role of endocannabinoids as a diffusible neuromodulator at synapses. First, cannabinoid receptors, the molecular targets of the active component (Δ^9-tetrahydrocannabinol) of marijuana and hashish, are distributed

---

[*]M. Kano, T. Ohno-Shosaku, T. Maejima and T. Yoshida, Department of Cellular Neurophysiology, Graduate School of Medical Science, Kanazawa University, Kanazawa 920-8640, Japan.

Excitatory-Inhibitory Balance: Synapses, Circuits, Systems
Edited by Hensch and Fagiolini, Kluwer Academic/Plenum Publishers, 2003

widely in the CNS [1-4]. They have amino acid sequences characteristic of G protein-coupled seven-transmembrane-domain receptors, and consist of type 1 (CB1) and type 2 (CB2) receptors with different distributions[5-7]. While the CB2 receptor is expressed mainly in the immune system of the periphery[8], the CB1 receptor exhibits a widespread distribution in the brain. In situ hybridization, autoradiographic and immunohistochemical studies on brain tissue show that the distribution of CB1 receptor is heterogeneous, with high levels in some areas including the hippocampus, associational cortical regions, the cerebellum, and the basal ganglia[1-4]. The CB1 receptor distribution correlates with the known effects of cannabinoids on memory, perception, and the control of movement. The distribution is also heterogeneous among different populations of neurons in a certain brain region. In the hippocampus, for example, very high levels of CB1 immunoreactivity and mRNA are associated with subpopulations of GABAergic neurons[9,10]. In addition to CB1 receptors, it is recently reported that the third type of cannabinoid receptors (CB3) is present in excitatory synapses of the mature hippocampus[11,12]. Its molecular identity, however, has not been clarified so far.

Second, potential endocannabinoids such as anandamide and 2-arachidonoylglycerol (2-AG) and pathways for their biosynthesis have been identified[13-16]. These molecules are widely distributed in the brain, with high levels in the hippocampus, striatum, cerebellum, and cortex showing a good correlation with the CB1 receptor distribution[17]. These are lipid in nature, are synthesized from membrane phospholipid in activity- and Ca^{2+}-dependent manners[18-22], and can diffuse out across the cell membrane due to its hydorphobicity. The released endocanabinoids are removed from the extracellular space through selective and saturable uptake system or enzymatic degradation. The degradating enzyme FAAH (fatty acid amide hydrolase) is strongly expressed in the CB1 receptor-rich regions[23-25], and is located complementally to CB1 receptors[26].

Third, the activation of cannabinoid receptors induces a variety of actions on neural functions[20,5]. Cannabinoid agonists inhibit adenylate cyclase[27] and N- and P/Q-type Ca^{2+} channels, and activate inwardly rectifying potassium channels[28-31]. They also suppress the release of neurotransmitters, such as glutamate, GABA, acetylcholine, and noradrenaline, presumably through an inhibition of presynaptic Ca^{2+} channels[32-35,9].

All these findings suggest that endocannabinoids can work as a diffusible and short-lived mediator that is released from activated neurons, binds to its receptors on neighboring neurons and modulate their functions.

3. DEPOLARIZATION-INDUCED SUPPRESSION OF INHIBITION (DSI)

In the hippocampus[36,37] and the cerebellum[38], action potential firing or depolarization of the postsynaptic neuron induces a transient suppression of inhibitory synaptic inputs to the depolarized neuron (Fig. 1A). This phenomenon is termed DSI (depolarization-induced suppression of inhibition). The postsynaptic depolarization triggers Ca^{2+} influx through voltage-gated Ca^{2+} channels[39], and induces a transient elevation of intracellular Ca^{2+} concentration ($[Ca^{2+}]_i$). When the Ca^{2+} influx is inhibited by perfusing the postsynaptic neuron with a Ca^{2+}-free solution during depolarization[36], or when the fast Ca^{2+} buffer BAPTA is injected into the postsynaptic neuron[37,40,41] (Fig. 1B), the depolarization fails to induce DSI. Thus, it is evident that the postsynaptic Ca^{2+} elevation is required for the induction of DSI. While the site of DSI induction is postsynaptic, the

ENDOCANNABINOID-MEDIATED SYNAPTIC MODULATION

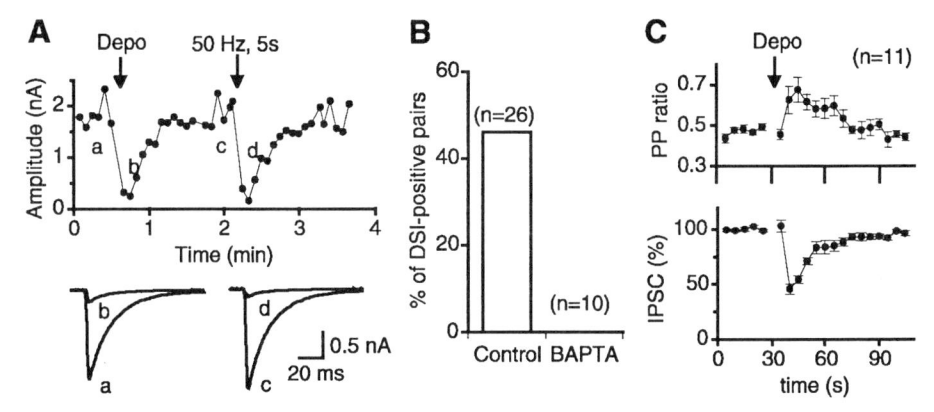

Fig. 1. Postsynaptic induction and presynaptic expression of DSI
(A) An example of DSI in a cultured hippocampal neuron pair. Inhibitory postsynaptic currents (IPSCs) were evoked by stimulating the presynaptic neuron at 0.2 Hz, and were measured from the postsynaptic neuron at -80 mV. DSI can be induced by either a continuous depolarization (0 mV, 5 s) or a train of action potentials (50 Hz, 5 s) in the postsynaptic neuron. The time course of the change in IPSC amplitudes (upper) and the IPSC traces (lower) acquired at the indicated time points are shown.
(B) DSI requires elevation of Ca^{2+} concentration in the postsynaptic neuron. DSI occurred in none of the 10 neuron pairs when the recording pipette for the postsynaptic neuron contained BAPTA (30 mM).
    DSI accompanies significant increase in the paired-pulse ratio.

site of its expression is apparently presynaptic. The postsynaptic response to the exogenously applied GABA is not suppressed significantly after the postsynaptic depolarization[40, 37]. DSI accompanies a clear increase in the paired-pulse ratio[36, 42] (Fig. 1C), a widely used indicator for presynaptic modulation[43]. These data indicate that the depolarization-induced elevation of $[Ca^{2+}]_i$ transiently suppresses the release of the inhibitory transmitter GABA from presynaptic terminals. Therefore, some retrograde signal must exist from the depolarized postsynaptic neurons to the presynaptic terminals.

The possibility that endocannabinoids may mediate the retrograde signaling of DSI has been evaluated recently, by using hippocampal cultures[40] and acute slices[42] from the hippocampus and the cerebellum[44-46]. In cultures, DSI occurs at about half of the inhibitory synapses[36]. At all the DSI-positive synapses, the synthetic cannabinoid agonist

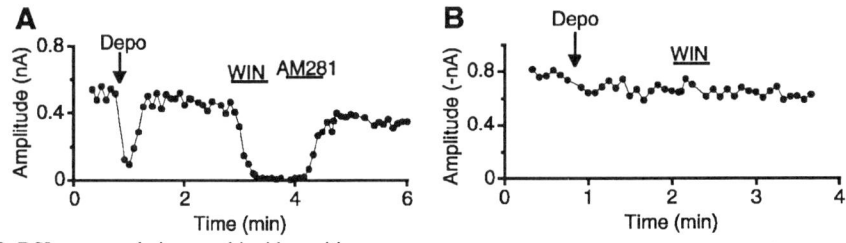

Fig. 2. DSI occurs only in cannabinoid-sensitive synapses.
(A) The cannabinoid agonist WIN55, 212-2 (WIN, 0.1µM) effectively suppressed IPSCs in a DSI-positive pair, which was reversed by the cannabinoid antagonist AM281 (0.3µM).
(B) WIN55, 212-2 (0.1µM) had no effect on a DSI-negative pair.

WIN55,212-2 suppresses the release of the inhibitory neurotransmitter GABA from presynaptic terminals[40] (Fig. 2A). In contrast, WIN55,212-2 is ineffective at most of the synapses that do not exhibit DSI[40] (Fig. 2B). Most importantly, DSI is completely eliminated by synthetic cannabinoid antagonists, AM 281 and SR141716A[40] (Fig. 3). In hippocampal slices, Wilson and Nicoll present essentially the same results as those in cultures[42]. In addition, they show that DSI is not inhibited by postsynaptically-applied botulinum toxin, making vesicular release from the postsynaptic cell highly improbable[42]. They also demonstrate that liberation of Ca^{2+} alone via flash photolysis of caged Ca^{2+} is sufficient to induce DSI[42]. All these data are consistent with the notion that endocannabinoids function as a retrograde messenger for DSI. The role of endocannabinoids as the major mediator of DSI has been confirmed also in the cerebellum[44-46] and the basal ganglia[47].

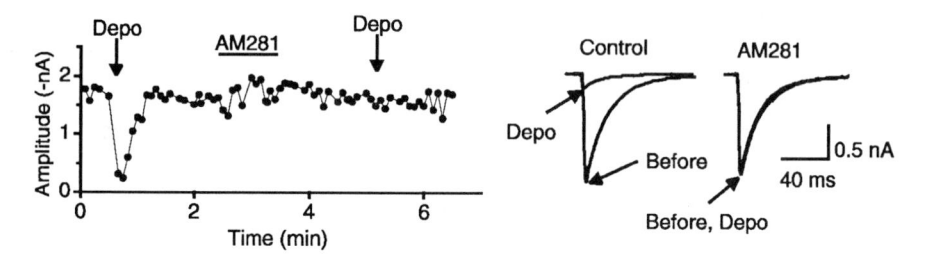

Fig. 3. DSI is eliminated completely by the cannabinoid antagonist AM281 (0.3 µM).

4. DEPOLARIZATION-INDUCED SUPPRESSION OF EXCITATION (DSE)

Because the CB1 receptors is rich in not only inhibitory but also excitatory presynaptic fibers[1], it is highly likely that endocannabinoids produced in postsynaptic neurons can reduce the release of excitatory transmitter. Kreitzer and Regehr have demonstrated that brief depolarization of cerebellar Purkinje cells in the rat induces a transient suppression of excitatory synaptic transmission at both climbing fiber and parallel fiber synapses[48]. This depolarization-induced suppression of excitation (DSE) lasts for tens of seconds, is blocked by the postsynaptic injection of a fast Ca^{2+} chelator BAPTA[48]. DSE is eliminated by the cannabinoid receptor antagonist[48]. We confirmed these results in mouse cerebellar Purkinje cells[49]. All these properties are quite similar to those of DSI, and indicate that endocannabinoids are released from depolarized Purkinje cells, and suppress the glutamate release from presynaptic terminals through the activation of the presynaptic CB1. In addition, Kreitzer and Regehr demonstrate that action potential-evoked rise of $[Ca^{2+}]_i$ in climbing fiber terminals is decreased during DSE[48]. This result strongly suggests that DSE results from the decreased Ca^{2+} influx to presynaptic terminals.

We have reported that DSE occurs also in the hippocampus[50]. We quantitatively examined the effects of postsynaptic depolarization and a cannabinoid agonist on excitatory and inhibitory synapses in the hippocampal slices and cultures. We found that both DSE and DSI could be induced, but DSE was much less prominent than DSI (Figure. 4). For the induction of DSE, depolarization with longer duration was required than for DSI. The magnitude of DSE was much smaller than that of DSI in hippocampal cultures

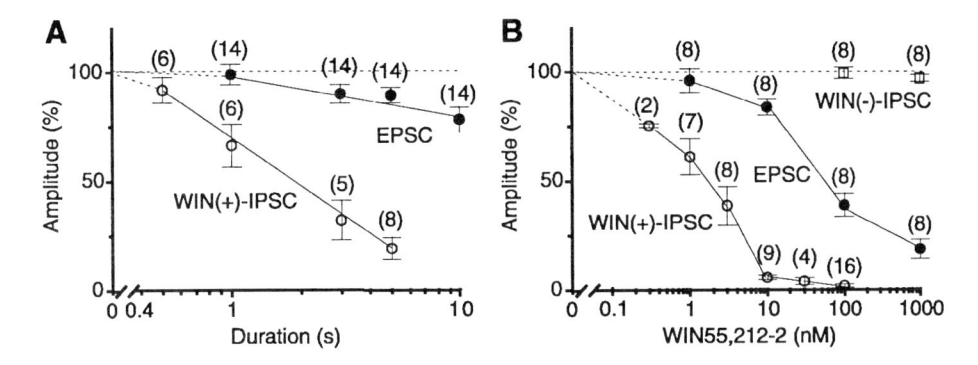

Fig. 4. DSE is much less prominent than DSI in the hippocampus.
(A) The relationships between the depolarizing pulse duration and the relative amplitude of WIN55,212-2-sensitive IPSCs (open circles) and EPSCs (closed circles) obtained 6 to 16 s after the end of depolarization. The amplitude was normalized to the averaged value before depolarization. Each symbol represents the averaged value obtained from the indicated number of neuron pairs.
(B) Amplitudes of WIN55,212-2-sensitive IPSCs (open circles), WIN55,212-2-insensitive IPSCs (squares) and EPSCs (closed circles) plotted against the concentration of WIN55,212-2. In each cell, current amplitudes in the presence of WIN55,212-2 were expressed as the percentages of the control values obtained before application of WIN55,212-2. Each symbol represents the average value from the indicated number of neuron pairs.

(Figure 4A). We tested the sensitivity of excitatory postsynaptic currents (EPSCs) and inhibitory postsynaptic currents (IPSCs) to a cannabinoid agonist, WIN55,212-2, in hippocampal cultures. IPSCs were dichotomized into two distinct populations, the one with a high sensitivity to WIN55,212-2 and the other with no sensitivity (Fig. 4B). In contrast, EPSCs were homogeneous and exhibited a low sensitivity to WIN55,212-2 (Fig. 4B). There was no difference between the excitatory and inhibitory synapses in terms of possible endocannabinoid concentration. These results indicate that the presynaptic cannabinoid sensitivity is a major factor that determines the extent of DSI and DSE in the hippocampus.

5. ENDOCANNABINOID SIGNALING TRIGGERED BY METABOTROPIC GLUTAMATE RECEPTOR ACTIVATION

We found a new type of synaptic modulation that involves retrograde signaling from postsynaptic metabotropic glutamate receptors (mGluR) to presynaptic cannabinoid receptors[49]. In mouse cerebellar slices, activation of group I mGluR in Purkinje cells (PCs) by its agonist, (RS)-3,5-dihydroxyphenylglycine (DHPG), reduced neurotransmitter release from excitatory climbing fibers. This inhibition involved activation of G proteins but did not require intracellular Ca^{2+} elevation in postsynaptic PCs. This effect was eliminated by specific cannabinoid receptor antagonists. The DHPG-induced suppression was absent in mutant mice lacking mGluR subtype 1 (mGluR1), a member of the group I mGluR that is abundantly expressed in PCs. Depolarization-induced Ca^{2+} transients in PCs caused robust DSE. Thus, endocannabinoid production in PCs can be initiated by two distinct stimuli. Furthermore, we demonstarted that activation of mGluR1 by repetitive stimulation of parallel fibers,

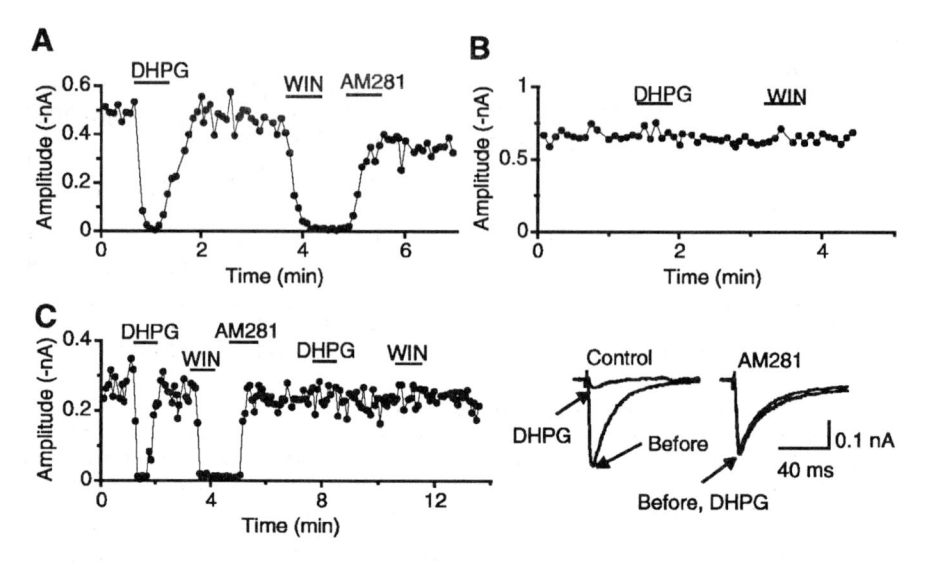

Fig. 5. Suppression of IPSCs by the group I mGluR agonist DHPG is mediated by cannabinoid receptors.
(A) DHPG (50 μM) effectively suppressed IPSCs in a cannaninpoid-sensitive pair. The IPSCs were suppressed by WIN55,212-2 (0.1 μM) and the effect was reversed by AM281 (0.3 μM).
(B) DHPG (50 μM) had no effect on IPSCs of a cannaninpoid-insensitive pair.
(C) Suppression of IPSCs by DHPG (50 μM) and WIN55,212-2 (0.1 μM) were both emiminated by AM281 (0.3 μM).

the other excitatory input to PCs, caused transient cannabinoid receptor-mediated depression of climbing fiber input. This result suggests that the mGluR1-mediated production of endocannabinoids is functional *in vivo* and may contribute to the control of excitatory synaptic activity.

Endocannabinoid production triggered by activation of group I mGluR was also found in the hippocampus[51-52]. We found that activation of group I mGluRs by DHPG suppressed IPSCs in about half of the neuron pairs (Fig. 5A). This effect was blocked by an antagonist specific to mGluR5, the other member of group I mGluR. A cannabinoid agonist, WIN55,212-2, suppressed IPSCs in all DHPG-sensitive pairs (Figure 5A) but not in most of DHPG-insensitive pairs (Fig. 5B). Effects of DHPG and WIN55,212-2 were both abolished by cannabinoid antagonists, AM281 and SR141716A (Fig. 5C), indicating that activation of group I mGluR releases endocannabinoids and suppress inhibitory neurotransmitter release through activation of presynaptic cannabinoid receptors. Importantly, DSI was significantly enhanced when group I mGluRs were activated simultaneously by DHPG[51] (Fig. 6). This enhancement was much more prominant than what was expected from the simple summation of depolarization-induced and group I mGluR-induced endocannabinoid release. DHPG caused no change in depolarization-induced Ca^{2+} transients, indicating that the enhanced DSI by DHPG was not due to the augmentation of Ca^{2+} influx. Enhancement of DSI by DHPG was also observed in hippocampal slices[51-52]. These results suggest that two pathways work in a cooperative manner to release endocannabinoids through a common intracellular cascade.

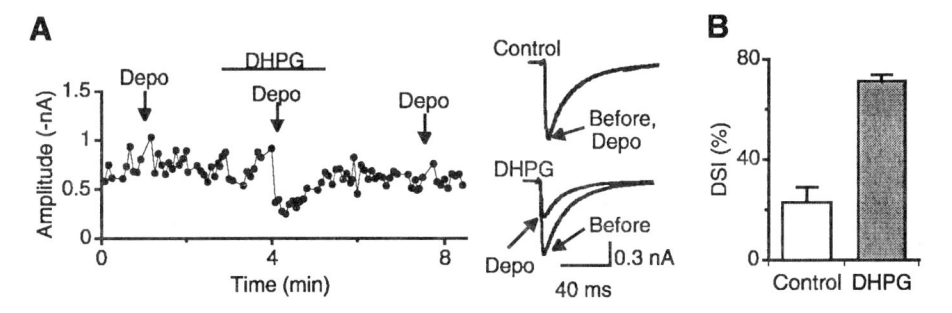

Fig. 6. DHPG significantly enhances DSI.
(A) In this neuron pair, depolarization alone (1 s, Depo) caused no detectable suppression of IPSCs. In the presence of DHPG (5 μM), the same depolarization induced clear suppression.
(B) Summary bar graph showing the enhancement of DSI (induced by 1 s depolarization) by DHPG (5 μM).

6. ENDOCANNABINOID SIGNALING TRIGGERED BY MUSCARINIC ACETYLCHOLINE RECEPTOR ACTIVATION

Cholinergic effects in the CNS are largely mediated by muscarinic acetylcholine receptors (mAChRs) that consist of five subtypes (M_1-M_5). Kim et al. have reported very recently that muscarinic activation induces endocannabinoid production in hippocampal slices[53]. They also reported that muscarinic activation enhanced DSI. We have confirmed their results in cultured hippocampal neurons. We made paired whole-cell recordings from cultured hippocampal neurons, and monitored IPSCs. The cholinergic agonist carbachol (CCh) markedly enhanced DSI at 0.01-0.3 μM without changing the presynaptic cannabinoid sensitivity. The facilitating effect of CCh on DSI was mimicked by the muscarinic agonist oxotremorine-M, whereas it was eliminated by the muscarinic antagonist atropine. It was also blocked by a non-hydrolizable analogue of GDP (GDP-β-S) that was applied intracellularly to postsynaptic neurons. CCh still enhanced DSI significantly under the blockade of postsynatpic K^+ conductance, and did not significantly influence the depolarization-induced Ca^{2+} transients. These results indicate that the activation of postsynaptic muscarinic receptors facilitates the depolarization-induced release of endocannabinoids.

7. DISCUSSION

How can DSI and DSE contribute to the excitatory-inhibitory balance of the hippocampal neurons? Previous studies show that endocannabinoids are synthesized "on demand" in stimulated neurons and released from them in a Ca^{2+}-dependent manner[14,20,21] We have shown that the postsynaptic elevation of $[Ca^{2+}]_i$ and DSI had similar time courses in cultured hippocampal neurons[40]. Therefore, the amount of released endocannabinoids can directly reflect the activity of postsynaptic neurons and the resultant elevation of $[Ca^{2+}]_i$. If endocannabinoids suppress excitatory and inhibitory inputs to the same extent, it will cause no change in the excitability of postsynaptic neurons. Our data, however, indicate that excitatory and inhibitory synapses of the hippocampus had different sensitivities to cannabinoids[50]. While excitatory synapses were homogeneous and had moderate sensitivities to WIN55,212-2, inhibitory synapses were dichotomized into two distinct populations, the one with a high sensitivity to WIN55,212-2 and the other with no sensitivity. Thus, endocannabinoids can control the balance between excitatory and inhibitory inputs depending on the local concentration around synapses (Fig. 7). When the activity of hippocampal neurons elevated moderately, endocannabinoids will be released, which can suppress DSI-positive inhibitory synapses but are not sufficient to affect excitatory synapses. Under this condition, the endocannabinoid system will exert net excitatory action. DSI may facilitate the induction of synaptic plasticity, as reported recently[54]. When the activity of hippocampal neurons increased further, even larger amount of endocannabinoids will be released, which may be sufficient to suppress not only DSI-positive inhibitory synapses but also excitatory synapses. This may lead to net inhibition of the neural circuitry.

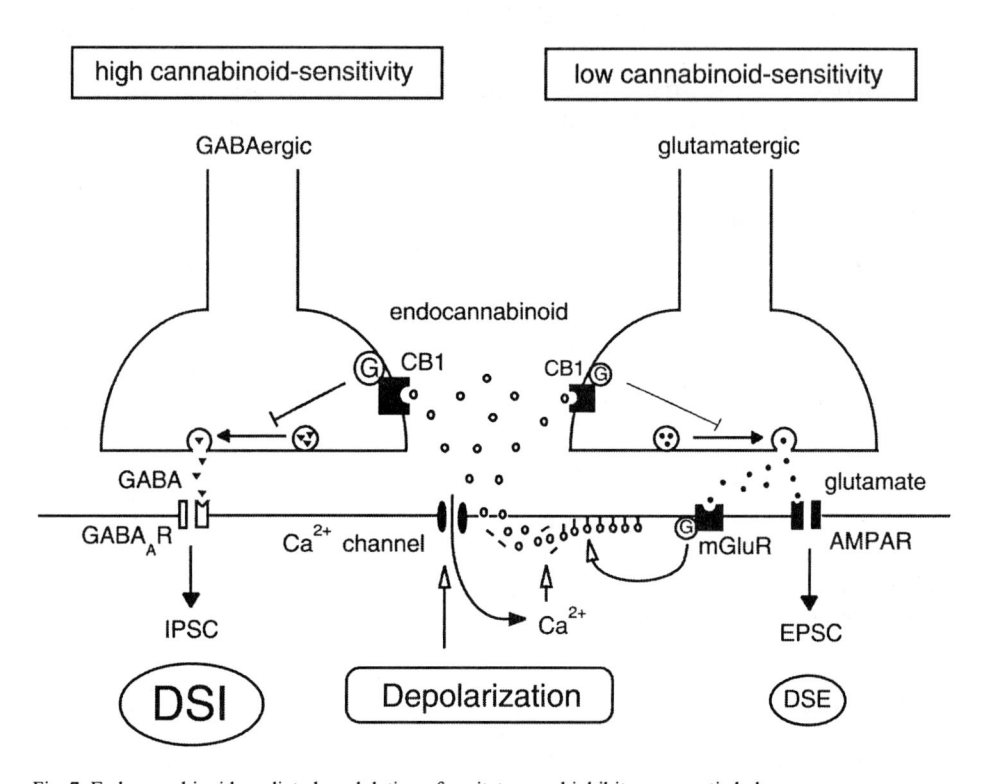

Fig. 7. Endocannabinoid-mediated modulation of excitatory and inhibitory synaptic balance

In addition to the modulation of synaptic transmission, cannabinoid agonists have been reported to exert variable effects on neurons[20, 5]. These include inhibition of adenylate cyclase[27], inhibition of voltage-gated Ca^{2+} channels[29, 31], activation of inwardly rectifying K^+ channels[30]. These effects can be produced by 1-100 nM WIN55,212-2. It is, therefore, likely that these effects can work in concert with DSI and DSE to regulate the net excitability of the postsynaptic neuron *in vivo*.

Experimental evidence indicates that activation of group I mGluR releases endocannabinoids and suppress GABA release through activation of presynaptic CB1 receptors[51,52]. In cultured hippocampal neurons, a low dose of DHPG, which alone suppressed IPSCs only slightly, significantly enhanced DSI. The magnitude of the enhanced DSI was much larger than the simple summation of the magnitude of the control DSI and that of DHPG-induced suppression[51]. Similar results were obtained in hippocampal slices. These results indicate that the postsynaptic depolarization and the activation of group I mGluRs work in a cooperative manner to release endocannabinoids.

The CB1 receptor are expressed at presynaptic terminals of various regions of the CNS including the cerebellum and the hippocampus[1, 55, 3]. Most neurons in the CNS express either mGluR1 or mGluR5 on the postsynaptic site[56-58, 25]. Therefore, modulation of synaptic transmission by the group I mGluR-induced endocannabinoid production may be of general importance in neural functions. Furthermore, endocannabinoid production can also be facilitated following activation of postsynaptic muscarinic receptors. The muscarinic receptor subtypes involved in this process are presumed to be M1, M3 or M5, because they are Gq/11-coupled similar to group I mGluRs. Thus, the enhancement of DSI, and presumably also DSE, by the facilitation of endocannabinoid production could be an important and widespread mechanism in the CNS through which the activation of postsynaptic Gq/11 coupled receptors could modulate synaptic transmission.

8. ACKNOWLEDGMENTS

This work was supported by grants-in aid for scientific research and the special coordination funds for promoting science and technology from the Ministry of Education, Science, Sports, Culture and Technology of Japan, by the Sumitomo Foundation, and by the Cell Science Research Foundation.

9. REFERENCES

1. Egertova, M., and Elphick, M. R., (2000) Localisation of cannabinoid receptors in the rat brain using antibodies to the intracellular C-terminal tail of CB. *J Comp Neurol* **422**: 159-171.
2. Herkenham, M., Lynn, A. B., Johnson, M. R., Melvin, L. S., de Costa, B. R., and Rice, K. C., (1991) Characterization and localization of cannabinoid receptors in rat brain: a quantitative in vitro autoradiographic study. *J Neurosci* **11**: 563-583.
3. Matsuda, L. A., Bonner, T. I., and Lolait, S. J., (1993) Localization of cannabinoid receptor mRNA in rat brain. *J Comp Neurol* **327**: 535-550.
4. Tsou, K., Brown, S., Sanudo-Pena, M. C., Mackie, K., and Walker, J. M., (1998) Immunohistochemical distribution of cannabinoid CB1 receptors in the rat central nervous system. *Neuroscience* **83**: 393-411.
5. Felder, C. C., and Glass, M., (1998) Cannabinoid receptors and their endogenous agonists. *Annu Rev Pharmacol Toxicol* **38**: 179-200.
6. Matsuda, L. A., Lolait, S. J., Brownstein, M. J., Young, A. C., and Bonner, T. I., (1990) Structure of a cannabinoid receptor and functional expression of the cloned cDNA. *Nature* **346**: 561-564.
7. Munro, S., Thomas, K. L., and Abu-Shaar, M., (1993) Molecular characterization of a peripheral receptor for cannabinoids. *Nature* **365**: 61-65.
8. Klein, T. W., Newton, C., and Friedman, H., (1998) Cannabinoid receptors and immunity. *Immunol Today* **19**: 373-381.
9. Katona, I., Sperlagh, B., Sik, A., Kafalvi, A., Vizi, E. S., Mackie, K., and Freund, T. F., (1999) Presynaptically located CB1 cannabinoid receptors regulate GABA release from axon terminals of specific hippocampal interneurons. *J Neurosci* **19**: 4544-4558.
10. Tsou, K., Mackie, K., Sanudo-Pena, M. C., and Walker, J. M., (1999) Cannabinoid CB1 receptors are localized primarily on cholecystokinin-containing GABAergic interneurons in the rat hippocampal formation. *Neuroscience* **93**: 969-975.

11. Hajos, N., and Freund, T. F., (2002) Pharmacological separation of cannabinoid sensitive receptors on hippocampal excitatory and inhibitory fibers. *Neuropharmacology* **43**: 503-510.
12. Hajos, N., Ledent, C., and Freund, T. F., (2001) Novel cannabinoid-sensitive receptor mediates inhibition of glutamatergic synaptic transmission in the hippocampus. *Neuroscience* **106**: 1-4.
13. Devane, W. A., Hanus, L., Breuer, A., Pertwee, R. G., Stevenson, L. A., Griffin, G., Gibson, D., Mandelbaum, A., Etinger, A., and Mechoulam, R., (1992) Isolation and structure of a brain constituent that binds to the cannabinoid receptor. *Science* **258**: 1946-1949.
14. Mechoulam, R., Ben-Shabat, S., Hanus, L., Ligumsky, M., Kaminski, N. E., Schatz, A. R., Gopher, A., Almog, S., Martin, B. R., Compton, D. R., and et al., (1995) Identification of an endogenous 2-monoglyceride, present in canine gut, that binds to cannabinoid receptors. *Biochem Pharmacol* **50**: 83-90.
15. Sugiura, T., Kodaka, T., Nakane, S., Miyashita, T., Kondo, S., Suhara, Y., Takayama, H., Waku, K., Seki, C., Baba, N., and Ishima, Y., (1999) Evidence that the cannabinoid CB1 receptor is a 2-arachidonoylglycerol receptor. Structure-activity relationship of 2-arachidonoylglycerol, ether-linked analogues, and related compounds. *J Biol Chem* **274**: 2794-2801.
16. Sugiura, T., Kondo, S., Sukagawa, A., Nakane, S., Shinoda, A., Itoh, K., Yamashita, A., and Waku, K., (1995) 2-Arachidonoylglycerol: a possible endogenous cannabinoid receptor ligand in brain. *Biochem Biophys Res Commun* **215**: 89-97.
17. Bisogno, T., Berrendero, F., Ambrosino, G., Cebeira, M., Ramos, J. A., Fernandez-Ruiz, J. J., and Di Marzo, V., (1999a) Brain regional distribution of endocannabinoids: implications for their biosynthesis and biological function. *Biochem Biophys Res Commun* **256**: 377-380.
18. Bisogno, T., Melck, D., De Petrocellis, L., and Di Marzo, V., (1999b) Phosphatidic acid as the biosynthetic precursor of the endocannabinoid 2-arachidonoylglycerol in intact mouse neuroblastoma cells stimulated with ionomycin. *J Neurochem* **72**: 2113-2119.
19. Cadas, H., Gaillet, S., Beltramo, M., Venance, L., and Piomelli, D., (1996) Biosynthesis of an endogenous cannabinoid precursor in neurons and its control by calcium and cAMP. *J Neurosci* **16**: 3934-3942.
20. Di Marzo, V., Melck, D., Bisogno, T., and De Petrocellis, L., (1998) Endocannabinoids: endogenous cannabinoid receptor ligands with neuromodulatory action. *Trends Neurosci* **21**: 521-528.
21. Piomelli, D., Giuffrida, A., Calignano, A., and Rodriguez de Fonseca, F., (2000) The endocannabinoid system as a target for therapeutic drugs. *Trends Pharmacol Sci* **21**: 218-224.
22. Stella, N., Schweitzer, P., and Piomelli, D., (1997) A second endogenous cannabinoid that modulates long-term potentiation. *Nature* **388**: 773-778.
23. Cravatt, B. F., Giang, D. K., Mayfield, S. P., Boger, D. L., Lerner, R. A., and Gilula, N. B., (1996) Molecular characterization of an enzyme that degrades neuromodulatory fatty-acid amides. *Nature* **384**: 83-87.
24. Goparaju, S. K., Ueda, N., Yamaguchi, H., and Yamamoto, S., (1998) Anandamide amidohydrolase reacting with 2-arachidonoylglycerol, another cannabinoid receptor ligand. *FEBS Lett* **422**: 69-73.
25. Thomas, E. A., Cravatt, B. F., Danielson, P. E., Gilula, N. B., and Sutcliffe, J. G., (1997), Fatty acid amide hydrolase, the degradative enzyme for anandamide and oleamide, has selective distribution in neurons within the rat central nervous system. *J Neurosci Res* **50**: 1047-1052.
26. Egertova, M., Giang, D. K., Cravatt, B. F., and Elphick, M. R., (1998) A new perspective on cannabinoid signalling: complementary localization of fatty acid amide hydrolase and the CB1 receptor in rat brain. *Proc R Soc Lond B Biol Sci* **265**: 2081-2085.
27. Howlett, A. C., and Fleming, R. M., (1984) Cannabinoid inhibition of adenylate cyclase. Pharmacology of the response in neuroblastoma cell membranes. *Mol Pharmacol* **26**: 532-538.
28. Felder, C. C., Joyce, K. E., Briley, E. M., Mansouri, J., Mackie, K., Blond, O., Lai, Y., Ma, A. L., and Mitchell, R. L., (1995) Comparison of the pharmacology and signal transduction of the human cannabinoid CB1 and CB2 receptors. *Mol Pharmacol* **48**: 443-450.
29. Mackie, K., and Hille, B., (1992) Cannabinoids inhibit N-type calcium channels in neuroblastoma-glioma cells. *Proc Natl Acad Sci U S A* **89**: 3825-3829.
30. Mackie, K., Lai, Y., Westenbroek, R., and Mitchell, R., (1995) Cannabinoids activate an inwardly rectifying potassium conductance and inhibit Q-type calcium currents in AtT20 cells transfected with rat brain cannabinoid receptor. *J Neurosci* **15**: 6552-6561.
31. Twitchell, W., Brown, S., and Mackie, K., (1997) Cannabinoids inhibit N- and P/Q-type calcium channels in cultured rat hippocampal neurons. *J Neurophysiol* **78**: 43-50.
32. Gifford, A. N., and Ashby, C. R., Jr., (1996) Electrically evoked acetylcholine release from hippocampal slices is inhibited by the cannabinoid receptor agonist, WIN 55212-2, and is potentiated by the cannabinoid antagonist, SR 141716A. *J Pharmacol Exp Ther* **277**: 1431-1436.
33. Hoffman, A. F., and Lupica, C. R., (2000) Mechanisms of cannabinoid inhibition of GABA(A) synaptic transmission in the hippocampus. *J Neurosci* **20**: 2470-2479.

34. Ishac, E. J., Jiang, L., Lake, K. D., Varga, K., Abood, M. E., and Kunos, G., (1996) Inhibition of exocytotic noradrenaline release by presynaptic cannabinoid CB1 receptors on peripheral sympathetic nerves. *Br J Pharmacol* **118**: 2023-2028.
35. Shen, M., Piser, T. M., Seybold, V. S., and Thayer, S. A., (1996) Cannabinoid receptor agonists inhibit glutamatergic synaptic transmission in rat hippocampal cultures. *J Neurosci* **16**: 4322-4334.
36. Ohno-Shosaku, T., Sawada, S., and Yamamoto, C., (1998) Properties of depolarization-induced suppression of inhibitory transmission in cultured rat hippocampal neurons. *Pflugers Arch* **435**: 273-279.
37. Pitler, T. A., and Alger, B. E., (1992) Postsynaptic spike firing reduces synaptic $GABA_A$ responses in hippocampal pyramidal cells. *J Neurosci* **12**: 4122-4132.
38. Llano, I., Leresche, N., and Marty, A., (1991) Calcium entry increases the sensitivity of cerebellar Purkinje cells to applied GABA and decreases inhibitory synaptic currents. *Neuron* **6**: 565-574.
39. Lenz, R. A., Wagner, J. J., and Alger, B. E., (1998) N- and L-type calcium channel involvement in depolarization-induced suppression of inhibition in rat hippocampal CA1 cells. *J Physiol* **512 (Pt 1)**: 61-73.
40. Ohno-Shosaku, T., Maejima, T., and Kano, M., (2001) Endogenous cannabinoids mediate retrograde signals from depolarized postsynaptic neurons to presynaptic terminals. *Neuron* **29**: 729-738.
41. Vincent, P., and Marty, A., (1993) Neighboring cerebellar Purkinje cells communicate via retrograde inhibition of common presynaptic interneurons. *Neuron* **11**: 885-893.
42. Wilson, R. I., and Nicoll, R. A., (2001) Endogenous cannabinoids mediate retrograde signalling at hippocampal synapses. *Nature* **410**: 588-592.
43. Zucker, R. S., and Regehr, W. G., (2002) Short-term synaptic plasticity. *Annu Rev Physiol* **64**: 355-405.
44. Diana, M. A., Levenes, C., Mackie, K., and Marty, A., (2002) Short-term retrograde inhibition of GABAergic synaptic currents in rat Purkinje cells is mediated by endogenous cannabinoids. *J Neurosci* **22**: 200-208.
45. Kreitzer, A. C., and Regehr, W. G., (2001a) Cerebellar depolarization-induced suppression of inhibition is mediated by endogenous cannabinoids. *J Neurosci* **21**: RC174.
46. Yoshida, T., Hashimoto, K., Zimmer, A., Maejima, T., Araishi, K., and Kano, M., (2002) The cannabinoid CB1 receptor mediates retrograde signals for depolarization-induced suppression of inhibition in cerebellar Purkinje cells. *J Neurosci* **22**: 1690-1697.
47. Wallmichrath, I., and Szabo, B., (2002) Cannabinoids inhibit striatonigral GABAergic neurotransmission in the mouse. *Neuroscience* **113**: 671-682.
48. Kreitzer, A. C., and Regehr, W. G., (2001b) Retrograde inhibition of presynaptic calcium influx by endogenous cannabinoids at excitatory synapses onto Purkinje cells. *Neuron* **29**: 717-727.
49. Maejima, T., Hashimoto, K., Yoshida, T., Aiba, A., and Kano, M., (2001) Presynaptic inhibition caused by retrograde signal from metabotropic glutamate to cannabinoid receptors. *Neuron* **31**: 463-475.
50. Ohno-Shosaku, T., Tsubokawa, H., Mizushima, I., Yoneda, N., Zimmer, A., and Kano, M., (2002b) Presynaptic cannabinoid sensitivity is a major determinant of depolarization-induced retrograde suppression at hippocampal synapses. *J Neurosci* **22**: 3864-3872.
51. Ohno-Shosaku, T., Shosaku, J., Tsubokawa, H., and Kano, M., (2002a) Cooperative endocannabinoid production by neuronal depolarization and group I metabotropic glutamate receptor activation. *Eur J Neurosci* **15**: 953-961.
52. Varma, N., Carlson, G. C., Ledent, C., and Alger, B. E., (2001) Metabotropic glutamate receptors drive the endocannabinoid system in hippocampus. *J Neurosci* **21**: RC188.
53. Kim, J., Isokawa, M., Ledent, C., and Alger, B. E., (2002) Activation of muscarinic acetylcholine receptors enhances the release of endogenous cannabinoids in the hippocampus. *J Neurosci* **22**: 10182-10191.
54. Carlson, G., Wang, Y., and Alger, B. E., (2002) Endocannabinoids facilitate the induction of LTP in the hippocampus. *Nat Neurosci* **5**: 723-724.
55. Herkenham, M., Lynn, A. B., Little, M. D., Johnson, M. R., Melvin, L. S., de Costa, B. R., and Rice, K. C., (1990) Cannabinoid receptor localization in brain. *Proc Natl Acad Sci U S A* **87**: 1932-1936.
56. Abe, T., Sugihara, H., Nawa, H., Shigemoto, R., Mizuno, N., and Nakanishi, S., (1992) Molecular characterization of a novel metabotropic glutamate receptor mGluR5 coupled to inositol phosphate/Ca2+ signal transduction. *J Biol Chem* **267**: 13361-13368.
57. Shigemoto, R., Kinoshita, A., Wada, E., Nomura, S., Ohishi, H., Takada, M., Flor, P. J., Neki, A., Abe, T., Nakanishi, S., and Mizuno, N., (1997) Differential presynaptic localization of metabotropic glutamate receptor subtypes in the rat hippocampus. *J Neurosci* **17**: 7503-7522.
58. Shigemoto, R., Nakanishi, S., and Mizuno, N., (1992) Distribution of the mRNA for a metabotropic glutamate receptor (mGluR1) in the central nervous system: an in situ hybridization study in adult and developing rat. *J Comp Neurol* **322**: 121-135.

Excitatory-Inhibitory Balance:

Circuits

7

BALANCED RECURRENT EXCITATION AND INHIBITION IN LOCAL CORTICAL NETWORKS

David A. McCormick, You-Sheng Shu and Andrea Hasenstaub [*]

1. INTRODUCTION

A basic feature of cortical networks is a large degree of divergence and convergence, both locally and between cortical areas. Each cortical pyramidal cell receives approximately 10,000 synaptic inputs, of which about 75% are excitatory[1-3]. The vast majority of these excitatory inputs arise from other cortical neurons, with each presynaptic cell donating only a few synapses, leading to a very large degree of convergence of synaptic input onto each cortical cell. Since cortical pyramidal cells typically give rise to a dense innervation of the local region in which they reside, a high degree of local, recurrent connectivity is generated. Pyramidal to pyramidal cell recurrent connectivity is expressed within a cortical lamina as well as between cortical laminae, resulting in a local horizontal and vertical interconnected network. Interposed between this recurrent excitatory network of pyramidal cells are a large variety of GABAergic inhibitory interneurons[4,5]. One prominent pattern of connectivity in the cerebral cortex is that local axons from pyramidal cells innervate not only other pyramidal cells, but also GABAergic interneurons, which in turn innervate local pyramidal cells. This pattern of connectivity places GABAergic neurons in the role of regulator of excitatory communication within the cortex. Indeed, reduction or loss of this inhibitory control quickly results in the generation of the abnormal and powerful synchronized cortical discharges of epilepsy[6].

This basic pattern of cortical connectivity suggests that there is a functional balance between recurrent excitation and inhibition in the cortex, although the precise nature of this balance is only recently coming to light. Computational modeling studies have predicted a strong relation between the amplitude of recurrent excitation and inhibition in cortical networks, resulting in a proportionality of these two feedback signals[7]. Such a proportionality may result in a relatively stable network in which large swings in

---

[*] D.A. McCormick, Y-S. Shu, A. Hasenstaub, Department of Neurobiology, Yale University School of Medicine, 333 Cedar St, New Haven, CT 06510 USA

neuronal activity (e.g. seizures) are prevented. So-called "balanced" networks have been proposed to account for a large variety of observations of cortical activity including the stochastic properties of discharge rates[8,9], the generation of persistent activity for seconds or longer[10-12], and the modulation of neuronal excitability with attention[13,14] or sensory-motor transformations[15].

Here we examine the nature of proportionality and balance between local recurrent excitation and inhibition in the cerebral cortex and how this may be used to rapidly control neuronal excitability.

2. CORTICAL NETWORKS GENERATE TWO ACTIVITY STATES IN VIVO AND IN VITRO

Extracellular and intracellular recordings in vivo in anesthetized animals revealed the presence of two states of cortical activity which alternated with a period of approximately 4 to 10 seconds (0.1 to 0.25 Hz)[17-23]. During the active or "UP" period of the activity, cortical neurons were depolarized (by about 10 mV on average) and received strong barrages of synaptic potentials.

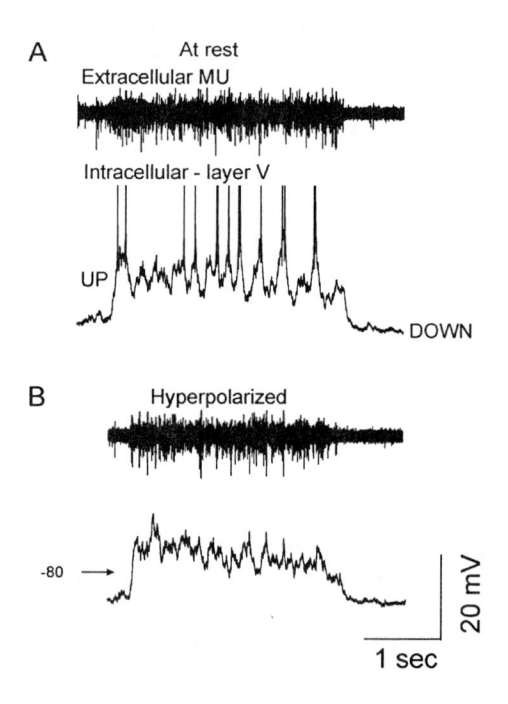

Fig. 1. The prefrontal cortical slice in vitro spontaneously generates prolonged periods of activity through synaptic bombardments. A. Simultaneous intracellular and extracellular recordings in layer 5 of the ferret prefrontal cortex. The network enters into the "UP" state for approximately 3 seconds prior to a rapid transition to the DOWN state. B. Hyperpolarization of the intracellularly recorded pyramidal cell reveals the barrage of PSPs arriving during the UP state and the reduction of PSPs during the DOWN state. Note that hyperpolarization or depolarization does not affect the duration of the UP state, which is the same as the duration of action potential discharge in the local network. From ref. 16.

These UP states were relatively stable in that the membrane potential did not deviate substantially from its mean value (SD of approximately 2-4 mV)[17, 23]. During UP states, cortical pyramidal cells discharge at a low to moderate rate (average rates of between 5 and 20 Hz). Between UP states, the network of cortical cells is relatively quiet with individual cells exhibiting a hyperpolarized membrane potential, a marked reduction in synaptic bombardment, and relatively little action potential activity. This period of inactivity became known as the DOWN state[17]. This recurrent pattern of UP and DOWN states is promoted by certain anesthetics (e.g. urethane) but can also be observed during natural sleep[17], indicating that is more than a curio of anesthesia.

While studying the mechanisms of synaptic depression, Maria Sanchez-Vives[24] observed that the so-called slow oscillation between UP and DOWN states occurred in cortical slices in which the composition of the ionic medium was changed to more closely match that found in situ (namely, extracellular Ca^{2+} and Mg^{2+} were reduced to 1 – 1.2 mM). In this medium, cortical networks undergo recurring transitions between hyperpolarized DOWN states and depolarized UP states approximately once every 4-10 seconds (Fig. 1). The occurrence of this activity in cortical slices indicates that it can be generated by a restricted local cortical network, and therefore does not require large scale corticocortical interactions.

Intracellular recordings in both pyramidal and non-pyramidal (fast spiking interneurons) cells in vitro revealed that the UP state was associated with strong barrages of synaptic potentials that caused the membrane potential to depolarize by an average of 8 mV[24,25] (Fig. 1). We hypothesize that the transition to the UP state is initiated by the progressive recruitment of neurons, through recurrent excitation initiated by spontaneously active neurons, following recovery from refractory mechanisms. The spontaneous transition to the DOWN state appears to result from the build up of a refractory mechanism[24]. For example, initiating an UP state with the local excitation of neurons by the application of glutamate results in UP states whose duration lengthens with increasing delays from the cessation of the previous UP state. Intracellular recordings in pyramidal cells reveal that the DOWN state is associated with an afterhyperpolarization, presumably generated by a Ca^{2+} or Na^+ activated K^+ current[24].

Therefore, we propose that the buildup in many networked cells of this intrinsic hyperpolarizing current, from the entry of Ca^{2+} and Na^+ during synaptic and action potential activity of the UP state, may be responsible for the spontaneous transition into the DOWN state (Fig. 2). During the DOWN period, the reduced level of synaptic and action potential activity results in a gradual dissipation of the afterhyperpolarization, priming the network for the generation of the next UP state (Fig. 2).

3. UP PERIOD ACTIVITY IS STRONGEST IN LAYER 5

Placement of several multiple unit recording electrodes between layer 1 and the white matter of the cerebral cortical slice revealed that the UP period activity occurred earliest and was strongest in layer 5[24]. Preparation of slices in which layers 2/3 were isolated from layer 5 revealed that the deeper layers still robustly generated the slow oscillation, while layers 2/3 did so much less frequently[24]. Thus, although supragranular layers could generate the UP state, the infragranular layers did so much more strongly, at

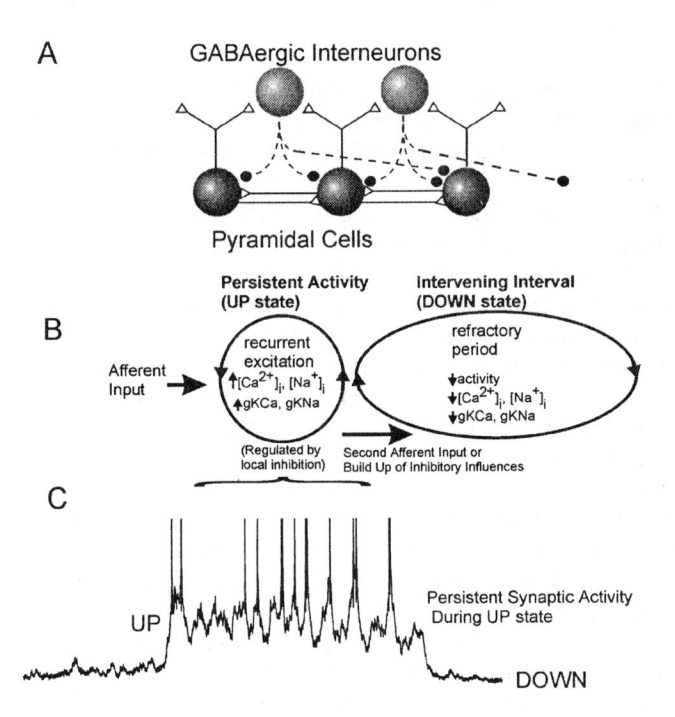

Fig. 2. Summary diagram of the proposed mechanisms for the spontaneous generation of the UP and DOWN states in cortical networks. Cortical pyramidal and local GABAergic interneurons are highly interconnected through local axonal connections such that activity in pyramidal cells excites both other pyramidal cells as well as local interneurons. The inhibitory feedback from the local interneurons controls the level of the UP state. Activation of an afferent input can trigger the transition from the DOWN to the UP state. Build up of intracellular levels of Ca^{2+} and Na^+ can activate K^+ currents, which eventually tip the balance back to the DOWN state, during which time the levels of Ca^{2+} and Na^+ inside the cell decrease, thus allowing the network to spontaneously generate another UP state. Another mechanism for making the transition from the UP to the DOWN state is the activation of afferent inputs[25].

the shortest latency, and with the longest discharge.

Placement of electrodes horizontally along the length of the slice within layer 5 revealed that the slow oscillation propagated, typically in a preferred direction, at a speed of approximately 11 mm/sec[24]. Recordings of the slow oscillation in vivo suggest that it may propagate at a significantly higher velocity, and in more complicated patterns, presumably owing to preserved connectivity[18]. Either way, the propagation of the slow oscillation is a natural consequence of the induction of anatomically connected cells (either locally or distally connected) into the UP state through recurrent excitation.

4. PERSISTENT ACTIVITY OF THE UP STATE IS GENERATED THROUGH A MIXTURE OF RECURRENT EXCITATION AND INHIBITION

To investigate the cellular mechanisms of generation of the persistent activity of the UP state, we examined the effect of changing membrane potential on the amplitude of

PSP barrages arriving during the UP state either with (Fig. 3) or without the reduction of voltage dependent Na$^+$ currents and K$^+$ currents through the inclusion of QX314 and CsAc in the intracellular recording electrode. The ionic currents arriving in cortical pyramidal cells during the generation of the UP state revealed a consistent reversal potential of approximately -35 mV throughout its duration (Fig. 3). Calculation of the changes in membrane conductance during the UP state revealed a steady increase followed by decrease (Fig. 3). Since the reversal potential is constant despite large changes in membrane conductance, the excitatory and inhibitory conductances must increase and decrease in precise proportion[25]. Plotting the inhibitory conductance as a function of excitatory conductance reveals a relatively linear relation, again indicating that these increase and decrease with each other in precise proportion during this type of

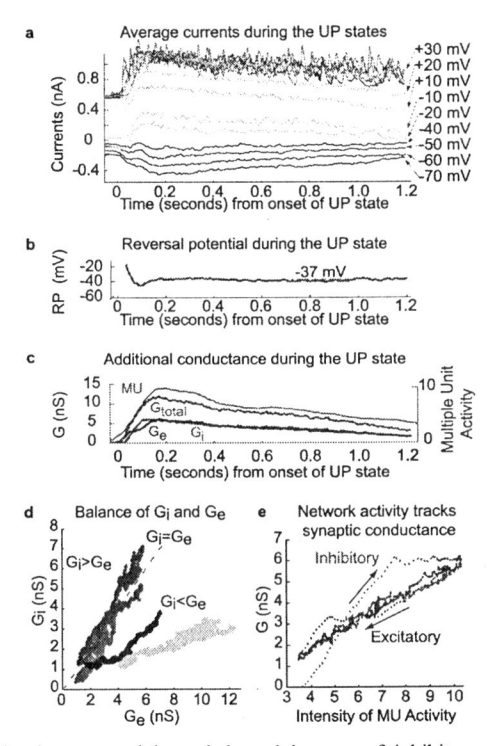

Fig. 3. Recurrent activity is generated by a balanced barrage of inhibitory and excitatory postsynaptic potentials. A. Average currents during the UP state under voltage clamp. Each trace is an average of 9-17 trials. Several raw traces at +30 mV are shown with the average for comparison. The average synaptic currents reverse around −37 mV in this cell. Electrode contained 2 M CsAc and 50 mM QX-314 to minimize the contribution from K$^+$ and Na$^+$ currents. Similar results were obtained with electrodes containing KAc only. B. Calculation of the reversal potential of the average synaptic currents over time. C. Illustration of the amplitude-time course during the UP state of the average multiple unit activity, the total increase in conductance (as measured by the change in slope of the IV plot) and the calculated conductance of excitatory and inhibitory currents. D. Plot of the inhibitory conductance as a function of the excitatory conductance in 5 different neurons. E. Relationship between the average intensity of neuronal activity in the local network, as measured by multiple unit activity, and the amplitude of the calculated excitatory and inhibitory conductances. In this cell, as the network transitioned into the UP state, the inhibitory conductance lagged the excitatory conductance. Following the onset of the UP state the excitatory and inhibitory conductances were proportional and strongly correlated with the intensity of activity in the nearby (< 100 μm) layer 5 multiple unit recording. From ref. 25.

cortical activity. This proportionality yields an impressive "reversal potential clamp" in which the reversal potential of the synaptic response is relatively constant (Fig. 3). Computational models of recurrent activity in the cortex have typically assumed such a proportionality between recurrent excitation and inhibition in the cortex[7-12], although it has not previously been directly demonstrated.

5. BARRAGES OF SYNAPTIC POTENTIALS ENHANCE NEURONAL RESPONSIVENESS

Cortical neurons are under the constant influence of barrages of synaptic potentials. How do so-called "background" barrages of synaptic potentials influence the neurons response to other synaptic inputs? We utilized the UP and DOWN states of the cortical slice in vitro to address this question. The UP state was associated with an increase in the probability that a synaptic input will cause action potentials in both pyramidal cells and fast spiking interneurons[25]. To address the mechanisms of this increase in responsiveness,

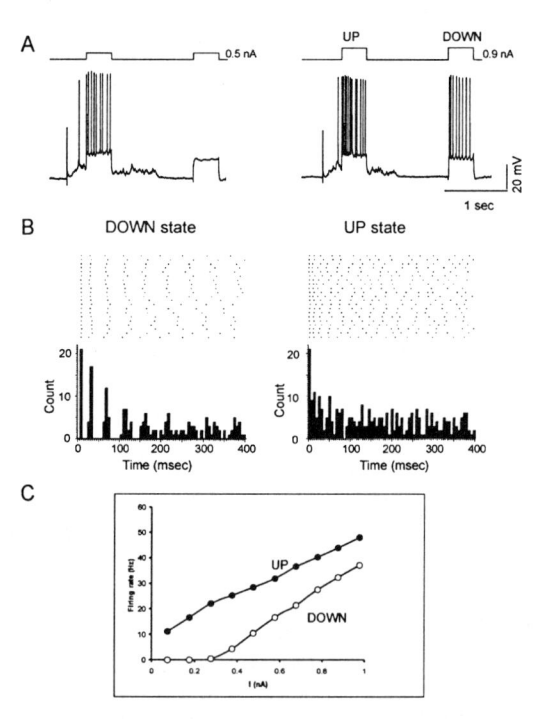

Fig. 4. The UP state is associated with a marked increase in neuronal excitability. A. Intracellular injection of depolarizing current pulses during the UP and DOWN states reveal a marked increase in neuronal responsiveness during the UP in comparison to the DOWN state. Two different amplitudes of current pulses are illustrated. B. Stimulus histograms of the action potential response of the cell to repeated injections of the same current pulse (0.7 nA) in the DOWN and UP states. The cell responds with more action potentials in the UP state, and the timing of these action potentials varies from pulse to pulse. C. Response of the neuron to different amplitude current pulses in the UP and DOWN states. Note the marked increase in neuronal responses, especially to smaller amplitude inputs. From ref. 16.

we used the dynamic clamp technique to inject into single pyramidal cells artificial synaptic conductances, or simply injected depolarizing current pulses of varying amplitudes (Fig. 4). During the UP state, the neuron's response to both of these inputs was facilitated considerably, such that previously subthreshold inputs could become suprathreshold (Fig. 4). There are three factors that may be involved in this enhancement associated with the UP state: depolarization of the membrane potential, increase in membrane conductance, and increase in membrane variance (noise). By varying each of these independently, we found that depolarization of the membrane potential during the UP state was largely responsible for the increase in neuronal responsiveness[16]. Increases in membrane conductance had the opposite effect, decreasing the number of action potentials generated by each current pulse or artificial synaptic conductance, while increases in membrane variance resulted in a decrease in the slope of the input-output relation (by facilitating the response to small inputs more than those to large inputs; see refs 13, 14). Interestingly, the UP state also had strong effects on spike timing, which depended upon the characteristics of the current or conductance waveform injected into the neuron studied. The latency to generate an action potential and jitter of spike timing to small, short duration inputs that mimicked single, small amplitude EPSPs were both decreased in response to the addition of moderate noise, or during the UP state. In contrast, the UP state or the addition of membrane potential variance resulted in an increase in the variance of action potential timing in response to constant current pulses (Fig. 4). Thus barrages of synaptic potentials that depolarize cortical cells during the UP state can have a profound facilitatory influence on the response of these cells to other depolarizing inputs and can increase spike precision to high frequency, while decreasing spike precision to low frequency, components of synaptic barrages.

6. IMPLICATIONS OF BALANCED RECURRENT EXCITATION AND INHIBITION: AROUSAL, ATTENTION, AND MEMORY

Our results indicate that a basic operation of the cerebral cortex is the generation of self-sustained periods of activity mediated by a proportional and balanced relation between recurrent excitation and inhibition. These intracortically generated UP states are most commonly observed during slow wave sleep or anesthesia[17-22]. However, the mechanisms by which they are generated has important implications for our understanding of the cellular basis not only of sleep rhythms, but also potentially of working memory, attention, and arousal.

Stimulation of ascending activating systems in the brainstem results in activation of the EEG (e.g. a pattern of EEG activity similar to waking), even in anesthetized animals. Intracellular recordings in cortical pyramidal cells during the slow oscillation reveal that brainstem stimulation results in an abolition of the DOWN state and a maintenance of the membrane potential at a level and noisiness that is similar to the UP state[20]. This result suggests that the UP state is similar to the resting state of the waking brain. However, intracellular recordings in vivo during natural sleep and waking reveal that the input resistance of pyramidal cells in the waking state my be significantly higher than during the UP state[17] and that the membrane potential may still exhibit significant depolarizing and hyperpolarizing jumps in the waking state. The mechanisms underlying the apparent increase in input resistance during the natural waking state in comparison to the UP state remain to be uncovered. This difference suggests that the UP state may be distinct from simply the resting state of the cell in the waking state. The UP state therefore has some, but not all, properties of several aspects of cortex function in the waking animal, including a maintained depolarization and persistent activity, an increase in neuronal excitability, and spontaneous low frequency discharge.

Recordings in multiple regions of the brain reveal that working memory tasks (in which a feature of a stimulus must be remembered for several seconds) result in the persistent discharge at rates of 5-40 Hz of selected cortical (and subcortical) neurons during the "memory" or delay period[26-34]. This persistent activity has been hypothesized to be generated either through recurrent excitatory and inhibitory interactions[11,12,35-39] or through the activation of intrinsic membrane mechanisms[40,41], although these mechanisms are not mutually exclusive. Our results demonstrate that local recurrent networks within the cortex are perfectly capable of generating second long periods of sustained activity mediated through the properties of recurrent excitation controlled by inhibition. We have not yet found a strong role for intrinsic membrane properties in the generation of the UP period. For example, hyperpolarization of the recorded neuron does not result in an abolition of the UP state nor does it change its duration or rate of recurrence. In addition, neuronal discharge during the delay period exhibits a broad distribution of interspike intervals, which is not seen when persistent activity is generated through intrinsic membrane mechanisms (Hasenstaub, Shu, Ghandi, McCormick, unpublished observations). Intracellular recordings in vivo, or the analysis of spike trains during delayed memory tasks, may help to resolve the issue of how this persistent activity is generated.

Possible Contribution of Neuronal Barrages to Attention Mechanisms

Attention to specific regions of sensory space, or specific features of an object, can be rapidly shifted at will or in response to a stimulus and results in a significant increase in signal detectability (e.g. salience) and decrease in reaction times. Extracellular recordings in the visual system often reveal attention to be associated with an increase in neuronal responsiveness, especially to less salient stimuli (such as a low contrast grating)[42,43]. Through what mechanisms could attention result in a rapid change in the excitability of cortical neurons? The release of neuromodulatory agents, such as acetylcholine or norepinephrine, is unlikely to be responsible since these have far too slow of a time course to underlie the rapid changes in excitability associated with shifts in attention. The leading hypothesis is that attentional mechanisms involve the rapid reconfiguration of neuronal networks through shifts in the synaptic bombardment of key elements of the network that corresponds to the stimulus region or feature that is being attended to. Our results suggest that increasing the synaptic bombardment of a cortical cell with a depolarizing barrage of EPSPs and IPSPs may result in enhancements of neuronal excitability that are similar to those observed in some attentional paradigms[43]. These shifts in responsiveness are naturally stronger for weak, versus strong stimuli (see Fig. 4).

Attention can also result in an increase in neuronal "gain" for all stimuli, meaning that the spike rate output for each stimulus is increased by the same percentage, regardless of the magnitude of the input[42]. How might such an increase in neuronal gain be achieved? It has been proposed that the rapid removal of barrages of synaptic potentials may underlie attentional changes in gain: if the incoming PSPs are perfectly balanced between excitation and inhibition then there will be no net change in membrane potentials, and the reduced neuronal conductance will make the cell more responsive to its other inputs[14]. However, this proposal has two unusual features. First, it implies that the neurons representing all of the large parameter space that is unattended are constantly bombarded with synaptic activity so as to keep their responsiveness low. Second, this model of attention requires a precise and ongoing balance of incoming EPSPs and IPSPs so that the membrane potential of the cell is unaffected by changes in background PSP rate (even though the membrane potential itself is constantly changing). It is difficult to imagine exactly how this precise balance between EPSPs, IPSPs, and membrane potential, even as it fluctuates, could be so precisely obtained. Rather, we propose that the network dynamics of neurons representing the attended object or attended spatial location are altered to facilitate a "pop-out" effect (Fig. 5). We envision a dynamic interaction of a facilitatory attentional template of the attended with the neurons that represent that object leading to an increase in response to the attended object. This results in an increase in its apparent salience in a manner similar to that recently proposed with computational models[44].

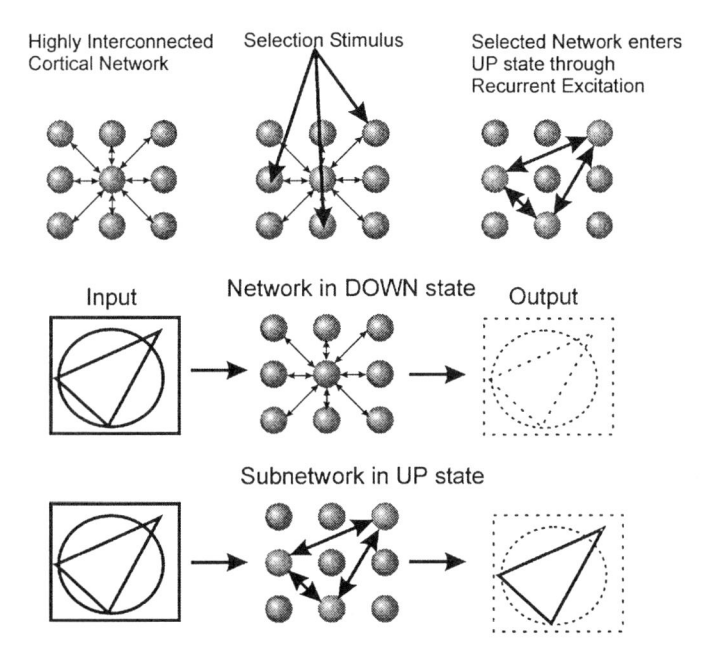

Fig. 5. Hypothetical involvement of synaptic barrages and recurrent excitation in the facilitatory effects of selective attention. In a highly interconnected network of cells, many possible configurations are possible. A "top down signal" may select a configuration of cells that roughly matches the form that is being attended. These cells may go into the UP state (or a similar depolarized state) through lateral interactions in response to the attentional signal. The arrival of a complex input to the network will then result in the selective transfer of neurons representing the attended signal if the subnetwork is in the UP state. In the DOWN state, cells are equally unresponsiveness (dashed lines in middle, right).

7. SUMMARY

Our results demonstrate that local cortical networks have the ability to generate recurrent periods of relatively low firing rate activity mediated by precisely balanced increases in both recurrent excitation and inhibition. This is exactly the type of activity that is expected from the basic architecture of the cortex: wide divergence and convergence with strong local recurrent excitation and disynaptic inhibition. The spontaneous recurrence of periods of locally generated UP states interspersed with periods of relative inactivity (DOWN) states resembles the transition between states observed in vivo during slow wave sleep and anesthesia. Although UP and DOWN states have not yet been observed in the waking brain, we hypothesize that similar mechanisms may underlie rapid changes in neuronal activity or excitability. The generation of local recurrent activity may be used to form temporary "memories" or to control neuronal responsiveness – computationally advantageous modifications that may serve multiple functions in the brain.

8. ACKNOWLEDGMENTS

Parts of this work were done in collaboration with Thierry Bal and Mathilde Badoual and we thank them for helpful suggestions. Supported by the NIH (DAM) and the Howard Hughes Foundation (AH).

9. REFERENCES

1. White, E.L. (1989) Cortical circuits. Boston, Birkhauser.
2. Abeles, M. (1991) *Corticonics. Neural Circuits of the Cerebral Cortex.* Cambridge University Press. Cambridge.
3. Braitenburg, V. and Shuz, A. (1998) Cortex: Statistics and geometry of Neuronal Connectivity. Springer.
4. Peters, A., Jones, E.G. (1984) Cerebral Cortex. Vol. 1. Cellular Components of the Cerebral Cortex. Plenum Press, New York.
5. Gupta, A., Wang, Y., Markram, H. Organizing principles for a diversity of GABAergic interneurons in the neocortex. *Science* **287**, 273-278 (2000).
6. McCormick, D.A. and Contreras, D. (2001) On the cellular and network bases of epileptic seizures. *Ann. Rev. Physiol.* **63**, 815-846.
7. Compte, A., Sanchez-Vives, M.V., McCormick, D.A., and Wang, X.J. (2003) Cellular and network mechanisms of slow oscillatory activity (< 1 Hz) in a cortical network model. *J. Neurophysiol.* In press.
8. van Vreeswijk, C. & Sompolinsky, H. Chaotic balanced state in a model of cortical circuits. *Neural Comp.* **10**, 1321-1371 (1998).
9. Shadlen, M.N., & Newsome, W.T. (1998) The variable discharge of cortical neurons: implications for connectivity, computation, and information coding. *J. Neurosci.* **18**, 3870-3896.
10. Brunel, N., and Wang, X.J. (2001) Effects of neuromodulation in a cortical network model of object working memory dominated by recurrent inhibition. *J. Comp. Neurosci.* **11**, 63-85.
11. Durstewitz, D., Seamans, J.K., and Sejnowski, T.J. (2000) Neurocomputational models of working memory. *Nature Neurosci.* **3**, 1184-1191.
12. Wang, X.J. (2001) Synaptic reverberation underlying mnemonic persistent activity. *Trends Neurosci.* **24**, 455-463.
13. Hô, N. and Destexhe, A. (2000). Synaptic background activity enhances the responsiveness of neocortical pyramidal neurons. *J. Neurophysiol.* **84**, 1488-1496.

14. Chance, F.S., Abbott, L.F., and Reyes, A.D. (2002) Gain modulation from background synaptic input. *Neuron* **35**, 773-782.
15. Andersen R.A., Essick G.K. and Siegel R.M. (1985). Encoding of spatial location by posterior parietal neurons. *Science* **230**, 450-458.
16. McCormick, D.A., Shu, Y-S., and Hasenstaub, A. (2003) Cortical persistent activity: mechanisms of generation and effects on neuronal excitability. *Cerebral Cortex*, in press.
17. Steriade, M., Timofeev, I., Grenier, F. (2001). Natural waking and sleep states, a view from inside neocortical neurons. *J. Neurophysiol.* **85**, 1969-1985.
18. Amzica, F. and Steriade, M. (1995) Short and long-range neuronal synchronization of the slow (< 1 Hz) oscillation. J. *Neurophysiol.* **73**, 20-38.
19. Contreras, D., and Steriade, M. (1995). Cellular basis of EEG slow rhythms, a study of dynamic corticothalamic relationships. *J. Neurosci.* **15**, 604–622.
20. Steriade, M., Amzica, F., and Nunez, A. (1993) Cholinergic and noradrenergic modulation of the slow (approximately 0.3 Hz) oscillation in neocortical cells. *J. Neurophysiol.* **70**, 1385-1400.
21. Cowan, R.L., Wilson, C.J. (1994) Spontaneous firing patterns and axonal projections of single corticostriatal neurons in the rat medial agranular cortex. *J. Neurophysiol.* **71**, 17-32.
22. Metherate, R. & Ashe, J.H. (1993) Ionic flux contributions to neocortical slow waves and nucleus basalis-mediated activation: whole-cell recordings *in vivo*. *J. Neurosci.* **13**, 5312-5323.
23. Stern, E.A., Kincaid, A.E., and Wilson, C.J. (1997) Spontaneous subthreshold membrane potential fluctuations and action potential variability of rat corticostriatal and striatal neurons in vivo. *J. Neurophysiol.* **77**, 1697-1715.
24. Sanchez-Vives, M.V., and McCormick, D.A. (2000) Cellular and network mechanisms of rhythmic recurrent activity in neocortex. *Nature Neurosci.* **3**, 1027-1034.
25. Shu, Y., Hasenstaub, A., and McCormick, D.A. (2003) Turning on and off recurrent balanced cortical activity. *Nature*, in press.
26. Fuster, J.M., and Alexander, G.E. (1973) Firing changes in cells of the nucleus medialis dorsalis associated with delayed response behavior. *Brain Res.* **61**, 69-91.
27. Fuster, J.M., and Jervey, J.P. (1981) Inferotemporal neurons distinguish and retain behaviorally relevant features of visual stimuli. *Science* **212**, 952-955.
28. Miyashita, Y., and Chang, H.S. (1988) Neuronal correlate of pictorial short-term memory in the primate temporal cortex. *Nature* **331**, 68-70.
29. Hikosaka, O., Sakamoto, M., Usui, S. (1989) Functional properties of monkey caudate neurons. III. Activities related to expectation of target and reward. *J. Neurophysiol.* **61**, 814-832.
30. Fuster, J.M. (1995) *Memory in the Cerebral Cortex*. MIT press, Cambridge.
31. Pesaran, B., Pezaris, J.S., Sahani, M., Mitra, P.P., and Andersen, R.A. (2002) Temporal structure in neuronal activity during working memory in macaque parietal cortex. *Nature Neurosci.* **5**, 805-811.
32. Goldman-Rakic, P.S. (1995) Cellular basis of working memory. *Neuron* **14**, 477-485.
33. Funahashi, S., Bruce, C.J., and Goldman-Rakic, P.S. (1989) Mnemonic coding of visual space in the monkey's dorsolateral prefrontal cortex. J. Neurophysiol. **61**, 331-349.
34. Miller, E.K., Erickson, C.A., and Desimone, R. (1996) Neural mechanisms of visual working memory in prefrontal cortex of the macaque. *J. Neurosci.* **16**, 5154-5167.
35. Camperi, M., and Wang, X.-J. (1998) A model of visuospatial memory in prefrontal cortex: recurrent network and cellular bistability. *J. Comp. Neurosci.* **5**, 383-405.
36. Amit, D.J., and Brunel, N. (1997) Model of global spontaneous activity and local structured activity during delay periods in the cerebral cortex. *Cerebral Cortex* **7**, 237-252.
37. Lisman, J.E., Fellous, J.-M., and Wang, X.-J. (1998) A role for NMDA-receptor channels in working memory. *Nature Neurosci.* **4**, 273-275.
38. Wang, X-J. (1999) Synaptic basis of cortical persistent activity: the importance of NMDA receptors to working memory. *J. Neurosci.* **19**, 9587-9603.
39. Compte, A., Brunel, N., Goldman-Rakic, P.S., and Wang, X.-J. (2000) Synaptic mechanisms and network dynamics underlying spatial working memory in a cortical network model. *Cerebral Cortex* **10**, 910-923.
40. Haj-Dahmane, S., and Andrade, R. (1998) Ionic mechanism of the slow afterdepolarization induced by muscarinic receptor activation in rat prefrontal cortex. *J. Neurophysiol.* **80**, 1197-1210.
41. Egorov, A.V., Hamam, B.N., Fransen, E., Hasselmo, M.E., and Alonso, A.A. (2002) Graded persistent activity in entorhinal cortex neurons. *Nature* **420**, 173-178.
42. McAdams, C.J., and Maunsell, J.H.R. (1999). Effects of attention on orientation-tuning functions of single neurons in macaque cortical area V4. J. Neurosci. *19*, 431-441.
43. Reynolds, J.H., Pasternak, T. and Desimone, R. (2000). Attention increases sensitivity of V4 neurons. *Neuron* **26**, 703-714.

44. Hahnloser, R.H.R., Douglas, R.J., Hepp, K. (2002) Attentional recruitment of inter-areal recurrent networks for selective gain control. *Neural Comp.* **14**, 1669-1689.

8

LOCAL CIRCUIT NEURONS IN THE FRONTAL CORTICO-STRIATAL SYSTEM

Yasuo Kawaguchi *

1. INTRODUCTION

The neostriatum is the principal recipient of afferents to basal ganglia from the cerebral cortex, especially, from the frontal cortex[1]. These two large forebrain structures, the neostriatum (caudate-putamen) and cortex, are both composed of morphologically diverse types of neurons[2-4]. Recently striatal and frontal cortical interneurons have been characterized from various points of view and they are now known to be not only morphologically but also physiologically and chemically very diverse[5-8]. Since the functional roles of each neuron type remain to be investigated, we do not understand the meaning of the neuronal diversity. In recent times it has been shown that striatal and cortical interneurons developmentally originate from the same distant sites then migrating to their location[9-11]. This suggests that the cortex and striatum may have a similar interneuronal organization. In fact, chemically similar interneurons exist in both structures. One chemical type shared in common contains the calcium-binding protein, parvalbumin (PV), and another the peptide, somatostatin (SOM) (somatotrophin-release inhibiting factor), these being expressed in separate interneuron populations. To facilitate understanding of the meaning of neuronal diversity and local circuit organization, we compare various characteristics of PV and SOM cells in the functionally intimately related frontal cortex and striatum here.

2. THE FRONTAL CORTICO-STRIATAL SYSTEM

The frontal cortico-striatal system is considered to be involved in the context-dependent release of various motor and cognitive circuits in the brainstem and frontal

---

* Y. Kawaguchi, Division of Cerebral Circuitry, National Institute for Physiological Sciences, Myodaiji, Okazaki 444-8585, Japan

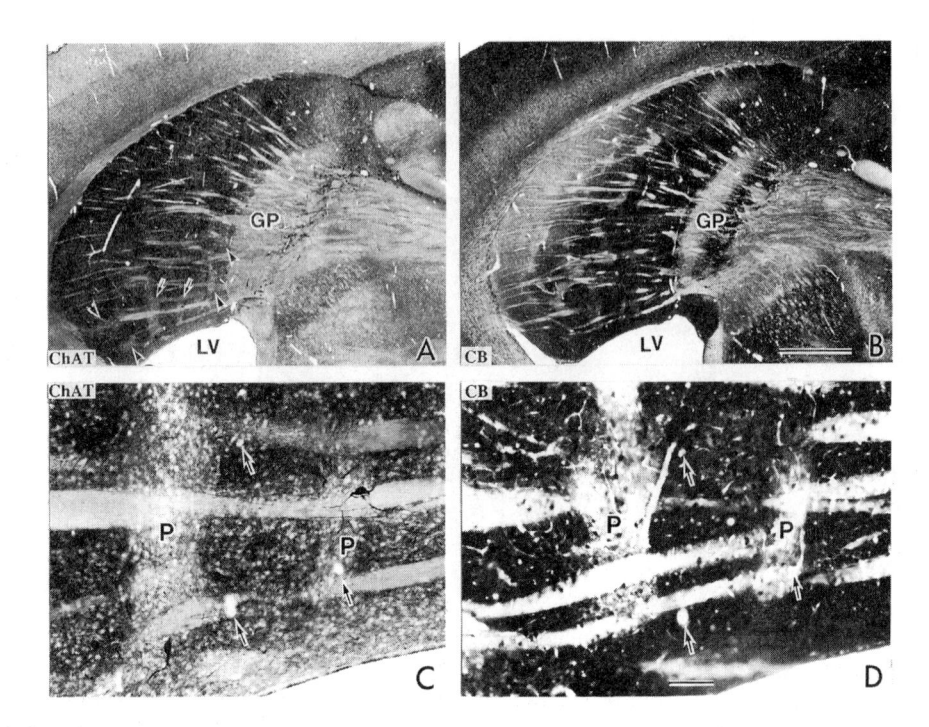

Fig. 1. Choline acetyltransferase (ChAT; a synthetic enzyme for acetylcholine) and calbindin D_{28k} (CB) immunoreactivities in the striatum (oblique horizontal sections). **A:** Note the ChAT-poor regions within striatum (arrows and arrowheads). The immunoreactivity is denser in the rostrolateral part, and weaker in regions close to the lateral ventricle (LV) and globus pallidus (GP). **B:** Calbindin D_{28k} (CB) immunoreactivity in a section serial to **A**. The rostrolateral region and the patch compartment show weak immunoreactivity. Note that the immunoreactivity of ChAT is spatially complementary to that of calbindin D_{28k}. **C:** ChAT poor-immunoreactive regions in part of the striatum close to the lateral ventricle (indicated by arrows in **A**). These correspond to calbindin D_{28k} poor patches (P) on a serial section (**D**). Arrows in C and D indicate capillaries as landmarks. Scale bars, 1 mm for **A** and **B**, and 100 μm for **C** and **D**. Modified from Ref. 167.

cortical areas[12,13]. Information transfer at the frontal cortico-striatal circuit is a crucial point for the forebrain neural loop through the cortex, basal ganglia, and thalamus. The cortex possesses layered structures, whereas the striatum shows cell clustering into two compartments, different in chemical characteristics, incoming afferents and projection sites, called the matrix (85% in area), and the patch (striosome) (15%) respectively[12,14]. Calbindin D_{28k} (calbindin) can be immunohistochemically demonstrated in matrix projection cells, although there is a medioventral to rostrocaudal gradient in staining intensity, but not in patch cells (Fig. 1). On the other hand, μ opioid receptors are expressed on patch cells, but not on the matrix counterparts. Intrinsic cholinergic fibers, mostly originating from cholinergic interneurons, mainly innervate the matrix (Fig. 1). It is interesting how cortical and striatal modular structures, the layered and compartmental arrangement, are synaptically connected and regulated by local circuit elements in each area.

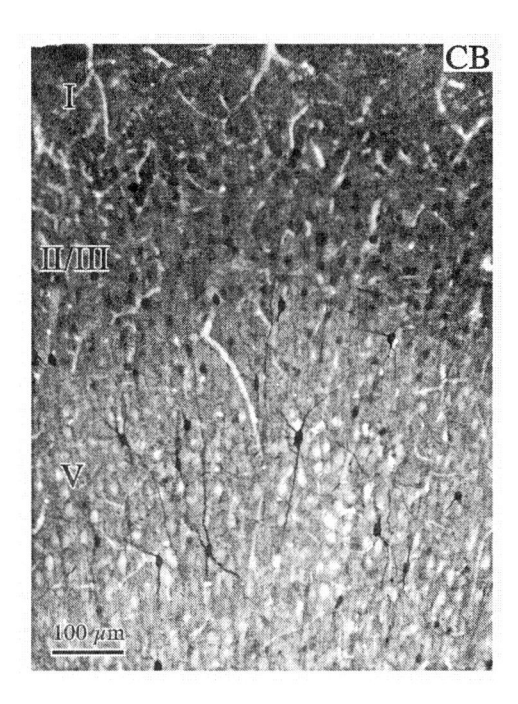

Fig. 2. Calbindin D_{28k} (CB) immunoreactivity in the frontal cortex. Roman numerals correspond to the cortical layers. In layers II/III both pyramidal and GABAergic non-pyramidal cells contain calbindin D_{28k}-positive cells, but in layer V calbindin D_{28k}-immunoreactivty is found in some non-pyramidal cells.

2.1. Corticostriatal pyramidal cells

Most cortical areas have projections to some part of the matrix, whereas medial frontal and orbital cortical areas are connected with both the patch and the matrix[14]. According to other projection sites than the striatum, corticostriatal cells in the frontal cortex are divided generally into two types (Fig. 3)[15-17]. Some pyramidal cells (crossed corticostriatal cells) send axons to the corpus callosum, and innervate the contralateral cortex and striatum in addition to the ipsilateral striatum. Other pyramidal cells (ipsilateral corticostriatal cells) in layer V innervate the ipsilateral striatum and project to the ipsilateral brain stem. Ipsilateral corticostriatal cells send axon collaterals to the thalamic nuclei, including the posterior thalamic group. The striatum can thus continuously receive both efferent signals from the frontal cortex to the ipsilateral thalamus/brainstem and those to the contralateral cortex/striatum. Corticostriatal axons show two intrastriatal arborization patterns[17,18]. Some axons provide a focal collateral innervation in a restricted area in one of the patch/matrix compartments. The other pattern is extended arborization which innervates both compartments[19]. These two arborization types are not correlated with projection types of corticostriatal cells (crossed and ipsilateral corticostriatal cells). These findings suggest a complicated wiring pattern from the frontal cortex to the striatum.

Frontal Cortex

GABAergic
non-pyramidal cell
chemical class
(1) parvalbumin
(2) somatostatin
(3) VIP, calretinin, CCK/VIP
(4) CCK
(5) others

Cortico-striatal
pyramidal cell
projection class
(1) ipsi-/cotralateral striatum
(2) ipsilateral striatum/thalamus/brainstem

Striatum

Aspiny interneuron
chemical class
(1) acetylcholine
(2) parvalbumin
(3) somatostatin
(4) calretinin
(5) others

Spiny projection cell
compartmental class
(chemical class)
(1) matrix (calbindin D_{28k})
(2) patch (μ opioid receptor)
projection class
(chemical class)
(1) GP (enkephalin)
(2) GP, SN (substance P)
(3) GP, EP, SN (substance P)
[(1), (2), (3): local intrastriatal arbor]
(4) GP [wide intrastriatal arbor]

GP: globus pallidus, external segment
EP: entopeduncular nucleus (globus pallidus, internal segment)
SN: substantia nigra

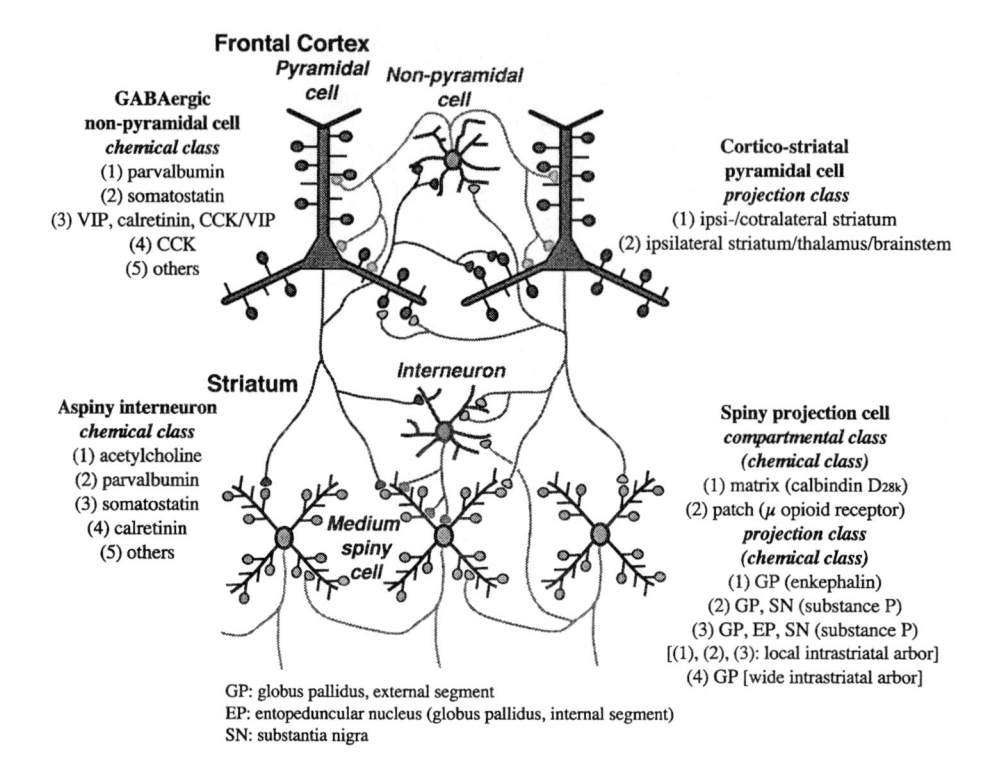

Fig. 3. Major neuronal types in the frontal cortico-striatal system. In the frontal cortex, GABAergic non-pyramidal cells are divided into several immunohistochemical classes (*chemical class*). Corticostriatal pyramidal cells in frontal cortex are separated into two main types according to their projecting sites (*projection class*). Striatal interneurons are also divided into several immunohistochemical classes. Striatal spiny projection cells belong to the matrix or patch region (*compartmental class*). The compartmental classes also differ in expression of calbindin or µ opioid receptors. Striatal projection cells are also separable into several types according to their projecting sites and intrastriatal innervation pattern.

2.2. Recurrent excitation of cortical pyramidal cells

Both local circuits, especially in the frontal cortex, contain many types of neurons and mutual connections. Projection neurons in both regions have spiny dendrites and receive excitatory inputs from the cortex and thalamus. Cortical projection cells, the pyramidal cells, are excitatory and recurrently connected with each other[20-23]. Recurrent excitatory interactions can induce slow rhythmic (<1Hz) depolarization (a hyperpolarized "down" state and a depolarized "up" state) generated intrinsically in the cortex during sleep or anaesthesia[24-27]. Sensory stimulation increases the probability of the up state of some visual cortical cells, suggesting this fluctuation may provide a substrate for encoding sensory information[28]. Reverberation by excitatory recurrent connections may play an important role in the computation of cortical circuits[29,30].

2.3. Recurrent connections of medium spiny cells in striatum

Striatal medium spiny cells are GABAeric inhibitory neurons projecting to the globus pallidus (external segment), entopeduncular nucleus (internal segment of globus pallidus) and/or substantia nigra (Fig. 3)[31]. Striatal spiny cells show up and down membrane potential fluctuation generated by cortical inputs[32-34]. Projection cells have local axonal arbors around the dendritic domain and induce GABAergic inhibition, although weak, on adjacent spiny cells[35,36]. Each spiny projection cell receives thousands of cortical inputs[18]. Each excitatory input on a spine is weak, and simultaneous activation of hundreds of inputs is needed to move to the up state for firing[37]. From these characteristics, the striatum is considered to use competitive learning to classify cortical inputs[38,39]. Although the competitive learning model requires input sharing among spiny projection cells, common cortical inputs between spiny cells are very few[19].

3. INTERNEURON ORGANIZATION IN THE STRIATUM AND FRONTAL CORTEX

Cortical pyramidal cells originate in the ventricular zone of the pallium and migrate radially into the cortex, whereas striatal projection cells are derived from the lateral ganglionic eminence[11,40]. Both cortical and striatal interneurons originate in the ganglionic eminence and migrate tangentially. Cortical and striatal interneurons are segregated for their destination during the migration, according to their expression patterns of neurophilins, receptors for semaphorin proteins (chemorepellents for growing axons)[41]. The same developmental origin for cortical and striatal interneurons suggests some similar characteristics.

In both the striatum and cerebral cortex, calcium binding proteins and peptides are expressed in some neuron types. Calbindin is found in striatal projection cells in the matrix compartment, but not in the patch (Fig. 1). In the frontal cortex, calbindin is expressed in pyramidal cells in the superficial layer, and in some GABAergic non-pyramidal cells in both the superficial and deep layers (Fig. 2). On the other hand, some other markers are expressed more selectively in cortical and striatal interneurons. Although the physiological roles of these marker substances remain to be elucidated, we assumed that chemically-differentiated neuron types have distinct functions in local circuits[6,7]. We have not been able to confirm this background assumption because chemical and physiological classes are enormously diverse in these regions[42-44]. Furthermore, the morphology, electrophysiology, and neurotransmitter receptor expression of interneurons do not coincide well in other cortical regions[45]. However, this idea is still helpful for us to dissect the complicated organization of the cortical local circuit[8].

3.1. Striatal interneuron types

In the striatum, projection cells express substance P, enkephalin, and/or calbindin[14]. Three main types of aspiny interneurons modulate the activity of the spiny projection cells, and can be histochemically distinguished (Fig. 3, 4)[5,46]. Striatal aspiny interneurons are chemically divided into three major groups: (1) cholinergic cells; (2) parvalbumin (PV) cells; and (3) somatostatin (SOM) cells, most of which also express nitric oxide

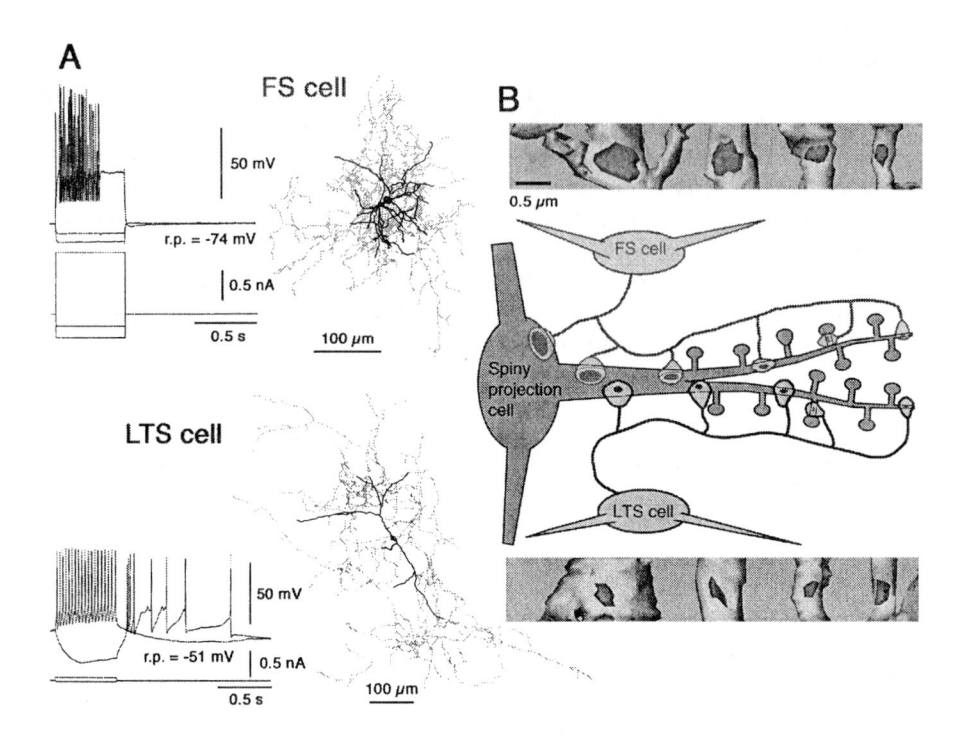

Fig. 4. FS cells and LTS cells in the striatum. **A:** Morphological and physiological characteristics of the two interneuron types. The somata and dendrites are shown in *black* and the axons in *gray*. FS cells have a hyperpolarized resting potential (r.p.), which rapidly, but transiently, fires short duration spikes with a constant interspike interval when depolarized with a current pulse. Firing sometimes resumes during the depolarization. LTS cells had depolarized resting potentials (r.p.) and high input resistances. Two potential responses are shown superposed for the injected currents shown in the lower traces. Note that the cell fired a low threshold spike after cessation of a hyperpolarization pulse. **B:** Relationships of synaptic junction area to target structure circumference for FS and LTS cells. The data were obtained from the 3-D reconstructions of serial ultrathin sections. 3-D reconstruction images of postsynaptic dendrites of FS (*above*) and LTS cells (*below*). The area of the synaptic junctions (*dark regions*) of the FS cells clearly increases with the size of the postsynaptic structures, whereas those of the LTS cells does not increase sharply. A schematic view of synaptic connections of GABAergic FS and LTS cells of the rat striatum is shown in the middle. Modified from Ref. 88.

synthase (NOS) and neuropeptide Y. These interneurons are distributed in both the patch and matrix compartments.

Cholinergic and dopaminergic interaction in the striatum has a crucial role in the control of voluntary movements[47,48]. Cholinergic innervation is mainly from the basal forebrain in the cortex. On the contrary, cholinergic fibers come mostly from the interneurons in the striatum[49-51]. Cholinergic neurons are identified by immunohistochemistry for choline acetyltransferase, an enzyme involved in acetylcholine synthesis. Striatal cholinergic cells are large aspiny cells with soma diameter larger than 20 μm which show tonic irregular firing at 2 ~ 10 Hz (tonically active neuron) (Table 1). They are innervated by dopaminergic fibers from the substantia nigra compacta[52].

Table 1. Choline acetyltransferase (ChAT) cells in the striatum and cortex

	Striatum	Cortex
cell type	large aspiny interneuron	small sparsely spiny or aspiny interneuron
GABA [1]	(-)	(+)
VIP [1]	(-)	(+)[a]
calretinin [2]	(-), rat ; (+), human	(+)[b]
firing pattern	large afterhyperpolarization cell (LA cell) tonically-active neuron	non-fast spiking cell (non-FS cell)
synaptic type	symmetrical	symmetrical
axonal arborization	wide (mostly in the matrix)	descending
transmitter action[§]		
nicotine		[5]
muscarine	hyperpolarization [3]	[6]
dopamine	depolarization (D1/D5) [4]	

[a] 30~50% of VIP cells (Ref. 73 ; unpublished observation)
[b] 30~60% of calretinin cells (Ref. 168; unpublished observation)
§ , recorded at soma; D1/D5, dopamine D1/D5 receptor.
[1] Ref. 71, 72, 73, 168; [2] Ref. 168, 169, 170; [3] Ref. 171; [4] Ref. 54.
[5] depolarization in VIP cells (Ref. 141); [6] depolarization in VIP cells (Ref. 172).
References for other items are in the text.

Tonically active neurons display firing changes in relation to sensory stimuli which trigger a rewarded movement, and the acquired sensory responsiveness depends on dopaminergic innervation[53]. Dopamine modulates striatal cholinergic cells through the postsynaptic receptors[54,55] or action on terminals onto cholinergic interneurons[56,57]. Striatal endogenous acetylcholine exerts a complex regulation of striatal synaptic transmission, which produces both short-term and long-term effects[47].

3.2. Cortical GABAergic interneuron types

GABAergic inhibition contribute to control of the input and output of cortical cells, and generation of rhythmic and synchronized activities[58,59]. Intracellular recordings in the visual cortex in vivo have revealed that diverse types of inhibitory connections are involved in generation of orientation selectivity, depending on the location of the orientation map or positioned layer[60,61]. Perforated-patch recordings from cortical pyramidal cells have shown that dendritic GABA responses are excitatory at somata regardless of timing, whereas somatic GABA responses are inhibitory when coincident with excitatory input but excitatory at earlier times[62]. Under certain circumstances GABA has an excitatory role in synaptic integration in the cortex. These recent observations indicate cortical GABAergic IPSPs regulate excitability of cortical cells in a spatially and temporally refined way.

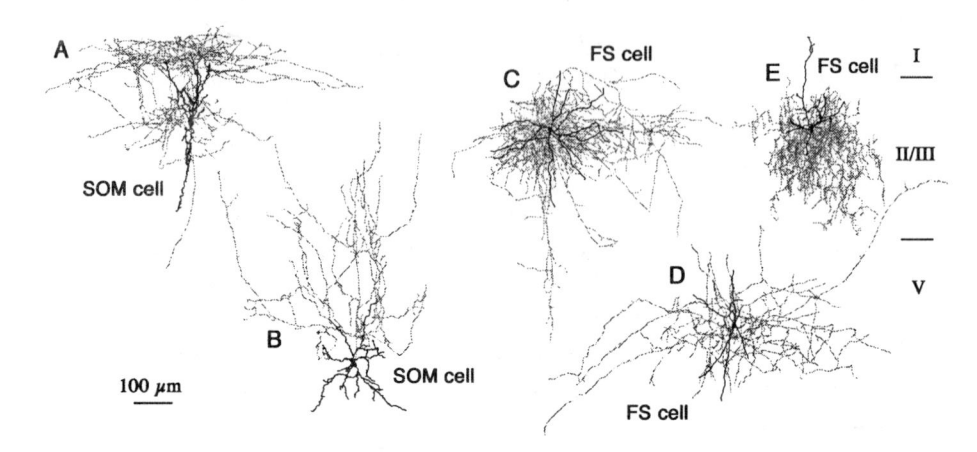

Fig. 5. Drawings of identified somatostatin and FS cells of the rat frontal cortex assembled from reconstructions of intracellularly labeled neurons in the slice preparations. The somata and dendrites are shown in *black*, and the axons in *gray*. Roman numerals correspond to cortical layers. **A:** A layer II/III somatostatin RSNP cell with ascending axonal arbors. Bouton percentage attaching to other somata (Btsoma) = 0.5 %. **B:** A layer V somatostatin BSNP (LTS) cell with ascending axonal arbors. Btsoma = 1.0 %. Cells in **A** and **B** correspond to Martinotti cells. **C:** A layer II/III FS basket cell. Btsoma = 23.1 %. **D:** A layer V FS basket cell. Btsoma = 25.0 %. **E:** A layer II/III FS chandelier cell. Btsoma = 2.0 %. Axonal boutons apposed to somata are apparent with differential interference contrast. The bouton proportion with contacts with other somata was calculated from 200 randomly-sampled boutons.

Compared to excitatory pyramidal cells, cortical inhibitory interneurons are highly diverse in morphological, chemical and physiological characteristics[2,42,43,63-67]. To understand the organizing principles of cortical circuits, it is necessary to define functional subtypes of inhibitory interneurons and reveal their synaptic connection rules within the columnar circuit[44,58 68-70].

In the frontal cortex, GABAergic non-pyramidal cells (interneurons) are segregated into several immunohistochemical groups (Fig. 3, 5). However, specific chemical markers have yet to be determined for some physiological and morphological types of cortical GABA cells yet[7,8]. Four main chemical types have been identified : (1) PV cells (some of them also containing calbindin); (2) SOM cells (most of them also containing calbindin and neuropeptide Y); (3) vasoactive intestinal polypeptide (VIP) cells [some of them also containing calretinin or cholecystokinin (CCK)]; and (4) large CCK cells. As in the striatum, PV and SOM cells belong to the major interneuron type. Each chemical group contains further morphological subgroups[8]. In contrast to the striatum, most cholinergic fibers in the cortex come from the basal forebrain. However, there are a few non-pyramidal cells immunohistochemically positive for choline acetyltransferase[71] (Table 1), which are supposed to be cholinergic interneurons, but are also positive for GABA[72,73]. In the frontal cortex, choline acetyltransferase-positive cells are immunoreactive for VIP and also positive for calretinin (unpublished observations). Choline acetyltransferase has been found in some VIP cells (41% in layers II/III, 54% in layer V, 38% in layer VI), and calretinin cells (65% in layers II/III, 35% in layer V, 32%

in layer VI) (unpublished observations). Morphological and physiological differences between corticopetal and intrinsic cholinergic innervations remain to be investigated.

3.3. Similar interneuron types in cortex and striatum

Although PV is an excellent marker for neuronal subpopulations in the central nervous system[74], its physiological function in the nervous system remains to be elucidated. In the cerebellar cortex and hippocampus PV modulates short-term plasticity of inhibitory synaptic transmission from interneurons to principal cells[75,76]. SOM, also found in both striatal and cortical interneurons, reduces the release of growth hormone from pituitary. In addition to its neuroendocrine role, SOM has diverse neurophysiological effects[77], reducing high-voltage activated Ca^{2+} currents in hippocampal pyramidal cells and striatal projection cells[78,79], and activating K^+ currents in cortical and hippocampal pyramidal cells[80,81].

The striatum and cortex possess common interneuron types in the local circuits. Although cholinergic interneurons are considered to be most important in the striatum from the physiological and pathological point of view, PV and SOM cells belong to the major interneuron types in both regions. Each interneuron type may have similar functional roles in the local circuit operation.

4. PARVALBUMIN AND SOMATOSATIN CELLS IN THE STRIATUM

4.1. Morphologies, transmitters and firing patterns

Striatal PV cells have dense innervation close to the dendritic field, with a resting potential hyperpolarized in a similar way to projection cells. In response to depolarizing current pulses, they fire a train of short duration spikes with little adaptation, so that the cells are called fast-spiking cells, FS cells (Fig. 4, Table 2). The repetitive firing ceases abruptly after a short time, and occasionally resumes. FS cells were further divided into two morphological types: those with local and those with extended dendritic fields[46]. SOM/NOS cells innervate wider areas than PV FS cells, and the resting potential is depolarized as in cholinergic cells. In contrast to cholinergic interneurons, calcium-dependent low-threshold spikes are induced from the hyperpolarized potential. These SOM/NOS cells are called low-threshold spike (LTS) cells (Fig. 4, Table 2). PV FS cells are GABAergic, showing immunoreactivity for GABA and glutamate decarboxylase (GAD)[82,83], and inducing GABA-A IPSPs[84]. An amino acid transmitter of SOM cells has not been definitely identified, but it may be GABA. GAD mRNAs are not normally detected in SOM cells[85,86], but on treatment with colchicine, 67 kD GAD immunoreactivity was found[87]. Post-embedding GABA immunohistochemistry using colloidal gold particles showed particles on PV boutons to be more numerous than on GABA-negative boutons. Particles on SOM boutons were more numerous than GABA-negative boutons, but significantly fewer than on PV boutons[88], indicating that somatostatin boutons express GABA with lower content than their parvalbumin counterparts.

The dendrites and axon collaterals of spiny cells generally do not cross the boundaries between the patch and matrix[89]. The separate distribution of the cortical

Table 2. Parvalbumin and somatostatin cells in the striatum and cortex

	Parvalbumin (PV) cells		Somatostatin (SOM) cells	
	Striatum	Cortex	Striatum	Cortex
cell type	aspiny or sparsely spiny interneuron	aspiny or sparsely spiny interneuron	aspiny or sparsely spiny interneuron	moderately spiny interneuron
developmental origin [1]	ganglionic eminence (medial)	ganglionic eminence	ganglionic eminence (medial)	ganglionic eminence
firing pattern	FS	FS	LTS	non-FS [RS in layers II/III] [RS and BS (LTS) in layer V]
transmitters [2]	GABA	GABA	NO NPY (80%) (GABA)	GABA NPY (30~40%) NO (10~30%)
GABAergic IPSP in target cell [3]	(+)	(+) (paired pulse depression)		(+) (paired pulsedepression)
synaptic type	symmetrical	symmetrical	symmetrical	symmetrical
synaptic target	soma dendritic shaft, spine	soma, axon dendritic shaft, spine	dendritic shaft, spine	dendritic shaft, spine
autaptic inhibition [4]		(+)		(-)
dependency of synaptic areas on target size	(+)	(+)*	(-)	(+)*
axonal arborization	local [basket cell]	local/horizontal wide (avoiding layer I) [basket cell] [chandelier cell]	wide (mostly in the matrix)	ascending wide (entering layer I) [Martinotti cell]
electrical coupling with the same type [5]	(+)	(+)		(+)
K^+ channel Kv3.1 [6]	(+)	(+)	(-)	(-)
Kv3.2 [6]				30% in layers V/VI
calbindin D_{28K} [2], [7]	(-)	10% +80%(weak) in layer II/III	20~40%	30~40% + 50%(weak) in layer II/III
		5% in layer V/VI		60~70% in layer V/VI
extrinsic GABA innervation [8]	(+) (from globus pallidus)	(+) (from basal forebrain)	(+) (from globus pallidus)	(+) (from basal forebrain)
transmitter action§ nicotine [9]	depo [firing]	n.d.		n.d. or depo [firing]
muscarine [9]	n.d.	n.d. or hyper		depo [firing]
noradrenaline [10]		depo (α)		depo (α) [firing]
dopamine [11]	depo (D1/D5) [firing]	depo (D1/D5)	depo (D1/D5) [firing]	

* unpublished observation.
FS, fast-spiking cell; LTS, low-threshold spike; RS, regular-spiking; BS, burst-spiking.
NPY, neuropeptide Y; NO, nitric oxide.
§, recorded at soma, n.d., not detected; depo, depolarization; hyper, hyperpolarization.
α, α-adrenoceptor; D1/D5, dopamine D1/D5 receptor.
[1] Ref.9, 10, 137, 173; [2] Ref. 174, 175, 176, 177; [3] Ref. 84, 105, 108, 114, 178; [4] Ref. 117;
[5] Ref. 84, 122, 147; [6] Ref. 179, 180; [7] Ref. 87, 167, 181; [8] Ref. 101, 132, 133;
[9] Ref. 94, 141, 142, 172; [10] Ref. 102; [11] Ref. 182, 183, 184. References for other items are in the text.

inputs into the two compartments is preserved in the synaptic input to individual spiny cells. On the other hand, dendrites of PV and SOM cells cross the boundaries, providing a basis for compartmental interactions. Somatostatinergic fibers show compartmental preference, mainly innervating the matrix.

4.2. Synaptic targets

The two interneuron types differ in synaptic targets and structures[88]. FS cells make symmetrical synapses on somata (28%) and dendrites (72%) including a few spines. The figures for LTS cells are somata (3%) and dendritic shafts (97%). Both cells innervate dendrites including spiny ones. Single FS cells innervated somata, dendritic shafts, and spines of striatal neurons including spiny cells[88]. Dendrites innervated by both types vary in thickness, the circumference of postsynaptic dendritic shafts and spines ranging from 0.94 to 5.15 µm (2.12 ± 1.25 µm, mean ± SD) in FS cells and from 0.84 to 4.25 µm (1.94 ± 0.8 µm) in LTS cells. These data suggest striatal PV and SOM cells innervated various domains of projection cells. It remains to be investigated whether these interneurons innervate a specific domain of each postsynaptic cell or several domains.

Since the postsynaptic junctional area may be related to the size of the synaptic current[90-92], we measured its dimensions from reconstructions. While the synaptic junctional areas of FS cells (0.024 - 0.435 µm^2, n = 28) sharply and linearly increased with the circumference of the postsynaptic dendrites or spines, the slope for the junctional area of LTS cells (0.02 - 0.103 µm^2, n = 29) against circumference was less steep, and a much weaker correlation was seen[88]. Since the circumference of the postsynaptic target is related to the input resistance and the synaptic junctional area to the number of receptors[90], the change of junctional area according to postsynaptic dimensions may be due to adjustment of GABAergic currents. The contrasting weak correlation between synaptic junctional area and postsynaptic size suggests that similar postsynaptic effects may be induced in LTS cells irrespective of the synaptic location. Peptides and NO released from the axon terminals of LTS cell, may act on G protein-coupled receptors or diffusely on extrasynaptic receptors, whose effects are not directly related to the junctional area.

4.3. Recurrent connections from spiny projection cells

It has not been investigated whether projection cells can induce IPSPs in PV and SOM cells in a feedback manner. In addition to GABA, spiny cells synthesize neuropeptides such as substance P, dynorphin and enkephalin. Some of these are selectively expressed in interneurons. Substance P is expressed in axons of spiny cells projecting to the substantia nigra. Intrastriatal stimulation induces slowly depolarizing potentials in cholinergic cells, which are blocked by a substance P receptor antagonist[93]. Bath-applied substance P also causes depolarization through non-selective cation channels at resting potentials in cholinergic and SOM/NOS cells. Substance P, probably released from the collaterals of cells projecting to the SN, excites cholinergic and SOM/NOS cells, but not projection or PV cells. This intrastriatal peptidergic pathway can be used for selective feedback excitation of cholinergic and somatostatinergic interneuron types. Acetylcholine depolarizes FS cells through nicotinic receptors[94]. This pathway may be used for feedback excitation of PV cells through excitation of cholinergic cells by substance P released from spiny cells projecting to the substantia nigra.

4.4. Excitatory input from the cortex and inhibitory input from the globus pallidus

PV and SOM cells receive cortical inputs like projection cells[95,96]. Corticostriatal axons innervate cell bodies and proximal dendrites of PV cells, and often multiple contacts on single cells within a small distance[97]. Multiple axonal appositions are not the case for spiny projection cells, suggesting that cortical innervation rules are different for projection and PV interneurons, with more selective innervation on the latter. In contrast, SOM cells may receive cortical inputs onto small dendritic spines or spine-like appendages[98]. The different cortical innervation patterns may relate to differences in firing induction: firing of PV cells may need strong synchronized excitatory inputs, but SOM cells may fire with single EPSPs like cholinergic interneurons[99]. Single PV cells receive convergent inputs from functionally distinct cortical regions[97] and are electrically connected with one another[84]. Thus, each PV cell receives excitation from various cortical regions by convergent synaptic inputs and shares depolarizations with other PV cells through gap junctions for exerting feed-forward inhibitions on spiny cells. Therefore, focal cortical excitation may activate a group of PV cells which are distributed more broadly than spiny cells[100]. In addition to extrinsic excitatory inputs, both PV and SOM neurons receive GABAergic inputs from the globus pallidus[101]. This pathway disinhibits projection cells by inhibiting GABAergic interneuron types, especially striatal PV cells, and may regulate the spread of cortical excitation within the striatum.

5. PARVALBUMIN AND SOMATOSATIN CELLS IN THE CORTEX

In the rat frontal cortex, GABAergic interneurons are divided mainly into three groups according to the intrinsic firing pattern in response to depolarizing current pulses[65,66,102]: fast-spiking (FS) cells, late-spiking (LS) cells and non-FS cells. FS cells show abrupt episodes of non-adapting repetitive discharges of short-duration spikes in response to depolarizing currents[103]. LS cells exhibit a ramp-like depolarizing response before spike firing during a square wave current injection of threshold intensity. Non-FS groups contain regular-spiking non-pyramidal (RSNP) cells and burst-spiking non-pyramidal (BSNP) or LTS cells inducing low threshold spikes (LTS) from hyperpolarized potentials. Some non-FS cells fire spikes phasically only in the initial portion of depolarizing current pulses. These data suggest that non-FS cells are much more heterogeneous in electrophysiological characteristics than their FS counterparts. There are some correlations between chemical and firing classes. A group of non-pyramidal cells with a particular combination of firing pattern and the expressed substances includes morphological types with unique axon branching patterns[7,66]. Some morphological types such as basket cells belong to several classes[8]. Here I compare two major types of cortical GABAergic non-pyramidal cells, PV and SOM cells, incorporating the outcomes obtained in cortical regions other than the frontal cortex (Table 2).

5.1. Morphology and synaptic targets

Cortical PV cells generally belong to the FS group, but SOM cells are non-FS cells including both RSNP and BSNP types. SOM cells in layers II/III are mostly RS, but those in layer V are RS or BS with prominent low-threshold spikes[65,104]. Most FS cells have dense local and horizontal axon arborizations, and make multiple axonal boutons on other somata (FS basket cells) (Fig. 5). FS basket cells also innervate dendritic spines as well as dendritic shafts with GABA-containing symmetrical synapses[65,66,105,106]. In addition to basket cells, FS cells include another morphological type, chandelier cells with vertical arrays of several axonal boutons innervating axon initial segments (Fig.

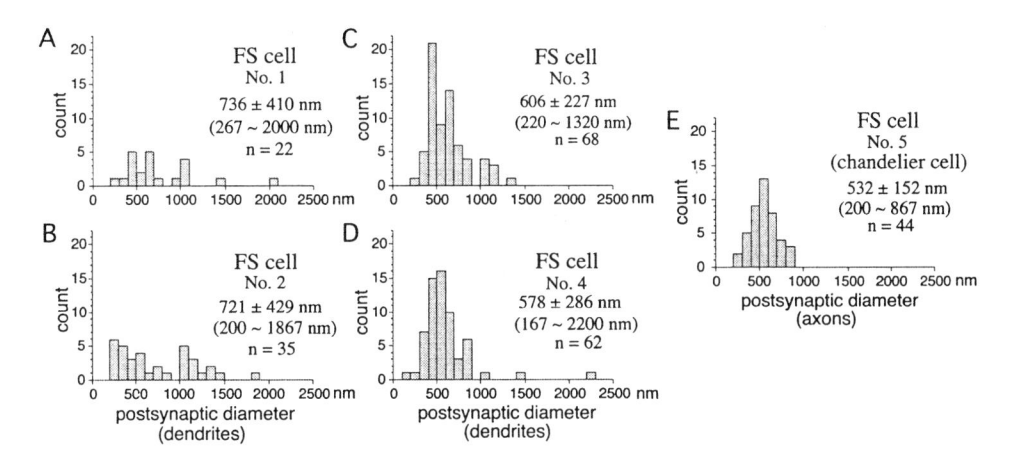

Fig. 6. Distributions of short-axis diameters of postsynaptic dendrites contacted by FS cells No.1-4, or axons contacted by FS cell No. 5 (chandelier cell) in the rat frontal cortex. Mean ± SD of diameters and number of observation (n) are written under the cell numbers. Modified from Ref. 66.

5)[66,107]. There are a few FS cells emitting axon collaterals extending to all directions without frequent branching (wide arbor FS cells). In contrast, most non-FS SOM Martinotti cells have ascending axonal arbors to layer I and show almost no basket terminals (Martinotti cells) (Fig. 5). There are some SOM cells with wide rather than ascending axonal arbors (wide arbor SOM cells). SOM Martinotti cells make symmetrical synapses immunoreactive for GABA, and innervate the dendritic shafts, and the heads and stalks of spines receiving other excitatory inputs simultaneously[65,108,109]. Like striatal FS cells, single cortical FS cells innervate various domains of the neuronal surface, and make synapses on dendrite or spines of differing thickness, except for chandelier cells (Fig. 6)[66], which innervate relatively homogeneous targets, axon initial segments (Fig. 6E)[110].

The synaptic junctional area of axon terminals from cortical FS cells linearly increases along with the circumference of postsynaptic dendrites or spines in a similar way to that of striatal FS cells. However, in contrast to striatal somatostatin cells the junctional area of axons from Martinotti cells linearly increases with the dendritic circumference or the surface area of the postsynaptic spine heads (unpublished observations). This suggests that striatal PV, cortical PV and SOM cells employ a similar kind of GABAergic transmission, but striatal SOM cells may utilize a different synaptic interaction, and that GABA may not be a main transmitter from the terminals of striatal SOM cells.

5.2. Synaptic actions

FS basket cells and non-FS Martinotti cells induce unitary inhibitory postsynaptic currents (IPSCs) with paired-pulse depression in short-term plasticity[105,111-114]. In layer V

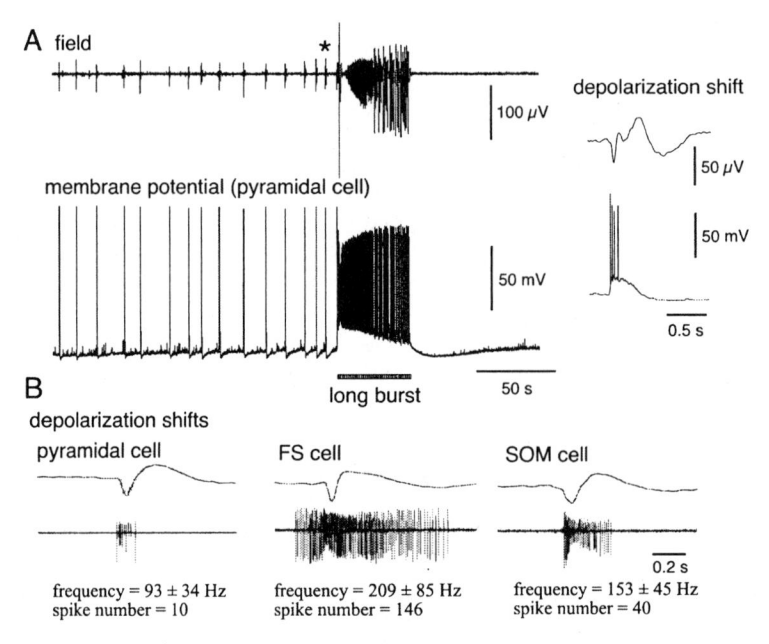

Fig. 7. **A:** Omission of Mg^{2+} from the external solution induces two types of spontaneous depolarizations in the rat frontal cortex in vitro. A membrane potential of a pyramidal cell is shown below, and a field potential recorded nearby above. Resting potential = -65 mV. In the Mg^{2+}-free solution, spontaneous depolarizations of pyramidal cells with spike firing (depolarization shift) appeared repeatedly, resulting in a long-lasting depolarization (long burst). **B:** Firing patterns of cortical neuron types during depolarization shifts. Field potentials and extracellularly recorded units are shown for pyramidal, fast spiking (FS), and somatostatin (SOM) cells during depolarization shifts before the transition to long bursts, corresponding to * in **A**. Average spike frequencies during these depolarization shifts are given as mean ± SD values. FS and somatostatin cells fire more spikes at higher frequency than pyramidal cells. Modified from Ref. 118.

of the visual cortex, IPSCs evoked by FS cell firing are larger in amplitude and faster in rise time than those with LTS cell firing[114]. Dopamine depresses inhibitory transmission between FS interneurons and pyramidal neurons but enhances inhibition between non-FS interneurons and pyramidal cells[115].

Some non-pyramidal cells make synaptic contacts with themselves (autapses)[105,116]. Autaptic connections are found in FS cells, but not in LTS cells in layer V of the sensorymotor cortex, and have significant inhibitory effects on repetitive firing[117]. FS cells show more prominent action potential accommodation when autaptic transmission is blocked[117]. Therefore, firing properties of FS cells may be controlled by both intrinsic membrane properties and autaptic transmission.

5.3. Firing characteristics

PV and somatostatin cells differ in firing patterns in response to artificially applied depolarizing current pulses, but also to cortical spontaneous synchronized depolarizations. On lowering extracellular Mg^{2+} in vitro to nominally zero, spontaneous depolarization starts periodically at about 0.1 Hz. After a 10 to 15 min, much stronger synchronized depolarizations we term the "long burst" occur (Fig. 7)[118]. These rhythmic depolarizations are synchronized among cortical cells, including pyramidal and non-pyramidal forms, and are composed of three phases (Fig. 7): initial strong depolarizations accompanying spikes with the highest frequency, termed "initial discharges"; rhythmic depolarizations at 6 ~ 10 Hz on relatively steady depolarizations, which were similar to fast runs observed in vivo[119,120], termed "fast run-like potentials (FR)"; and several slow-rhythmic strong depolarizations, termed "afterdischarges". The FR starts at about 10 Hz and ends at 6 Hz. In each phase of the synchronized activity, cortical neuron types exhibit distinct firing frequencies.

The frequency of spike discharges and the spike number have been measured for each neuron type at the depolarization shift preceding the long burst[118]. FS and SOM cells fire more vigorously than pyramidal cells. FS cells and SOM cells fire more spikes at a higher frequency than pyramidal cells (Fig. 7) [mean spike number = 84.5 (range, 4 - 300) for FS cells, 33.7 (20 - 53) for somatostatin cells, 6.8 (2 - 13) for pyramidal cells; mean spike frequency = 180 Hz (100 - 270 Hz) for FS cells, 120 Hz (60 - 170 Hz) for somatostatin cells, 70 Hz (30 - 110 Hz) for pyramidal cells at 30°C]. In the initial discharges, pyramidal cells discharge spikes at 150 Hz (110 - 190 Hz). The maximum frequencies in the initial discharge of FS cells (250 - 410 Hz; mean = 330 Hz) and SOM cells (170 -350 Hz; mean = 250 Hz) are larger than those for pyramidal cells. Some FS cells continue to discharge spikes for several seconds at frequencies higher than 200 Hz. During the FR, pyramidal cells increase the firing frequency periodically up to 25 - 55 Hz (mean = 35 Hz), whereas the firing frequency of FS cells rhythmically increases up to 150 Hz (90 - 190 Hz). SOM cells fire very similarly to pyramidal cells in the FR (35 - 75 Hz; mean = 40 Hz). During the FR, many FS cells and some SOM cells tend to be inactivated due to the strong depolarization. More depolarizing currents are necessary for induction of spike firing in FS than in other non-pyramidal cells. In synchronized activities, FS cells show the highest frequency of spike discharge, but tend to be inactivated more easily than other types. Thus, FS cells may show high-frequency firing only in the case of an appropriate excitation level in the cortex.

5.4. Recurrent excitation from pyramidal cells

Excitatory inputs from pyramidal cell to somatostatin LTS cells (or RSNP cells) show paired-pulse facilitation[108,111,121,122]. Excitatory connections from pyramidal cells to FS cells show either paired-pulse facilitation or depression[111,122,123]. The short term plasticity of EPSPs from pyramidal to FS cells is dependent on stimulus frequency of presynaptic cells and may change during maturation or depend on the presynaptic pyramidal cell subtype[124]. In the frontal cortex, excitations from pyramidal to pyramidal cells and to FS cells are differentially regulated by dopamine[125].

In the rat visual cortex, layer II/III FS cells receive strong excitatory input from the middle cortical layers. In contrast, adapting inhibitory interneurons, possibly including SOM cells, receive their strongest excitatory input either from deep layers or laterally from within layers II/III[126]. In layer V of the visual cortex, pyramidal cells projecting to the superior colliculus mainly excite LTS cells in a certain layer position, but not FS cells[127]. These observations suggest that separate types of pyramidal cells in different intracortical positions may excite PV and SOM cells. In visual cortical areas, PV cells receive both forward and feedback interareal excitatory synapses, but at different subcellular locations and with different synapse morphologies[128]. These asymmetries of excitatory synaptic contacts on PV cells may reflect differences in the strength of disynaptic inhibition evoked by interareal forward and feedback inputs.

5.5. Excitation by thalamocortical inputs and inhibition from the basal forebrain

Thalamocortical inputs and axon collaterals of pyramidal cells excite PV and SOM cells. In layer IV of the mouse barrel cortex, stimulation of thalamocortical axons excites both PV FS cells and calbindin RSNP cells, which are considered to be somatostatin cells[129]. EPSPs induced by single thalamocortical axons are larger in FS than in LTS cells[122]. Therefore, both classes of GABA cells can mediate thalamocortical feed-forward inhibition, but with different strengths. Thalamocortical EPSPs to layer IV FS cells show paired pulse depression[122]. In the rabbit barrel cortex, putative FS cells respond to vibrissa displacement with very high sensitivity and temporal fidelity, but lack directional specificity. These cells receive potent and highly convergent and divergent synchronous inputs from thalamocortical neurons[130]. This feed-forward inhibition is suited to suppress spike generation in spiny neurons following all but the most optimal feed-forward excitatory inputs[131].

GABAergic projection cells in basal forebrain innervate cortical PV and SOM cells as do those in the globus pallidus in the striatum case[132,133]. Basal forebrain GABA cells innervating cortical interneurons seem to express PV[134,135] and receive excitatory cortical inputs[136]. By inhibiting cortical GABAergic cells, PV cells in the basal forebrain may activate the activity of pyramidal cells. Ganglionic eminence is the common developmental origin of GABAergic projection cells in the globus pallidus and basal forebrain as well as the interneurons in striatum and cortex[137]. By this developmental process, PV and SOM interneurons may selectively interact with GABAergic projections cells in the basal telencephalon.

5.6. Noradrenergic and cholinergic modulation

In the neocortex, noradrenaline and acetylcholine are respectively released from afferent fibers originating in noradrenergic cells in the locus coeruleus and cholinergic cells in the basal forebrain. Both systems are related to the control of arousal and attention[138-140]. Noradrenaline and acetylcholine affect membrane potentials of cortical PV or SOM cells more than pyramidal cells (Table 2)[102,141,142]. Noradrenaline induces an increase of IPSCs in pyramidal cells, and excites GABAergic cells via α-adrenergic receptors (Fig. 8)[102]. FS cells are depolarized, but none demonstrate spike firings. In contrast, SOM cells are depolarized, accompanied by spike firing. These findings suggest that the excitability of cortical GABAergic cell types is differentially regulated by noradrenaline.

In the cortex, inhibitory as well as excitatory circuits generate synchronized periodical activity[143-145]. Cholinergic inputs from the basal forebrain exert profound effects on cortical activities such as rhythmic synchronization. In the rat frontal cortex, both carbachol and muscarine cause two temporally different patterns of inhibitory

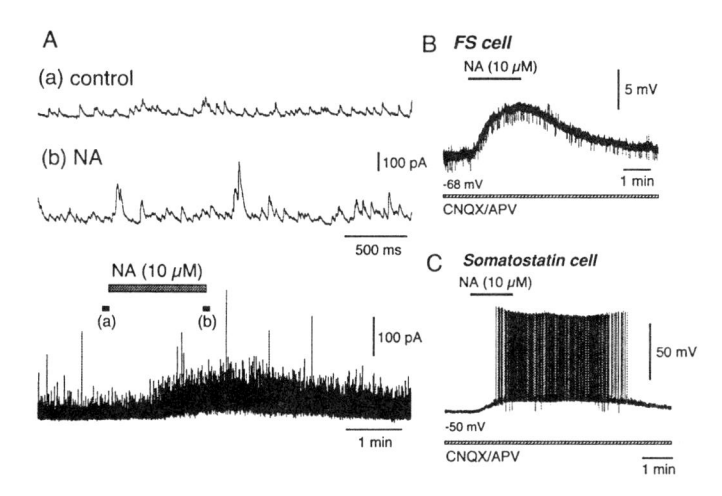

Fig. 8. **A:** The amplitude and frequency of spontaneous IPSCs are increased by application of noradrenaline (NA) in a solution containing 10 μM CNQX and 50 μM APV (CNQX/APV; blockers for ionotropic glutamate receptors). The currents at the black bars (a) and (b) are shown above. **B:** An FS cell depolarized by application of NA (10 μM) in a solution containing 20 μM CNQX and 50 μM APV. FS cells do not fire with NA application alone. **C:** A somatostatin cell depolarized with spike discharges by NA (10 μM) application in a solution containing CNQX/APV. Modified from Ref. 102.

postsynaptic current modulation in both pyramidal cells and inhibitory interneurons: tonic and periodic increase of GABA-A receptor-mediated currents[146]. The tonic pattern features continuous increase of inhibitory postsynaptic current frequency, while the periodic increase manifests itself as a rhythmic (0.1~0.3 Hz, mean 0.2 Hz) burst of inhibitory postsynaptic current (mean frequency, 24 Hz; mean burst duration, 2.2 s). Muscarinic receptor antagonists suppress both types of IPSC increase, but antagonists of ionotropic glutamate receptors do not affect the periodical inhibitory current bursts. In nearby cells these are synchronized as a whole, but individual inhibitory events within the bursts are not always temporally linked, suggesting synchronized depolarization of several presynaptic interneurons. The excitability of cortical GABAergic cell subtypes is differentially regulated by acetylcholine. Carbachol and muscarine affect the activities of peptide-containing non-FS cells, but not those of FS or LS cells[141]. We have found SOM cells to be depolarized, with accompanying spike firing. Following sufficient depolarization by muscarine in the solution containing tetorodotoxin, SOM Martinotti cells in layers II/III start slow rhythmic depolarizations[8]. These are considered to correspond to the muscarine-induced slow rhythm of IPSCs[145].

5.7. Electrical coupling and synchronized activities

Like striatal FS cells, cortical FS cells and SOM LTS cells are connected by electrical synapses[122,147,148]. This electrical coupling occurs frequently in the same class: between PV FS cells, or between SOM cells, but not between FS and SOM cells[122]. Connexin 36 is likely to be a component of electrical synapses between cortical interneurons[149]. Metabotropic glutamate or acetylcholine agonists induce rhythmic (3 - 6 Hz) synchronized excitation of layer IV LTS cells in the somatosensory cortex through electrical coupling[145]. The synchrony of these rhythms is weaker and more spatially more restricted in mice lacking connexin 36[150]. This suggests the electrical synapses are involved in synchronization of some cortical rhythmic activities. Using the electrical coupling combined with the mutual GABAergic inhibition and the temporal characteristics of spike transmission, groups of FS cells may fire in synchrony when receiving coincident excitatory inputs[151].

6. INTERNEURON CONNECTION TO MULTIPLE PROJECTION CELL TYPES

Spiny projection cells receive a mixture of inputs with different origins on diverse portions of dendrites. These inputs include recurrent collaterals from projection cells themselves. In addition to excitatory input, certain spines receive GABAergic synapses[152-154]. The specific or several combination of glutamatergic inputs may excite interneuron subtypes, which differentially regulate these diverse inputs on several domains of projection cells.

6.1. Multiple projection cell types in focal territories in the cortex and striatum

The hippocampus is a well-characterized circuit in the forebrain[58,155]. In each region of the hippocampal formation, spiny projections cells are relatively homogeneous in physiological and morphological aspects, and serially connected along the connection stream of hippocampal formation. In comparison with striatal and cortical interneurons, hippocampal interneuron types innervate more selectively the specific surface domains of spiny projection cells[156,157]. In addition, hippocampal interneuron types seem to receive more specific combinations of excitatory inputs[45]. The post- and presynaptic high selectivity of hippocampal interneurons may reflect both homogeneity of pyramidal cells in each region and serial regional processing in one direction.

In the cortex, pyramidal cells projecting to the same target are aggregated according to the layer structure, but each layer contains several projection types[158]. In the second somatosensory area, layer V pyramidal cells closely located in the same laminar position show different axon innervation patterns with or without branches to the striatum, thalamus, zona incerta, and/or substantia nigra[16]. This situation suggests that even the cortical micro-region in the same layer contains several projection cell types. In the striatum, spiny cell subtypes innervating different targets are intermingled, although those with the same projection sites may be aggregated into small clusters[31,159]. Their dendrites and local axons are overlapped within the striatum except for the patch/matrix compartmental segregation. Thus, in contrast to the hippocampus, the striatum and cortex have multiple projection systems in the local region. These mutual interactions among several projection cell types seem more important for function in the striatum and cortex than in the hippocampus. Striatal and cortical interneurons appear required to manage recurrent connections on various surface domains among multiple projection neurons types. This compound situation may make the connection selectivity seemingly less discriminating.

LOCAL CIRCUIT NEURONS IN THE FRONTAL CORTICO-STRIATAL SYSTEM 143

Possible connection patterns of interneurons and projection cell subtypes
I. output divergence pattern of interneurons innervating various domains
(2) *cell type-selective* (*e.g.* to P2, but not to P1)

Interneuron

Projection cell

(1) *cell type domain-selective*
(*e.g.* to dendrites of P1 and somata of P2)

(3) *non-selective*

II. input convergence pattern from projection cell subtypes
(1) *cell type-selective* (*e.g.* from P2, but not from P1)

(3) *non-selective*

(2) *domain-segregated*
(*e.g.* to different domains from P1 and P2)

P1, P2
spiny projection cell subtypes

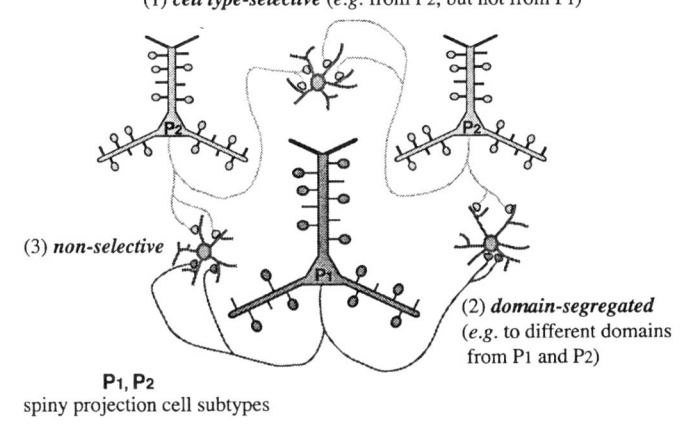

Fig. 9. Possible connection patterns of interneurons and projection cell subtypes. **I:** Selectivity levels of divergent innervation from interneurons making synapses on various domains of spiny projection cells. **II:** Selectivity levels of input convergence from recurrent collaterals of projection cell subtypes to interneurons.

6.2. Connection rules between interneurons and projection cell types

Cortical and striatal interneuron types preferentially innervate some surface domains of pyramidal cells, but also act on others[160,161]. Multiple projection cell and interneuron types may be connected according to strictly determined rules of each local circuit. Interneurons making synapses on various domains may possibly show several selectivity levels of divergent innervation (Fig. 9, I): (1) interneurons innervate a specific postsynaptic portion of one projection cell type, and another specific portion of another projection type (cell type domain-selective output divergence); (2) interneurons innervate a specific type of projection cells (cell type-selective divergence); (3) interneurons

Possible spine innervation patterns of interneurons
(1) *spine input-selective*   (2) *non-selective*

* to spines receiving inputs from P1, not from P2
** to spines receiving inputs from P2, not from P1

P1, P2 spiny projection cell subtypes

terminal boutons
○ from projection cell 1
● from projection cell 2

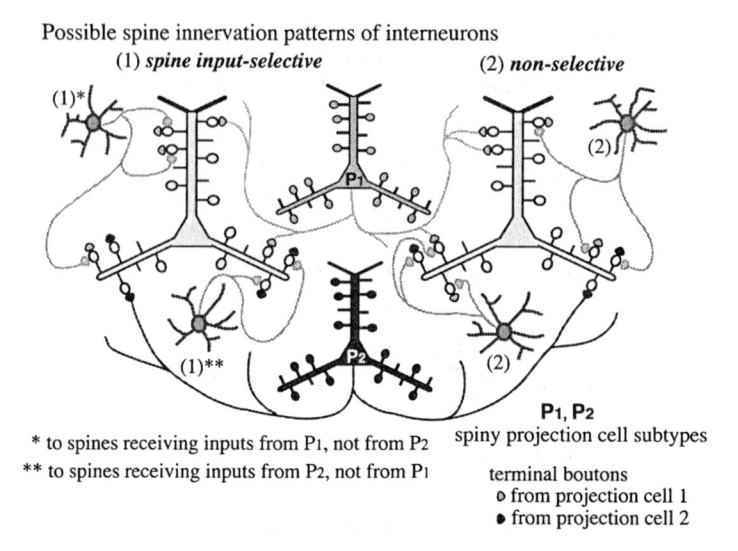

Fig. 10. Possible spine innervation patterns from interneurons and projection cell subtypes. Some interneuron types may innervate selectively spines receiving excitatory inputs from the same origin (spine-input selective).

innervate several portions of multiple types of projection cells (non-selective divergence). Recurrent innervation of projection cells on interneurons may also possibly show several selectivity levels of input convergence (Fig. 9, II). (1) Interneurons selectively receive inputs from specific types of projection cells (cell type-selective input convergence). (2) Interneurons receive inputs from several projection types on the separate surface domains (domain-segregated convergence). (3) Interneurons receive inputs from several projection types on similar domains (non-selective convergence). In the cat visual cortex, layer IV pyramidal cells innervate preferentially innervate the distal dendrites of layer IV basket cells[162].

Each spine of spiny projection cells usually is subjected to a single excitatory input, but the origin of spiny inputs is diverse. Pyramidal cells receive recurrent excitatory inputs on the spines from other pyramidal cells[163]. Inhibitory regulation of spine inputs is critical in the recurrent network in the cortex. Spine innervation by cortical interneurons may also possibly show several selectivity levels of divergence to spine types defined by the input origin and dendritic location (Fig. 10). (1) Interneurons innervate spines receiving inputs exclusively from the projection cell type (spine input-selective divergence; selective for presynaptic pyramidal cell type). (2) Interneurons innervate spines randomly (non-selective divergence). In the cat visual cortex, only a few spines receiving asymmetrical synapses of geniculate afferents have another symmetrical synapse[154].

7. CONCLUDING REMARKS

The cortex and striatum share PV and SOM interneurons in common. The two types are similar in several characteristics, but differ in others (Table 2). These two major interneuron types have also been identified in the hippocampus and amygdala[155,164].

It has not been well elucidated how precisely the local circuit connections are determined genetically in the cortex and striatum[165,166]. In the striatum, what each spiny cell encodes depends on the combination of excitatory inputs from 1000 to 5000 pyramidal cells[18]. It remains to be investigated what degree of selectivity striatal interneurons express in connections with cortical pyramidal cell types, and how divergent or convergent are the cortical inputs to interneuron types. In the cortex, excitatory pyramidal cells are recurrently connected on their spines. It is proposed that inhibition acts to control the gain of the recurrent amplification inherent in recurrent excitatory pathways[29]. The connection selectivity of PV and SOM interneuron types is critical for understanding their functional roles in intracortical and corticostriatal circuitry.

8. ACKNOWLEDGMENTS

The author thanks Fuyuki Karube, Yoshiyuki Kubota and Satoru Kondo for the intimate collaboration, and Kazuko Kawaguchi for continuous help and encouragements at Memphis, Wako, Nagoya and Okazaki.

9. REFERENCES

1. G.E. Alexander, M.R. DeLong and P.L. Strick (1986) *Annu. Rev. Neurosci.* **9**, 357.
2. S. Ramón y Cajal (1911) *Histology of the Nervous System, Vol. 2*, translated by N. Swanson and L.W. Swanson (Oxford UP, New York).
3. R. Lorente de Nó, (1949) Iin: *Physiology of the Nervous System*, J.F. Fulton, ed. (Oxford UP, New York), p 288.
4. J. Szentagóthai (1978) *Proc. R. Soc. Lond. B Biol. Sci.* **201**, 219.
5. Y. Kawaguchi, C.J. Wilson, S. Augood and P. Emson (1995) *Trends Neurosci.* **18**, 527.
6. Y. Kawaguchi (1997) *Neurosci Res.* **27**, 1.
7. Y. Kawaguchi and Y. Kubota (1997) *Cereb. Cortex* **7**, 476.
8. Y. Kawaguchi and S. Kondo (2003) *J. Neurocytol.*, in press.
9. S.A. Anderson, D.D. Eisenstat, L. Shi and J.L. Rubenstein (1997) *Science* **278**, 474.
10. N. Tamamaki, K.E. Fujimori and R. Takauji (1997) *J. Neurosci.* **17**, 8313.
11. O. Marín and J.L. Rubenstein(2001) *Nature Rev. Neurosci.* **2**, 780.
12. A.M. Graybiel, T. Aosaki, A.W. Flaherty and M. Kimura (1994) *Science* **265**, 1826.
13. O. Hikosaka, K. Nakamura, K. Sakai and H. Nakahara (2002) *Curr. Opin. Neurobiol.* **12**, 217.
14. C.R. Gerfen (1992) *Annu. Rev. Neurosci.* **15**, 285.
15. R.L. Cowan and C.J. Wilson (1994) *J. Neurophysiol.* **71**, 17.
16. M. Lévesque, S. Gagnon, A. Parent and M. Deschenes (1996) *Cereb. Cortex* **6**, 759.
17. M. Lévesque and A. Parent (1998) *Cereb. Cortex* **8**, 602.
18. A.E. Kincaid, T. Zheng and C.J. Wilson (1998) *J. Neurosci.* **18**, 4722.
19. T. Zheng and C.J. Wilson (2002) *J. Neurophysiol.* **87**, 1007.
20. H. Markram, J. Lübke, M. Frotscher, A. Roth and B. Sakmann (1997) *J. Physiol. (Lond.)* **500**, 409.
21. H. Markram (1997) *Cereb. Cortex.* **7**, 523.
22. A.M. Thomson and J. Deuchars (1997) *Cereb. Cortex* **7**, 510.
23. W.-J. Gao, L.S. Krimer and P.S. Goldman-Rakic (2001) *Proc. Natl. Acad. Sci. USA* **98**, 295-300.
24. M. Steriade, A. Nunez and F. Amzica (1993) *J. Neurosci.* **13**, 3266.
25. R. Metherate and J.H. Ashe (1993) *J. Neurosci.* **13**, 5312.
26. E.A. Stern, A.E. Kincaid and C.J. Wilson (1997) *J. Neurophysiol.* **77**, 1697.
27. M.V. Sanchez-Vives and D.A. McCormick (2000) *Nature Neurosci.* **3**, 1027.
28. J. Anderson, I. Lampl, I. Reichova, M. Carandini and D. Ferster(2000) *Nature Neurosci.* **3**, 617.

29. R. Douglas, C. Koch, M. Mahowald and K. Martin (1999) in: *Cerebral Cortex, Vol. 13, Models of Cortical Circuits*, P.S. Ulinski, E.G. Jones and A. Peters, eds. (Kluwer Academic/Plenum, New York), p 251.
30. X.J. Wang(2001) *Trends Neurosci.* **24**, 455.
31. Y. Kawaguchi, C.J. Wilson and P.C. Emson (1990) *J. Neurosci.* **10**, 3421.
32. C.J. Wilson and P.M. Groves (1981) *Brain Res.* **220**, 67.
33. C.J. Wilson and Y. Kawaguchi (1996) *J. Neurosci.* **16**, 2397.
34. E.A. Stern, D. Jaeger and C.J. Wilson (1998) *Nature* **394**, 475.
35. U. Czubayko and D. Plenz (2002) *Proc. Natl. Acad. Sci. USA* **99**, 15764.
36. M.J. Tunstall, D.E. Oorschot, A. Kean and J.R. Wickens (2002) *J. Neurophysiol.* **88**, 1263.
37. C.J. Wilson (1995) in: *Models of Information Processing in the Basal Ganglia*, J.C. Houk, J.L. Davis and D.G. Beiser, eds. (MIT Press, Cambridge), p 29.
38. D. Plenz and S.T. Kitai (2000) in: *Brain Dynamics and the Striatal Complex*, R. Miller and J.R. Wickens, eds. (Harwood Academic Publishers, Amsterdam), p. 165.
39. J.R. Wickens and D.E. Oorshcot (2000) in: *Brain Dynamics and the Striatal Complex*, R. Miller and J.R. Wickens, eds. (Harwood Academic Publishers, Amsterdam), p 141.
40. J.G. Parnavelas (2000) *Trends Neurosci.* **23**, 126.
41. O. Marín, A. Yaron, A. Bagri, M. Tessier-Lavigne and J.L. Rubenstein (2001) *Science* **293**, 872.
42. B. Cauli, J.T. Porter, K. Tsuzuki, B. Lambolez, J. Rossier, B. Quenet and E. Audinat (2000) *Proc. Natl. Acad. Sci. USA* **97**, 6144.
43. A. Gupta, Y. Wang and H. Markram (2000) *Science* **287**, 273.
44. Y. Wang, A. Gupta, M. Toledo-Rodriguez, C.Z. Wu and H. Markram (2002) *Cereb. Cortex* **12**, 395.
45. P. Parra, A.I. Gulyás and R. Miles (1998) *Neuron* **20**, 983.
46. Y. Kawaguchi (1993) *J. Neurosci.* **13**, 4908.
47. P. Calabresi, D. Centonze, P. Gubellini, A. Pisani and G. Bernardi (2000) *Trends Neurosci.* **23**, 120.
48. F.M. Zhou, C.J. Wilson and J.A. Dani (2002) *J. Neurobiol.* **53**, 590.
49. P. Bolam, B.H. Wainer and A.D. Smith (1984) *Neuroscience* **12**, 711.
50. P.E. Phelps, C.R. Houser and J.E. Vaughn (1985) *J. Comp. Neurol.* **238**, 286.
51. K. Semba (2000) *Behav. Brain Res.* **115**, 117.
52. Y. Kubota, S. Inagaki, S. Shimada, S. Kito, F. Eckenstein and M. Tohyama (1987) *Brain Res* **413**, 179.
53. T. Aosaki, A.M. Graybiel and M. Kimura (1994) *Science* **265**, 412.
54. T. Aosaki, K. Kiuchi and Y. Kawaguchi (1998) *J. Neurosci.* **18**, 5180.
55. Z. Yan, W.-J. Song and D.J. Surmeier (1997) *J Neurophysiol* **77**, 1003.
56. A. Pisani, P. Bonsi, D. Centonze, P. Calabresi and G. Bernardi(2000) *J. Neurosci.* **20**, RC69 (1-6).
57. T. Momiyama and E. Koga (2001) *J. Physiol. (Lond.)* **533**, 479.
58. P. Somogyi, G. Tamás, R. Luján and E.H. Buhl (1998) *Brain Res. Rev.* **26**, 113.
59. R.D. Traub, J.G.R. Jefferys and M.A. Whittington (1999) *Fast Oscillations in Cortical Circuits* (MIT Press, Cambridge).
60. L.M. Martinez, J.M. Alonso, R.C. Reid and J.A. Hirsch (2002) *J. Physiol. (Lond.)* **540**, 321.
61. C. Monier, F. Chavane, P. Baudot, L.J. Graham and Y. Frégnac (2003) *Neuron* **37**, 663.
62. A.T. Gulledge and G.J. Stuart (2003) *Neuron* **37**, 299.
63. Y. Kawaguchi (1993) *J. Neurophysiol.* **69**, 416.
64. J. DeFelipe (1997) *J. Chem. Neuroanat.* **14**, 1.
65. Y. Kawaguchi and Y. Kubota (1996) *J. Neurosci.* **16**, 2701.
66. Y. Kawaguchi and Y. Kubota (1998) *Neuroscience* **85**, 677.
67. L.S. Krimer and P.S. Goldman-Rakic (2001) *J. Neurosci.* **21**, 3788.
68. E.L. White (1989) *Cortical Circuits: Synaptic Organization of the Cerebral Cortex Structure, Function, and Theory*, E.L. White and A. Keller, eds. (Birkhäuser, Boston)
69. Y. Amitai and B.W. Connors (1995) in: *Cerebral Cortex, Vol. 11, The Barrel Cortex of Rodents*, E.G. Jones and I.T. Diamond, eds. (Plenum, New York), p 299.
70. E.G. Jones (2000) *Proc. Natl. Acad. Sci. USA* **97**, 5019.
71. F. Eckenstein and R.W. Baughman (1984) *Nature* **309**, 153.
72. T. Kosaka, M. Tauchi and J.L. Dahl (1988) *Exp. Brain Res.* **70**, 605.
73. T. Bayraktar, J.F. Staiger, L. Acsady, C. Cozzari, T.F. Freund and K. Zilles (1997) *Brain Res.* **757**, 209.
74. M.R. Celio (1990) *Neuroscience* **35**, 375.
75. O. Caillard, H. Moreno, B. Schwaller, I. Llano, M.R. Celio and A. Marty (2000) *Proc. Natl. Acad. Sci. USA* **97**, 13372.
76. M. Vreugdenhil, J. Jefferys, M. Celio and B. Schwaller (2003) *J. Neurophysiol.* **89**, 1414.
77. I. Selmer, M. Schindler, J.P. Allen, P.P. Humphrey and P.C. Emson (2000) *Regul. Pept.* **90**, 1.
78. H. Ishibashi and N. Akaike (1995) *J. Neurophysiol.* **74**, 1028.
79. C. Vilchis, J. Bargas, T. Perez-Rosello, H. Salgado and E. Galarraga (2002) *Neuroscience* **109**, 555.

80. G.A. Hicks, W. Feniuk and P.P. Humphrey (1998) *Br. J. Pharmacol.* **124**, 252.
81. P. Schweitzer, S.G. Madamba and G.R. Siggins (1998) *J. Neurophysiol.* **79**, 1230.
82. H. Kita, T. Kosaka and C.W. Heizmann (1990) *Brain Res.* **536**, 1.
83. R.L. Cowan, C.J. Wilson, P.C. Emson and C.W. Heizmann(1990) *J. Comp. Neurol.* **302**, 197.
84. T. Koós and J.M. Tepper (1999) *Nature Neurosci.* **2**, 467.
85. M.F. Chesselet and E. Robbins (1989) *Brain Res.* **492**, 237.
86. M.V. Catania, T.R. Tolle and H. Monyer (1995) *J. Neurosci.* **15**, 7046.
87. Y. Kubota, S. Mikawa and Y. Kawaguchi (1993) *NeuroReport* **5**, 205.
88. Y. Kubota and Y. Kawaguchi (2000) *J. Neurosci.* **20**, 375.
89. Y. Kawaguchi, C.J. Wilson and P.C. Emson (1989) *J. Neurophysiol.* **62**, 1052.
90. Z. Nusser, S. Cull Candy and M. Farrant(1997) *Neuron* **19**, 697.
91. Z. Nusser, N. Hájos, P. Somogyi and I. Mody (1998) *Nature* **395**, 172.
92. P.J. Mackenzie, G.S. Kenner, O. Prange, H. Shayan, M. Umemiya and T.H. Murphy (1999) *J. Neurosci.* **19**, RC13 (1-7).
93. T. Aosaki and Y. Kawaguchi (1996) *J. Neurosci.* **16**, 5141.
94. T. Koós and J.M. Tepper (2002) *J. Neurosci.* **22**, 529.
95. J.P. Bolam, B.D. Bennet (1995) in: *Molecular and Cellular Mechanism of Neostriatal Function*, M.A. Arino and D.J. Surmeier, eds. (R.G. Landes Company, Austin), p. 1.
96. S.R. Lapper, Y. Smith, A.F. Sadikot, A. Parent and J.P. Bolam (1992) *Brain Res.* **580**, 215.
97. S. Ramanathan, J.J. Hanley, J.-M. Deniau and J.P. Bolam (2002) *J. Neurosci.* **22**, 8158.
98. T.M. Thomas, Y. Smith, A.I. Levey and S.M. Hersch (2000) *Synapse* **37**, 252.
99. C.J. Wilson, H.T. Chang and S.T. Kitai (1990) *J. Neurosci.* **10**, 508.
100. H.B. Parthasarathy and A.M. Graybiel (1997) *J. Neurosci.* **17**, 2477.
101. M.D. Bevan, P.A. Booth, S.A. Eaton and J.P. Bolam (1998) *J. Neurosci.* **18**, 9438.
102. Y. Kawaguchi and T. Shindou (1998) *J. Neurosci.* **18**, 6963.
103. D.A. McCormick, B.W. Connors, J.W. Lighthall and D.A. Prince (1985) *J. Neurophysiol.* **54**, 782.
104. Y. Kawaguchi and Y. Kubota (1993) *J. Neurophysiol.* **70**, 387.
105. A.M. Thomson, D.C. West, J. Hahn and J. Deuchars (1996) *J. Physiol. (Lond.)* **496**, 81.
106. G. Tamás, E.H. Buhl and P. Somogyi (1997) *J. Physiol. (Lond.)* **500**, 715.
107. P. Somogyi (1977) *Brain Res.* **136**, 345.
108. J. Deuchars and A.M. Thomson (1995) *Neuroscience* **65**, 935.
109. Y. Kubota, F. Karube, K. Suzuki and Y. Kawaguchi (2000) *Soc. Neurosci. Abst.* **26**, 37.18.
110. P. Somogyi, T.F. Freund and A. Cowey (1982) *Neuroscience* **7**, 2577.
111. A. Reyes, R. Luján, A. Rozov, N. Burnashev, P. Somogyi and B. Sakmann (1998) *Nature Neurosci.* **1**, 279.
112. K. Tarczy-Hornoch, K.A. Martin, J.J. Jack and K.J. Stratford (1998) *J. Physiol. (Lond.)* **508**, 351.
113. M. Galarreta and S. Hestrin (1998) *Nature Neurosci.* **1**, 587.
114. Z. Xiang, J.R. Huguenard and D.A. Prince (2002) *J. Neurophysiol.* **88**, 740.
115. W.-J. Gao, Y. Wang and P.S. Goldman-Rakic (2003) *J. Neurosci.* **23**, 1622.
116. G. Tamás, E.H. Buhl and P. Somogyi (1997) *J. Neurosci.* **17**, 6352.
117. A. Bacci, R. Huguenard and D.A. Prince (2003) *J. Neurosci.* **23**, 859.
118. Y. Kawaguchi (2001) *J. Neurosci.* **21**, 7261.
119. M. Steriade, F. Amzica, D. Neckelmann and I. Timofeev (1998) *J. Neurophysiol.* **80**, 1456.
120. M.A. Castro-Alamancos and P. Rigas (2002) *J. Physiol. (Lond.)* **542**, 567.
121. A.M. Thomson, D.C. West and J. Deuchars (1995) *Neuroscience* **69**, 727.
122. J.R. Gibson, M. Beierlein and B.W. Connors (1999) *Nature* **402**, 75.
123. A.M. Thomson, J. Deuchars and D.C. West (1993) *Neuroscience* **54**, 347.
124. M.C. Angulo, J.F. Staiger, J. Rossier and E. Audinat (2003) *J. Neurophysiol.* **89**, 943.
125. W.-J. Gao and P.S. Goldman-Rakic (2003) *Proc. Natl. Acad. Sci. USA* **100**, 2836.
126. J.L. Dantzker and E.M. Callaway (2000) *Nature Neurosci.* **3**, 701.
127. J. Kozloski, F. Hamzei-Sichani and R. Yuste (2001) *Science* **293**, 868.
128. Y. Gonchar and A. Burkhalter (1999) *J. Comp. Neurol.* **406**, 346.
129. J.T. Porter, C.K. Johnson and A. Agmon (2001) *J. Neurosci.* **21**, 2699.
130. H.A. Swadlow (2002) *Nature Neurosci.* **5**, 403.
131. H.A. Swadlow (2003) *Cereb. Cortex* **13**, 25.
132. T.F. Freund and A.I. Gulyás (1991) *J. Comp. Neurol.* **314**, 187.
133. T.F. Freund and V. Meskenaite (1992) *Proc. Natl. Acad. Sci. USA* **89**, 738.
134. L. Zaborszky, K. Pang, J. Somogyi, Z. Nadasdy and I. Kallo (1999) *Ann. N.Y. Acad. Sci.* **877**, 339.
135. I. Gritti, I.D. Manns, L. Mainville and B.E. Jones (2003) *J. Comp. Neurol.* **458**, 11.
136. L. Zaborszky, R.P. Gaykema, D.J. Swanson and W.E. Cullinan (1997) *Neuroscience* **79**, 1051.
137. O. Marín, S.A. Anderson and J.L. Rubenstein (2000) *J. Neurosci.* **20**, 6063.

138. G. Aston-Jones, C. Chiang and T. Alexinsky (1991) in: *Progress in Brain Research, Vol. 88, Neurobiology of the Locus Coeruleus*, C.D. Barnes and O. Pompeiano, eds. (Elsevier, Amsterdam), p 501.
139. S.L. Foote, C.W. Berridge, L.M. Adams and J.A. Pineda (1991) in: *Progress in Brain Research, Vol 88, Neurobiology of the Locus Coeruleus*, C.D. Barnes and O. Pompeiano, eds. (Elsevier, Amsterdam), p 521.
140. B.E. Jones (1993) in: *Progress in Brain Research, Vol. 98, Cholinergic Function and Dysfunction*, A.C. Cuello ed. (Elsevier, Amsterdam), p 61.
141. Y. Kawaguchi (1997) *J. Neurophysiol.* **78**, 1743.
142. Z. Xiang, J.R. Huguenard and D.A. Prince (1998) *Science* **281**, 985.
143. J.G.R. Jefferys, R.D. Traub and M.A. Whittington (1996) *Trends Neurosci.* **19**, 202.
144. E.H. Buhl, G. Tamás and A. Fisahn (1998) *J. Physiol. (Lond.)* **513**, 117.
145. M. Beierlein, J.R. Gibson and B.W. Connors (2000) *Nature Neurosci.* **3**, 904.
146. S. Kondo and Y. Kawaguchi (2001) *Neuroscience* **107**, 551.
147. M. Galarreta and S. Hestrin (1999) *Nature* **402**, 72.
148. G. Tamás, E.H. Buhl, A. Lorincz and P. Somogyi (2000) *Nature Neurosci.* **3**, 366.
149. L. Venance, A. Rozov, M. Blatow, N. Burnashev, D. Feldmeyer and H. Monyer (2000) *Proc. Natl. Acad. Sci. USA* **97**, 10260.
150. M.R. Deans, J.R. Gibson, C. Sellitto, B.W. Connors and D.L. Paul (2001) *Neuron* **31**, 477.
151. M. Galarreta and S. Hestrin (2001) *Science* **292**, 2295.
152. E.G. Jones and T.P. Powell (1969) *J. Cell. Sci.* **5**, 509.
153. P. Somogyi and I. Soltesz (1986) *Neuroscience* **19**, 1051.
154. C. Dehay, R.J. Douglas, K.A. Martin and C. Nelson (1991) *J. Physiol. (Lond.)* **440**, 723.
155. T.F. Freund and G. Buzsaki (1 996) *Hippocampus* **6**, 347.
156. E.H. Buhl, K. Halasy and P. Somogyi (1994) *Nature* **368**, 823.
157. R. Miles, K. Tóth, A.I. Gulyás, N. Hájos and T.F. Freund (1996) *Neuron* **16**, 815.
158. E.G. Jones (1984) in: *Cerebral Cortex, Vol. 1, Cellular Components of the Cerebral Cortex*, A. Peters and E.G. Jones, eds. (Plenum, New York), p 40.
159. L.D. Loopuijt and D. van der Kooy (1985) *Brain Res.* **348**, 86.
160. P. Somogyi (1989) in: *Neural Mechanisms of Visual Perception*, D.K.-T. Lam and C.D. Gilbert, eds. (Portfolio Pub. Co., Texas), p 35.
161. Z.F. Kisvàrday (1992) in: *Progress in Brain Research, Vol. 90, Mechanisms of GABA in the Visual System*, R.R. Mize, R.E. Marc and A.M. Sillito, eds. (Elsevier, Amsterdam), p 385.
162. B. Ahmed, J.C. Anderson, K.A. Martin and J.C. Nelson (1997) *J. Comp. Neurol.* **380**, 230.
163. J. DeFelipe and I. Fariñas (1992) *Prog. Neurobiol.* **39**, 563.
164. A.J. McDonald and F. Mascagni (2002) *Brain Res.* **943**, 237.
165. G. Silberberg, A. Gupta and H. Markram (2002) *Trends Neurosci.* **25**, 227.
166. S.B. Nelson (2002) *Neuron* **36**, 19.
167. Y. Kubota and Y. Kawaguchi (1993) *J. Comp. Neurol.* **332**, 499.
168. B. Cauli, E. Audinat, B. Lambolez, M.C. Angulo, N. Ropert, K. Tsuzuki, S. Hestrin and J. Rossier (1997) *J. Neurosci.* **17**, 3894.
169. F. Cicchetti, T.G. Beach and A. Parent (2000) *Synapse* **30**, 284.
170. D.J. Holt, M.M. Herman, T.M. Hyde, J.E. Kleinman, C.M. Sinton, D.C. German, L.B. Hersh, A.M. Graybiel and C.B. Saper (1999) *Neuroscience* **94**, 21.
171. P. Calabresi, D. Centonze, A. Pisani, G. Sancesario, R.A. North and G. Bernardi (1998) *J. Physiol. (Lond.)* **510**, 421.
172. J.T. Porter, B. Cauli, K. Tsuzuki, B. Lambolez, J. Rossier and E. Audinat (1999) *J. Neurosci.* **19**, 5228.
173. S.A. Anderson, O. Marín, C. Horn, K. Jennings and J.L. Rubenstein (2001) *Development* **128**, 353.
174. Y. Kubota, R. Hattori and Y. Yui (1994) *Brain. Res.* **649**, 159.
175. W. Rushlow, B.A. Flumerfelt and C.C. Naus, *J. Comp. Neurol.* **351**, 499 (1995).
176. G. Figueredo-Cardenas, M. Morello, G. Sancesario, G. Bernardi and A. Reiner (1996) *Brain Res.* **735**, 317.
177. Y. Gonchar and A. Burkhalter (1997) *Cereb. Cortex* **7**, 347.
178. G. Tamás, P. Somogyi and E.H. Buhl (1998) *J. Neurosci.* **18**, 4255.
179. S. Lenz, T.M. Perney, Y. Qin, E. Robbins and M.F. Chesselet (1994) *Synapse* **18**, 55.
180. A. Chow, A. Erisir, C. Farb, M.S. Nadal, A. Ozaita, D. Lau, E. Welker and B. Rudy (1999) *J. Neurosci.* **19**, 9332.
181. B.D. Bennett and J.P. Bolam (1993), *Brain Res* **610**, 305.
182. E. Bracci, D. Centonze, G. Bernardi and P. Calabresi (2002) *J. Neurophysiol.* **87**, 2190.
183. D. Centonze, E. Bracci, A. Pisani, P. Gubellini, G. Bernardi and P. Calabresi (2002) *Eur. J. Neurosci.* **15**, 2049.
184. N. Gorelova, J.K. Seamans and C.R. Yang (2002) *J. Neurophysiol.* **88**, 3150.

9

INTERNEURON HETEROGENEITY IN NEOCORTEX

Anirudh Gupta, Maria Toledo-Rodriguez, Gilad Silberberg and Henry Markram *

1. INTRODUCTION

Complex behavior and cognitive functions, such as perception, attention and memory, are believed to be intimately associated with the ~1-2 mm thin neocortical sheet, that forms the outer envelope of mammalian brains[1, 2]. Neocortical function is believed to ultimately arise from the communication among the intricately interconnected constituent excitatory and inhibitory neurons[3-9]. In the late 1950's and early 1960's, experiments revealed small *repeating functional units* of several thousand cells, that are vertically arranged into cylinders/slabs (traversing all cortical layers) of about half a millimeter in diameter[10,11]. Functional modules, as exemplified by these *cortical columns*, have been found in many neocortical areas, establishing the notion that the neocortical sheet may be viewed as a mosaic of small repeating modules[1,7,8,12-18]. Understanding the structure and function of these modules, therefore, provides the foundation for understanding neocortical function.

Despite more than 100 years of research on the microstructure of the neocortex, a unifying view is still missing. The major obstacle to a unifying circuit diagram has been the daunting diversity of neuron types, each of which may employ diverse synapse patterns in the formation of neocortical microcircuits[17,19-28]. In particular, it is the small fraction of inhibitory neurons that display the greatest diversity and therefore impede attempts at simplifying neocortical microcircuit structure. Here we attempt to summarize our latest understanding of the extent of this diversity in terms of anatomy, electrophysiology, combined anatomy-electrophysiology and microcircuit connectivity.

---

* A. Gupta, M. Toledo-Rodriguez, G. Silberberg, H. Markram. Brain Mind Institute, EPFL, Lausanne, Switzerland

Excitatory-Inhibitory Balance: Synapses, Circuits, Systems
Edited by Hensch and Fagiolini, Kluwer Academic/Plenum Publishers, 2003

2. NEOCORTICAL NEURONS

Neocortical neurons are not randomly distributed in the cortical sheet, but arranged in layers (layers I-VI), each of which is characterized by different sets of (sub-) cortical in- and outputs[17,19,20]. In rodents a neocortical column of ~ 0.3mm in diameter contains roughly 7,500 neurons[29,30] (100 neurons in Layer I; 2150 in Layer II/II; 1500 in Layer IV; 1250 in Layer V and 2500 in Layer VI). The majority of neocortical neurons (70-80%) are excitatory cells (mainly pyramidal cells)[17,19,21], which are rather homogenous in many of their anatomical, physiological and molecular properties[25]. The remaining 20-30% of neocortical neurons are inhibitory interneurons, which contrary to their excitatory counterparts, are extremely diverse in their morphological, physiological and molecular characteristics[17,21-25,31,32]. Despite their tremendous diversity, inhibitory neurons display some common microcircuit features that allow them to be easily distinguished from excitatory neurons[17,21,33]. Firstly, inhibitory neurons either lack or only weakly form dendritic spines, and are, therefore, also known as aspiny (smooth) or sparsely spiny cells[9,17,34]. Secondly, in contrast to spiny neurons, inhibitory cells receive both excitatory and inhibitory inputs onto their somata[9,17,26]. Thirdly, the axonal arborization of the great majority of inhibitory neurons is generally restricted to their immediate vicinity and does not project to more distant regions. Two inhibitory cell types (Large Basket Cells and Martinotti Cells) are however specialized to project their axons laterally within the cortical sheet to inhibit distant microcircuits.

3. ANATOMICAL DIVERSITY OF INTERNEURONS

GABAergic interneurons constitute a morphologically highly heterogeneous population of neocortical cells[9,17,21,35] (Figure 1). Common to all is the lack of an apical dendrite, low spine densities (or complete lack of spines), and beaded dendrites. Instead of an apical dendrite toward the pia, many interneurons have a main dendrite with more branches extending towards the white matter. Moreover, the initial course of their axons is also often towards the pia. Contrary to excitatory cells, inhibitory interneurons are found throughout layers I-VI. Therefore, layer I is functionally unique in that it solely contains inhibitory cells[36]. Secondly, interneurons are apparently designed to innervate specific membrane domains of their target cells. Accordingly, they may be functionally subclassified into (i) axon-targeting, (ii) soma- and proximal dendritic-targeting, (iii) dendritic-targeting, and (iv) dendritic- and tuft-targeting cells[27,37] (see Figure 1). Thirdly, the axonal arborizations of interneurons may be either confined to or traverse layers and/or columns. Therefore, interneurons may also be functionally subclassified according to their "field of influence" into neurons of intralaminar-intracolumnar, interlaminar-intracolumnar, intralaminar-intercolumnar and interlaminar-intercolumnar impact. Finally, and of particular relevance for the microcircuit composition, each interneuron type seems to be differentially distributed across different layers and areas. It appears that basket cells are the most frequent interneuron type of the neocortex (about 50%), clearly predominating layers II/III, layer IV as well as layer V[38-40].

Following descriptions of interneuron anatomy emphasize the most characteristic features of each of the main inhibitory neuron types found in layers II-V of rat somatosensory cortex[25], although the basic features equally apply to higher species[3,17,32,41-50] Neocortical neurons of layers I and VI have not been studied as extensively.

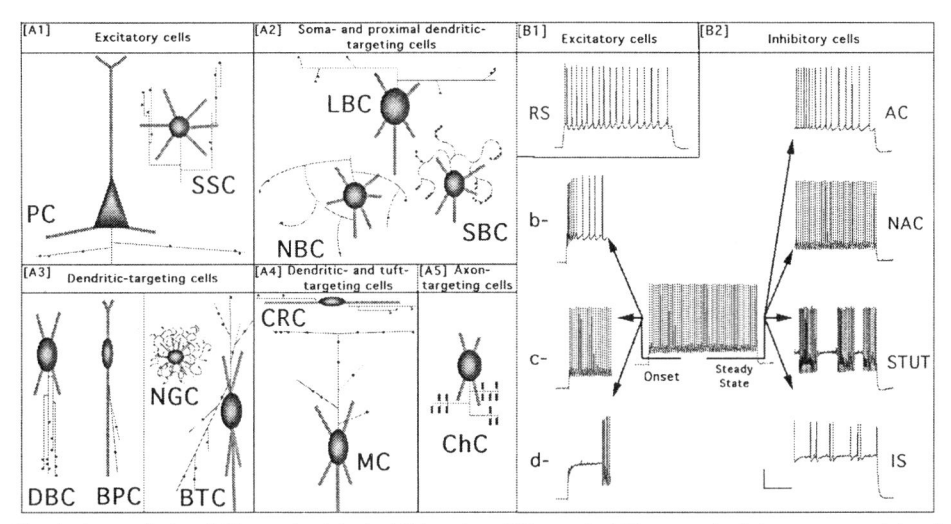

Fig. 1. Anatomical and Electrophysiological Diversity of Neocortical Neurons. **A**: Scheme summarizing the main anatomical properties of neocortical excitatory (A1) and inhibitory (A2-5) neurons. Each neuron type labeled by 3-letter abbreviation (for explanation, see text). Dendrites: thick, light gray; axon: thin, black lines; black dots: axonal boutons. Spines omitted for clarity. Neurons oriented with pia facing upwards and white matter (WM) downwards. Note the presence of a prominent, vertical dendrite directed towards WM on some interneurons (A2-4). Inhibitory interneurons (A2-5) are mainly distinguished by the structure of their axonal arbor (see text) and typically innervate selective domains (A2: (peri-) somatic; A3,4 dendritic; A5: axonal) of their target cells. **B**: Representative samples of the most common discharge responses of neocortical excitatory (B1) and inhibitory (B2) neurons to standardized intrasomatic step-current injections. B1: Excitatory cells typically display regular-spiking (RS) discharge behavior. B2: Inhibitory interneurons display a vast repertoire of discharge responses, displaying either bursts (b-), delays (d-) or neither burst/delay (classical, c-) at step-onset, and accommodation (AC), non-accommodation (NAC), stuttering (STUT) or irregular spiking (IS) at steady-state. Scale bar (20 mV; 500ms) applies to all traces. Figure reproduced with permission from ref. 25.

Basket Cells (BCs)

Basket cells are probably the most frequently encountered cortical interneurons[3,17,35,38,51-54]. BCs are distinguished by their preferential innervation of somata (20-40%) and proximal dendrites (onto shafts and spines)[40]. BCs are therefore soma- and proximal dendritic-targeting cells. Soma shape and dendritic morphology varies widely for all BCs (ovoid, triangular, inverted pyramidal or spindle shaped somata; multipolar, bitufted or bipolar dendritic morphologies)[38]. BCs are composed of three main subclasses each of which appears to be differentially distributed throughout layers II-VI:

- *Large Basket Cell (LBC):* LBC somata give rise to several smooth or sparsely spiny, beaded dendrites and emit axons with a low density of boutons that characteristically generate conspicuous long-range horizontal collaterals traversing multiple columns[3,9,17,35,38,40,51]. The axon usually arises from the pial aspect and produces a characteristic *sparse* cluster of axonal collaterals made up of a few, long and straight branches within the columns and layer as well as some vertically projecting collaterals that may cross all layers.

- *Small Basket Cell (SBC):* SBC somata give rise to 2-4 aspiny, beaded dendrites and a *dense* local cluster of highly varicose axons[3,35,38,40,51,55]. This characteristic cluster is formed by frequent, short, and curvy axonal branches, densely studded with boutons, which remain in the same column and layer. Occasionally SBCs may generate a few far-reaching collaterals projecting across layers and columns. A special subtype of SBC, termed *Clutch Cell*, has been observed in layer IV of the visual cortices of cat/monkey[40]. These cells are medium sized, multipolar cells that typically produce large bulbous terminals, which often "clutch" somata of their target cells[56].
- *Nest Basket Cell (NBC):* NBCs somata give rise to nearly aspiny, radially projecting, beaded dendrites and a *sparse to dense* local axonal cluster around somata and may generate long horizontal axons. This characteristic cluster is formed by infrequent, long and smoothly bending axonal branches, sparsely studded with boutons, which mainly remains in the same column and layer. NBCs also exhibit a characteristically simple dendritic arbor with few short and infrequently branching dendrites[28,38].

Chandelier Cell (ChC)

Chandelier cells mostly display ovoid or irregular somata, multipolar or bitufted dendritic morphologies, mostly aspiny and beaded dendrites that branch infrequently and a peculiar "chandelier-like" axonal arbor[57,58]. The dendrites may span one or several layers. The chandelier-like appearance is due to characteristic terminal portions forming short vertical rows of boutons, resembling candlesticks. Their axonal local clusters are formed by collaterals that branch with a high frequency at shallow angles and that often ramify around, above or below their somata with a high density of boutons. ChCs axons selectively innervate the axon initial segments of target cells and are therefore axon-targeting cells[27,37]. ChCs have been found in layers II-V.

Martinotti Cell (MC)

Martinotti cells display irregular somata, bitufted or multipolar dendritic morphologies, sparsely to medium spiny and beaded dendrites with a more elaborate dendritic arbor than most interneurons and a characteristic axonal projection which ramifies in layer I spreading across many columns[17,25,32,59,60]. A local and quite dense axonal cluster is formed by axons emerging either from the pial aspect of the soma or from an ascending dendrite, that branches with shallow angles to ramify mostly within the column before projecting up to layer I, where the segments form spiny boutons. MCs typically innervate dendrites (shafts and seldom spines; distal and tuft dendrites). MCs are therefore dendritic- and tuft-targeting cells. MCs are found in layers II-VI.

Bitufted Cell (BTC)

Bitufted cells display ovoid somata that emit two tufts of aspiny or sparsely spiny, beaded dendrites from opposite poles that are preferentially vertically oriented and may emit an additional oblique dendrite. Their axonal arborizations are characterized by long, vertically oriented, weakly varicose collaterals that may extend through all layers and mainly branch in a bifurcating manner. Their axonal ramification is mostly intracolumnar, although in some cases they may extend into neighboring columns. BTCs primarily innervate target cell dendrites[35] and are therefore dendritic-targeting cells. BTCs occur in layers II-VI.

Bipolar Cell (BPC)

Bipolar cells display small ovoid or spindle-shaped somata, that emit two long, vertically oriented, aspiny or sparsely spiny, beaded dendrites from opposite poles. Their characteristically simple and very sparsely varicose axonal arborization ramifies vertically across all layers[17,61-63]. The axon typically emerges from a primary dendrite. BPC dendritic arborization seems the simplest of all interneurons with infrequent, shallow branching producing a very narrow dendritic tree that may span all cortical layers occasionally forming a dendritic tuft in layer I. They preferentially form synapses with dendrites (shafts) and are therefore dendritic-targeting cells. BPCs occur in layers II-VI.

Double Bouquet Cell (DBC)

Double bouquet cells display ovoid and spindle-shaped somata, that emit several aspiny or sparsely spiny, beaded dendritic trees which are mainly bitufted or multipolar, and a thin axon that bifurcates to give rise to a characteristic, mainly descending, "horsetail-like", tight fascicular axonal cylinder[17,64-66]. The highly varicose collaterals forming these narrow columnar bundles are typically much thicker than the axonal main stem and may extend across all layers. A local ramification of different densities may occasionally be formed. They primarily innervate dendrites (spines and shafts) and are therefore dendritic-targeting cells. DBCs occur in layers II-V, although they appear to be preferentially located in the supragranular layers.

Neurogliaform Cell (NGC)

Neurogliaform cells display very small round or irregular somata, that give rise to a large number of fine, radiating dendrites that are short, aspiny, finely beaded and rarely branched[17,35,67]. They form a highly symmetrical, spherical dendritic field of less than columnar dimensions. Their ultra-thin axon originates from any part of the soma or from the base of a dendrite, shortly after which it breaks up into a very dense, highly intertwined arborization resembling a spider web. This axonal arborization is the most restricted among interneurons. NGCs target mainly dendrites (shafts) and are therefore dendritic-targeting cells[35]. NGCs occur in layers I-VI.

Neurons exclusive to Layer I

- *Cajal-Retzius Cells (CRC)* display large somata of varying shapes, long horizontal dendrites, and axons. Their horizontally projecting axons characteristically give rise to numerous short ascending and some descending, terminal fibrils[36,38].
- *Small layer I cells* are neurons with short processes that constitute a heterogeneous group of multipolar interneurons with varying axonal arborizations. These have been subdivided into small neurons with a rich axonal plexus and small neurons with a poor axonal plexus[36].

Neurons that have remained ill defined

Most of the neurons outlined above have been studied most extensively in layers II-V. Whereas some of these cells have been observed also in *layer VI*, this lamina is characterized by a multitude of local circuit neurons (~8-12 types) that still await precise description[69].

4. ELECTROPHYSIOLOGICAL DIVERSITY OF INTERNEURONS

Classification of interneuronal discharge responses has experienced a gradual refinement over the last decade. Initially, smooth or sparsely spiny cells recorded throughout layers II-VI were found to discharge APs of very brief duration repetitively without accommodation in response to sustained suprathreshold current injections[70,71]. This distinct fast-spiking (FS), non-accommodating discharge behavior was observed for "*all*" recorded cells, leading to the mistaken notion that inhibitory interneurons represent an electrophysiologically homogenous population in contrast to excitatory cells, that were shown to discharge repetitively either by spiking regularly or bursting initially (RS- and IB-cells, respectively). It became increasingly clear, however, that interneurons can exhibit a tremendous diversity of discharge responses and are in fact much more electrophysiologically diverse than pyramidal neurons (Figure 1).

The anatomical diversity of interneurons suggested to many that a morphologically defined interneuron would display a particular discharge behavior. This led to the gradual emergence of several new descriptors of interneuronal discharges. Whereas the original non-accommodating fast-spiking behavior was observed for chandelier cells and members of the basket cell family (mainly LBCs), other fast-spiking interneurons were shown to display accommodation[72,73], with some cells discharging repetitively with spike train adaptation that closely resembled the RS-behaviour of pyramidal neurons. These regular spiking non-pyramidal (RSNP) cells were found in layers II/III and V, some of which could be identified as DBCs, MCs and BPCs[23,74,75]. Other cells with bipolar morphology in these layers typically discharge with an initial burst of action potentials followed by APs at an irregular frequency (irregular spiking, or IS-cells)[24,76]. IS-cells have been further divided into two subclasses (IS1 and IS2) according to the duration of the initial burst[76], whose characteristics are quite different from the bursts observed for pyramidal IB and chattering cells[77,78]. Some interneurons located in layer V were also shown to display initial burst-like discharges, typically probed by delivering hyperpolarizing prepulses. Of these low-threshold spiking/ burst spiking non-pyramidal (LTS/BSNP) cells, some could be identified as MCs and DBCs[23,79,80]. Finally, some interneurons characteristically "avoid" discharging initially to sustained current injections. These late-spiking cells (LS-cells), that discharge only late after a slowly depolarizing ramp were found to be NGCs located in layers II/III and V[23,74], as well as in layer I[68].

In addition to the realization, that interneurons represent an electrophysiologically heterogenous group, these studies also made clear, that a particular discharge pattern may be found in more than a single interneuron type. Conversely, some morphologically identified interneuron types can display more than one discharge pattern. Recent studies demonstrated an even greater diversity of discharge behaviors and showed that each type of interneuron generally expresses a more or less limited set of discharge patterns[25,28,38] (Table 1).

These studies - aimed at understanding the functional diversity of large number of morphologically identified interneurons - adopted a simple classification scheme that encompasses previous schemes and considers both the *onset* of the discharge response and the *steady-state* response to a wide range of sustained somatic current injections (upto several times threshold current intensities) of variable durations (1-5 sec). According to this classification scheme, neocortical interneurons may display 15 distinct discharge behaviors (five main classes with three subclasses each). Although this (sub-) classification is based on the properties of repetitive discharge responses, additional differences in many of their detailed electrophysiological properties corroborate the validity of this classification scheme (see ref. 38).

Traditional Classification Schemes

Fast Spiking Cells (FS)
FS cells usually fire a single AP at threshold current pulses. Their AP waveform is characterized by a fast rise and fall rate. The fast falling rate causes their distinctive brief AP and leads to their characteristically fast afterhyperpolarization potential (fAHP)[71,81]. Sustained supra-threshold currents cause these cells to fire repetitively at high frequencies with little or negligible accommodation. The term "fast spiking", therefore, encompasses both the brevity of a single spike as well as the resulting fast discharge rate[77]. This firing pattern seems to mostly represent basket cells (mainly LBCs) and ChCs[23,73].

Burst Spiking Non-Pyramidal Cells (BSNP)
BSNP cells, originally described as *Low-Threshold Spiking Cells* (LTS), typically display burst-like discharges after a hyperpolarizing pre-pulse. These cells were found in a fraction of interneurons in layers II/III and V, some of which could be identified as MCs and DBCs[23,74,75].

Regular Spiking Non-Pyramidal Cells (RSNP)
RSNP discharge patterns are similar to the RS response of PCs. These cells were found in layers II/III and V, some of which could be identified as DBCs, MCs and BPCs[23,74,75].

Late-Spiking Cells (LS)
LS cells respond with a slowly ramp depolarization and a late onset discharge after a step current pulse. These cells were found in layers II/III and V, some of which could be identified as NGCs[23,68,74].

Irregular Spiking cells (IS)
IS cells typically discharge with an initial burst of APs followed by an irregular spiking response[24,76]. They are found in a small fraction of interneurons with bipolar morphology in layers II/III and V[24,76]. IS cells have been further divided into two subclasses (IS1 and IS2) according to the duration of the initial burst[76]. The characteristics of these bursts are quite different from the bursts observed for IB pyramidal cells and chattering cells[77,78].

Refined Classification Schemes

Interneuron discharge patterns, however, display an even richer diversity of behaviors that requires systematic classification. A unified classification scheme based on both the *onset* of the discharge response and the *steady-state* response to sustained somatic current injections was suggested[28] (Figure 1). This refined classification scheme encompasses and extends previous schemes (see Table 1 in ref. 25 for a comparison of traditional and refined classification schemes), but most importantly allows predictions about synaptic and microcircuit properties, which have not been able with the cruder schemes (see below).

Non-Accommodating Cells (NAC)
NACs fire repetitively without frequency adaptation for a wide range of sustained, long duration somatic current injections and can reach very high firing frequencies. The interspike intervals (ISIs) of consecutive APs during steady-state do not change or only change minimally. The steady-state discharge frequency increases steeply as a function of the injected current amplitude. APs are very brief and characteristically display a deep fAHP. NACs are the most frequently encountered cells in all layers. They can show different responses during the onset of a depolarization;

- *b-NACs* discharge initially a cluster of three or more APs. This **burst** persists even after the cell is tonically depolarized near to firing threshold and is, therefore, *qualitatively different* from the classical burst reported for IB cells. The NAC steady-state phase also *seamlessly* follows this initial high frequency burst. b-NAC behavior includes FS cells and is found in LBCs and NBCs.
- *d-NACs* display a distinct *delay* before discharging to a current pulse. The delay progressively decreases with increasing current injections. Delayed discharging cells characteristically show significantly higher action potential thresholds than the b- and c-subclasses[38]. d-NAC behavior includes FS and LS-cells and is found in BTCs, LBCs, NBCs SBCs, ChCs and NGCs[25,38].
- *c-NACs* neither display bursts nor delays at the onset of a depolarizing pulse, and are referred to as *classical* responses. The "onset" phase is therefore indistinguishable from the "steady state" phase[28]. c-NAC behavior includes FS cells and is found in BTCs, LBCs, MCs, NBCs and SBCs[25,38].

Accommodating Cells (AC)
ACs fire repetitively with a decrease in discharge frequency and do not reach high firing rates. ACs have been observed in layers II-VI and are the second most common discharge pattern[25,28]. All three subclasses may occur;

- *b-AC* behavior includes BSNP cells and is found in BTCs, MCs, NBCs and ChCs.
- *d-AC* behavior includes FS and LS-cells and is found in LBCs, NBCs and ChCs.
- *c-AC* behavior includes RSNP-cells and is found in BTCs, LBCs, MCs, NBCs, SBCs, ChCs, BPCs and DBCs.

Stuttering Cells (STUT)
STUT cells fire high frequency clusters of APs intermingled with unpredictable periods of silence ("morse-code"-like discharges) for a wide range of sustained, long duration somatic current injections[28]. The APs within clusters hardly accommodate. Stuttering is not just a behavior seen at threshold, but is stable for current injections several times threshold amplitude[38]. Cells displaying stuttering near threshold and fast spiking at slightly higher depolarizations, are not considered STUTs. STUTs have been observed in layers II-VI. All three subclasses may occur;

- **b-STUT** behavior is found in BTCs, MCs and NBCs.
- **d-STUT** behavior includes FS cells and is found in LBCs and NBCs.

- **c-STUT** behavior includes FS cells and is found in LBCS, NBCs and BPCs.

Irregular Spiking Cells (IS)
IS cells discharge single APs in a random manner throughout a depolarizing pulse, but do not form distinct clusters of APs. IS cells tend to show marked accommodation. IS cells have only occasionally been observed in layers II-V[24,76,82]. These cells are potentially sub-classified into;
- **b-IS** behavior includes IS1 & IS2 cells and is found in BPCs.
- **d-IS** have not been observed thus far.
- **c-IS** behavior is found in BPCs and MCs.

Bursting cells (BST)
BST cells characteristically fire a cluster of three to five APs riding on a slow *depolarizing wave* followed by a strong *slow afterhyperpolarization* (sAHP). This sAHP causes a *clear separation* of the initial burst response from the consecutive steady-state responses, even at high current injections. The peak amplitudes of these APs decrease during the bursts in most cells. The burst in BST cells is therefore similar to the classical burst found in pyramidal cells (see IB cells above) and not the same as the b-subclass (see above). BSTs have been observed in layers II-V. All three subclasses may occur;
- **r-BST** cells are characterized by *repetitive* bursting. r-BST behavior is similar to chattering[78] and is found in ChCs.
- **s-BST** cells are characterized by a *single* burst followed by complete cessation of spiking due to a complex of powerful AHPs. s-BST behavior includes BSNP cells and is found in BPCs and DBCs.
- **i-BST** cells are characterized by an *initial* burst, followed by a slow accommodating discharge response. i-BST behavior includes BSNP cells and in found in BPCs.

5. ANATOMO-ELECTROPHYSIOLOGICAL DIVERSITY OF INTERNEURONS

Understanding the structural and functional properties of neocortical microcircuits critically depends on the reliable characterization of homogenous populations of the constituent neuron types. A detailed analysis of the combined anatomical-electrophysiological properties of neocortical interneurons (Table 1) reveals that any anatomically distinct type of interneuron *typically* displays a set of discharge responses. Conversely, a given discharge behavior is *typically* found in several anatomically distinct types of interneurons. The high degrees of "redundancy" highlight the necessity of obtaining combined structural-physiological properties of neuron types as a necessary prerequisite for studying neocortical microcircuit function. Indeed, whereas it may appear that this "expansion" of interneuron types makes an understanding of the neocortical microcircuitry more difficult, this multidimensional classification has actually been crucial in unraveling many structural and functional principles of neocortical connectivity[28] (see below) and has helped resolving some of the "confusion" pertaining to the field of interneuron classification[25,38].

	AC b-AC	c-AC	d-AC	NAC b-NAC	c-NAC	d-NAC	STUT b-STUT	c-STUT	d-STUT	IS b-IS	c-IS	BST r-BST
BPC	x	x	-	x	-	-	-	x	-	x	x	-
BTC	x	x	-	x	x	x	x	-	-	x	x	-
ChC	x	x	x	-	x	x	x	-	-	-	-	x
DBC	x	x	-	x	-	-	-	-	-	x	x	-
LBC	x	x	x	x	x	x	x	x	x	x	x	-
MC	x	x	-	-	x	-	x	x	-	-	x	-
NBC	x	x	x	x	x	x	x	x	x	-	-	-
SBC	x	x	-	-	x	x	-	-	-	-	x	-
NGC	-	-	-	-	-	x	-	-	-	-	-	-

Table 1: Anatomical-Electrophysiological Diversity of Neocortical Inhibitory Neurons. Electrophysiological classification of interneurons: main classes and subclasses defined according to discharge responses at steady-state and onset-phase to intrasomatic current injections, respectively. Abbreviations explained in text; "-": not detected so far. Note that any given anatomically defined interneuron in general displays several distinct discharge behaviors and that conversely a given discharge behavior may be found in several anatomically defined interneuron types.

Importantly, the anatomical and physiological diversity displayed by inhibitory neurons is highly relevant for their functional role within neocortical microcircuits. Anatomical diversity provides a means for differentially sampling similar sets of (sub-) cortical in- and outputs, even more so as different neuron types appear to be differentially distributed throughout layers I-VI (see above). The characteristics of a neuron's dendritic and axonal arborizations as well as the spaces traversed by these arborizations determine the spatial context in which a neuron samples synaptic inputs from its sets of source neurons and affects its own sets of target neurons by synaptic outputs. The target-domain specificity displayed by different types of neocortical interneurons[17,27,35,40,83,84] in addition guarantees that the output will engage distinct compartments of the target cells, each of which may perform its own sets of functional transformations[85-88]. On top of these structural "constraints", the electrophysiological diversity displayed by neocortical neurons, indicates that identical spatio-temporal patterns of synaptic inputs will be differentially integrated and transformed into fundamentally different AP-patterns (and hence different synaptic outputs), thus potentially profoundly increasing the computational repertoire of the neocortical microcircuitry.

Molecular Diversity of Interneurons
Whereas excitatory cells (co-) express only a limited set of the commonly probed calcium binding proteins and neuropeptides, inhibitory interneurons display a much greater diversity in (co-) expression profiles[22-25,82]. They may express several combinations of CaBP and neuropeptides, co-expressing upto five different markers[38]. Importantly, it has become clear, that a particular anatomical or electrophysiological interneuron type cannot be exclusively described by the expression of its own *single* molecular marker. Whereas basket cells are commonly held to be fast-spiking, PV-positive cells[23,73,75,89], it has become clear, that basket cells can display a multitude of discharge behaviors, and a

considerable fraction of basket cells (nearly 50%) does not express PV[38]. The lack of a single anatomical or electrophysiological correlate is also evident for the other most commonly employed interneuronal markers VIP and SOM[38,60,74,75] and even certain combinations that were believed to be representative of distinct interneurons subtypes have been shown not to hold (for details see ref. 25). Yet, the molecular expression profiles as determined in terms of particular *arrays* of molecules that are *preferentially* expressed ("fingerprint") can be used to corroborate/validate anatomical and electrophysiological interneuron types[38]. These (co-) expression studies clearly indicate that the particular array of presence/absence of marker molecules is a manifestation of a cellular phenotype, and not merely an expression of transient differences in levels of peptides potentially present in all cortical cells (see Table 2 in ref. 25 for a comprehensive molecular characterization of different interneuron types).

6. DIVERSITY OF INHIBITORY CIRCUITS

Neocortical microcircuits are formed by intricately interconnecting heterogenous types of neurons with highly diverse types of synapses. Diversification of both neurons and synapses - ultimately caused by genetic variations as well as differences of environmental conditions – tremendously increases the number of possible interactions between these basic circuit elements, and the potential number of microcircuits (diversification of connectivity) is therefore enormous.

Anatomical Diversity of Excitatory Connections onto Interneurons

Glutamatergic neurons employ multiple synapses (range 1-12 synapses/connection) to innervate their target interneurons. These synapses are typically only formed onto a small fraction of dendrites[38], sometimes in a clustered fashion[90,91], i.e. they tend to be rather localized, which stands in contrast to the highly distributed manner in which glutamatergic synapses are distributed on excitatory cells. By far the majority of synapses are formed onto dendritic shafts (most interneurons display no or only few spines), although a small fraction may be formed onto dendritic spines. Importantly, glutamatergic synapses may be formed onto interneuron somata[17,33].

At present, synaptic innervation patterns have most thoroughly been described for basket cells, probably because they represent the most frequently encountered neocortical interneuron. Initial studies on the innervation of layer V, sparsely spiny, burst firing interneurons (most likely LBCs) by pyramidal cells (PCs) in rat neocortex showed that synaptic responses were mediated between 3-6 EM verified synapses[92,93]. The low sample size in this study, however, precluded any statements about synaptic innervation patterns. Later studies in supragranular layers of cat neocortex have found somewhat lower numbers of PC to LBC synapses (1-2 synapses/connection)[91], the discrepancy potentially being attributable to species differences. A recent systematic study of the synaptic innervation profiles of different types of basket cells by PCs in layers II-IV of rat neocortex yielded an average of 3.4 ± 1.1 synapses in connections onto NBCs[38]. These synaptic contacts were mainly formed onto interneuronal "basal" dendrites (59%), but a significant fraction was formed onto the main dendrite (32%) with the remainder onto the interneurons somata (9%). The number of synapses employed in this type of connection is significantly lower compared to innervation of LBCs (7 ± 3) in the same cortical area and species (Gupta A., Wang Y. and Markram H., unpublished). These synapses, the number of which is in good agreement with previous studies, are mainly distributed onto the main dendrite (68%) with only a small fraction on the "basal" dendrites (23%) and interneurons somata (9%). These results indicate that both synapse numbers and distributions may correlate with the type of interneuron targeted. Evidence for a correlation between synapse numbers and interneuron types have been observed in both cat and ferret neocortex[91,94], and preliminary data of a large number of connections from supragranular PCs onto the most common types of interneurons supports the notion of interneuron-specific patterns of synaptic innervation (Gupta A., Wang Y., Wu C.Z. and Markram H., unpublished)

Except for individual cases of synaptic innervation of DBCs (7 synapses)[91] and BPCs (4 synapses)[95], information pertaining to the intralaminar connectivity patterns for the other types of interneurons by PCs as well as by SSCs are lacking at present. Interlaminar connections from PCs onto interneurons have not been studied systematically. We are aware of only individual cases (i) in which a layer II/III PC innervated a dendritic targeting cell located in layer IV (5 synapses)[91] and (ii) in which a layer VI PC innervated a burst-firing LBC located in layer V (LM:12 synapses; EM:6;)[93]. The latter study deserves special mention, because it indicates that PCs may innervate specific types of interneurons in a target-cell specific manner. Layer V burst-firing (corresponding to BSNP) LBCs appear to be selectively targeted by layer V PCs, whose apical dendrite do not reach layer I (RS-discharge behavior; slender layer V PCs), but not by thick-tufted layer V PCs that produce a prominent tuft in layer I (with IB-discharge behavior). It is presently not known, if this target-selectivity also applies to other glutamatergic connections onto interneurons within and across cortical layers and columns.

Anatomical Diversity of Inhibitory Connections onto Pyramidal Neurons

Each distinct type of interneuron innervates its target cells by distributing multiple synapses in a highly stereotypical manner onto selected membrane domains (axon initial segments (AIS), somata, proximal and distal dendritic shafts and spines, dendritic tufts refs. 17,27,37; see above). Neurons that preferentially target the AIS are optimally positioned to "edit" a neuron's output (precise temporal sequences of APs; AP-patterning) by affecting both generation and timing of APs. The preferential innervation of the (peri-) somatic domain allows the respective neurons to effectively control the gain of summated potentials and thereby the AP-discharge of their target cells[38,96,97]. Indeed, it has been shown, that these cells play a role in phasing neuronal activity[39,98] and synchronizing the activity of a large number of cells[39,98,99]. Neurons that preferentially innervate the dendritic domain are strategically positioned (i) to influence dendritic processing and integration of synaptic inputs[88,100-103], (ii) to influence synaptic plasticity either locally or by interacting with backpropagating APs[104, 105] and (iii) to affect the generation and propagation of dendritic calcium spikes[97,100,106]. Finally, the preferential innervation of distal dendritic and tuft regions allows the respective neurons to mainly affect local dendritic integration.

In general, inhibitory neurons form a much larger number of synapses onto their target cells (2-22 synapses/connection) compared to excitatory neurons. The functional significance of utilizing such a large number of contacts is not understood, but likely contributes to maximizing the reliability of synaptic inhibition[28,38,107,108] and may be required to partly balance the excitation that comes from 4-5 times more pyramidal neurons. Axo-dendritic inhibitory synapses are typically highly distributed across the dendritic surface of target cells and are mainly formed onto dendritic shafts. In only rare cases have tendencies of synaptic clustering on target cell dendrites been observed[107]. Whenever formed onto spines, they provide an additional input (mainly "displaced" towards the spine neck region) to the excitatory synapse, which invariably impinges onto the spine head[37]. In other words, spines with sole inhibitory inputs are non-existent in the neocortex, indicating that inhibitory axo-spinous inputs may be primarily involved in "gating" excitation arriving at individual synapses (see also refs. 27,109).

As for the other types of connections, most information pertaining to anatomical properties of connectivity from inhibitory onto excitatory neurons comes from studies within the same layer and column. Inhibitory neocortical connections have been most extensively studied for basket cells onto PCs. Whereas fast-spiking LBCs in layers V/VI have been shown to employ a rather small number of synapses (range 1-5 synapses/connection), layer II/III SBCs[107] and NBCs in layers II-IV[38] utilize a considerably larger number of synapses when innervating neighboring PCs (SBC: 15 synapses; NBC: 15±4.1, 9-22 synapses/connection; see ref. 38). These differences may either allude to layer-specific or cell-type specific differences in the anatomical properties of connectivity, although layer-specific differences were not observed for NBCs located in layers II/III and IV, respectively[38]. However, each of these studies were carried out in different cortical areas, species and developmental ages, precluding definitive conclusions. Synaptic innervation patterns by other types of interneurons onto PCs have been reported only for individual cases of (i) a layer II/III DBC (10 synapses)[107], (ii) layer IV dendritic targeting cells (8 and 17 synapses, respectively)[107] and (iii) a layer II/III BPC (3 synapses)[95], precluding generalizations for these types of connections at present.

Anatomical Diversity of Inhibitory Connections onto Interneurons

With around 50 types of interneurons (see Table 1) and three types of inhibitory synapses[28], the potential number of connections between inhibitory neurons is enormous. Understanding interneuron interconnectivity is made even more difficult by the fact that the connection probabilities are much lower than between pyramidal cells and interneurons[28,38]. Despite these drawbacks, some general statements can be derived from the few studies carried out thus far. Connections between interneurons are mediated by multiple synapses (1-20 synapses/connection) that are apparently placed in a target-domain selective manner, distributed preferentially on different target domains, paralleling those onto PCs. The number of synapses deployed in a connection also appear to be determined by both pre- and postsynaptic interneurons. The axo-dendritic synapses tend to be formed onto a rather large fraction of dendrites (distributed innervation; Gupta A., Wu C.Z. and Markram H., unpublished). Finally, these axo-dendritic synapses are typically formed onto dendritic shafts, with only very few synapses formed onto dendritic spines.

Correlated anatomical and physiological studies of connections between interneurons have so far only been obtained for interneurons within the same layer and column, mainly involving basket cells. Supragranular basket cells (including LBCs and NBCs) have been shown to innervate neighboring BCs, dendritic-targeting cells (DTCs) and DBCs[108]. Interestingly, whereas the synapses in these connections were always located (peri-)somatically, clear differences were observed in the number of contacts deployed: both DTCs and DBCs received only a small number of synapses (~2 synapses), whereas postsynaptic BCs received a considerably higher number of synapses (7-20 synapses). This may suggest that whereas the *distributions* of synapses within connections among interneurons are primarily determined presynaptically (target-domain selectivity), the *numbers* of synapses deployed critically depend on both pre- and postsynaptic interneurons. Indeed, results obtained from studies of connections from (i) supragranular NBCs onto BPCs and BTCs (4 and 8 (peri-) somatic synapses, respectively)[38]; (ii) supragranular NGCs and BTCs onto NBCs (10 and 6 axo-dendritic synapses, respectively)[38] and (iii) supragranular DBCs onto BCs (1-4 axo-dendritic synapses)[108], corroborate this notion.

Differences in the numbers of synapses utilized at different types of interneuron connections have been interpreted to indicate both selectivity and preference within inhibitory neocortical microcircuits[27]. Hence, the utilization of large numbers of synapses between BCs[108] may underlie the formation of extensive BC-networks that have been implicated in long-range lateral disinhibition[110]. The coordination of activity within these BC-networks may not arise purely from chemical transmission, as recent studies have indicated additional parallel engagement of electrical transmission[111,112]. The functional significance of gap junctional coupling among neocortical interneurons[113-116] as well as to what extent this mode of communication may be preferentially utilized by neocortical BCs is currently not fully resolved[116-118].

Physiological Diversity of Excitatory Connections onto Interneurons

Initial studies of connections from PCs onto interneurons revealed that the synaptic properties of glutamatergic synapses onto inhibitory cells differed fundamentally from those onto excitatory neurons: Glutamatergic synaptic transmission at connections onto interneurons utilized different AMPA receptor subunits[119], lacked a significant NMDA-component[120] and displayed frequency-dependent facilitation[92,120,121]. Importantly, frequency-dependent facilitation was observed for interneurons with strikingly different electrophysiological behaviors (classical FS and LTS/BSNP cells) in both supra- and infragranular layers[92,121,122]. These findings suggested a functional dichotomy of the glutamatergic system, with different modes of recruitment for PCs and interneurons[73,120]. Furthermore, these studies indicated that glutamatergic synapses onto different target neurons may transmit the same AP-train differentially in a target-cell dependent manner[73,122]. Direct evidence for differential synaptic transmission was obtained by simultaneously recording from a common presynaptic PC forming depressing synapses onto other PCs and facilitating synapses onto interneurons[123]. Interestingly, static and dynamic properties of facilitating synapses from single PCs onto interneurons could vary considerable, causing the postsynaptic inhibitory cell to discharge APs when located in deep cortical layers, but not in supragranular layers[124]. This indicates, that there may be layer-specific differences in the recruitment of interneurons and hence for intracortically generated disynaptic (dis-) inhibition (see also ref. 125).

Although most of the connections from PCs onto interneurons in neocortical layers II-V displayed frequency-dependent facilitation[73,123,124], some interneurons were shown to receive depressing synapses from PCs[73]. Simultaneous recordings from the same presynaptic PC and different types of interneurons in layers II/III directly demonstrated differential synaptic transmission across different types of interneurons: accommodating interneurons with bitufted dendritic morphology received facilitating synapses, whereas non-accommodating interneurons with multipolar dendritic morphology (presumably BCs) received depressing synapses[89]. Depressing synapses have also been observed for intralaminar connections from PCs onto DBCs in layers II/III[91], onto fast-spiking BCs in layer V[126,127], onto irregular-spiking BPCs in layers II/III and V[76], and onto supragranular BPCs with a discharge behaviour resembling b-AC cells[95], indicating that these synapses are much more common than previously estimated. All these studies favor the notion that the physiological properties of glutamatergic synapses are determined in a purely target-cell dependent manner[89,95,128]. Whereas this is an attractive hypothesis, it is currently not resolved which properties determine the type of synapse formed onto distinct types of interneurons. For example, some interneurons with multipolar dendritic morphology, as well as FS-cells (presumably BCs) have been shown to receive facilitating synapses[73,124], indicating that gross anatomical or physiological properties are insufficient to predict the type of glutamatergic synapse formed (synapse mapping). Nevertheless, in a recent study, a perfect synapse mapping was obtained, when the interneuron types were characterized in both their detailed anatomical (quantitative) and electrophysiological properties, suggesting that the combined properties decide which type of synapse will be formed[38]. Seemingly homogenous populations of irregular-spiking BPCs, however, have been in some cases shown to receive facilitating instead of the expected depressing inputs[76]. This may be either due to presently unknown molecular differences of anatomically homogenous interneurons or more likely due to some anatomical heterogeneity of the sampled cells (anatomical properties were not studied in detail and quantitatively).

Systematic studies of interlaminar connections from PCs onto interneurons have not been carried out so far. According to our knowledge, only individual cases have been reported in the literature for connections (i) from layer VI PCs onto layer V BF (paired-pulse facilitation (PPF))[92], (ii) from supragranular PCs onto layer V FS-cells (paired-pulse depression (PPD))[73,122,129] and (iii) a supragranular PCs onto a layer IV dendritic targeting cell (no information about synaptic dynamics available)[91].

The physiological properties of glutamatergic synapses formed by spiny stellate cells (SSCs) onto interneurons have been studied only in rare cases. Basket cells in layer IV have been shown to receive depressing inputs[39], although facilitating inputs have also been observed in some cases (Gupta A., Wu C.Z. and Markram H., unpublished). Although the properties of innervation for other types of interneurons by SSCs in the same and in other layers have not been studied thus far, it seems reasonable to speculate that the physiological properties at these connections will parallel the synaptic heterogeneity as observed for connections from PCs onto interneurons. Interestingly, recent studies indicate that depressing PC to FS-basket cell connections[126] may express synaptic facilitation to different degrees during development. This "developmental heterogeneity" may either suggest that different PC to interneuron subcircuits are embedded within a seemingly homogenous synapse population, or that specific subcircuits are generated ("de novo") during development of the animal. In any case, these results again allude to fundamental differences between the recruitment properties of excitatory and inhibitory cells within the neocortical microcircuitry.

Physiological Diversity of Inhibitory Connections onto Pyramidal Neurons

Given the high degree of interneuron diversity, it is somewhat surprising that the physiological properties of GABAergic transmission were initially found to be rather uniform: whenever encountered, different types of neocortical interneurons (LTS/BSNP, RSNP, bitufted accommodating and multipolar FS/NAC) were found to invariably form depressing synapses onto PCs, both in supragranular[89,107,127,130] and infragranular layers[72,127]. A major advance in understanding the physiological properties of neocortical inhibitory circuits came with the systematic study of connections formed by a large number of anatomically (SBCs, NBCs, LBCs, MCs and BTCs) and electrophysiologically defined interneurons (NACs, ACs and STUTs, each with their respective b-, c-, d-subclasses) onto PCs in layers II-IV[28]. This study, which confirmed and extended previous findings, could show that GABA-A receptor mediated synaptic transmission was physiologically much more diverse than previously reported: inhibitory synapses displayed synaptic depression as well as facilitation to different degrees, yielding three distinct classes of GABAergic synapses (defined as F1-, F2- and F3-type, respectively, according to the ratio of the time constants of recovery from synaptic facilitation/depression). Importantly, each type of interneuron formed only a particular synapse class onto a given target neuron depending on the anatomical and physiological properties of both pre- and postsynaptic cells (synapse mapping principle). Furthermore, a distinct feature of GABAergic transmission was revealed by simultanously recording the synaptic responses from multiple postsynaptic neurons: target neurons of the same anatomical and electrophysiological type received synapses that displayed virtually identical temporal dynamics of synaptic transmission (homogeneity principle). This homogeneity of synaptic dynamics endows interneurons with the unique property of selecting and functionally affecting groups of neurons in an identical temporal manner (GABA groups), which stands in striking contrast to excitatory cells that innervate the same anatomical-electrophysiological type of target neuron with synapses that vary considerably in their synaptic properties. Hence, whereas excitatory cells effect each target cell in a potentially unique manner, interneurons produce identical temporal impacts onto selected, functionally related groups of neurons. Although this particular study[28] only investigated the most commonly occurring types of interneurons, the principles are likely to apply to the less frequent types of interneurons as well (ChC, NGC, DBC, BPC). Indeed, synaptic diversity has recently been observed for connections from VIPergic BPCs (b-AC like discharge response) onto PCs in layers II/III, with some displaying PPD and others PPF[95]. In how far these "types" of synapses "map" onto the previously reported diversity of BPCs[76] remains to be determined.

The properties of GABAergic transmission from interneurons onto SSCs have been studied much less frequently. Nonetheless, SSCs have been reported to receive depressing inhibitory inputs from dendrite-targeting cells[107] as well as LBCs[39]. Although currently not known, it seems reasonable to assume that GABAergic transmission onto SSCs is equally diverse. It will be of great interest to see if this presumed synaptic diversity follows similar or different organizational principles.

In addition to fast GABA-A receptor mediated inhibition, slow inhibitory synaptic responses have been recorded in neocortical neurons. These slow responses, mediated by metabotropic GABA-B receptors, have mainly been detected after strong extracellular stimulation or alternatively by repetitive, high frequency stimulation of interneurons[131,132]. This has led to the notion, that the relevant GABA-B receptors were likely located extrasynaptically and activated by "residual" GABA "spillover" from the synaptic cleft. Alternatively, it has been proposed that neocortical microcircuits may be composed of two separate populations of interneurons, each responsible for exclusively GABA-A or GABA-B receptor mediated responses, respectively[133]. Evidence for the segregation of interneuron populations has recently been obtained for few connections from FS and RSNP cells onto PCs in layer V[132]. The functional significance of these rare GABA-B responses is at present not clear.

Physiological Diversity of Inhibitory Connections onto Interneurons
Fast-spiking BCs in layers II-IV have been shown to form depressing synapses onto other BCs, DBCs and dendritic-targeting cells[108]. Differences in the degree of depression

have been observed for divergent connections formed onto non-accommodating multipolar interneurons (presumably BCs) and accommodating interneurons with bitufted dendritic morphology[89], indicating that GABAergic transmission between interneurons may be more diverse than previously expected. Indeed, it was subsequently shown that GABAergic transmission between interneurons is highly diverse with three distinct types of synapses (F1-, F2- and F3-type)[28]. As for the GABAergic innervation of PCs, distinct synapse types were formed according to the precise anatomical-electrophysiological properties of both pre- and postsynaptic interneurons. For example, NBCs in layers II-IV have been shown to receive F1- and F2-type synapses from BTCs, NGCs and NBCs and form F1- and F2-type synapses onto BTCs, LBCs, BPCs and other NBCs depending on the electrophysiological properties of both pre- and postsynaptic interneurons[38].

Although these recent studies of inter-interneuron connections may indicate that each type of interneuron potentially forms and receives synaptic innervation onto and from all other types of interneurons, this is rather unlikely to be the case. Firstly, ChC have been shown to selectively innervate excitatory cells, but not inhibitory cells (see ref. 37). Secondly, anatomical and physiological data suggest that some types of interneurons, such as LBCs[39,40,110], are highly interconnected, whereas other types of interneurons such as DBCs, appear to be much less interconnected, if indeed at all. This is suggested by their low occurrence, layer-specific distributions and narrow dendritic and axonal fields.

7. SPECULATIONS ON THE PURPOSE OF INTERNEURON DIVERSITY

A single pyramidal neuron in layers II/III receives approximately 16% inhibitory synapses[19,26] which is contributed by about 40 interneurons from the same layer and column, 5 interneurons from the same column but different layer and about 5 interneurons from the same layer but different columns (Gupta A. & Markram H., unpublished). In total about 50-60 interneurons each placing around 15 synapses will pack onto a pyramidal neuron. The diversity of the anatomical-electrophysiological types of interneurons is so large that it may even be possible that each interneuron which a pyramidal neuron receives is unique. Preliminary single cell genetics strongly support such diversity and further suggest that the electrophysiological diversity does not simply lie on a continuum, but rather distinct classes exist that are probably determined by combinations of only a few transcription factors (Toledo-Rodriguez M. & Markram H., unpublished). It seems unlikely that such an intricate design is arbitrary. The question is then whether this design principle buys computational power for the microcircuit.

There are several issues that may help understand the need for interneuron diversity. Recent *in vivo* patch-clamp studies indicate that the net excitatory and inhibitory conductances are essentially balanced even during intense activity. How is balance achieved with only 16% of the synapses on a pyramidal neuron being inhibitory? The time course of inhibitory synaptic currents is about 2 fold slower (e.g. refs. 28,134), but this is still not enough to deliver matching inhibition across a large dynamic range. Other "soft parameters" in the microcircuitry are required to generate such a balance dynamically. There are cellular and synaptic "soft parameters". Interneurons can generally discharge 2-3 times faster than pyramidal neurons which could deliver sufficient inhibition for a dynamic balance. It would however, be crucial for the inhibitory synapses to be able to reliably transmit at such higher frequencies and indeed, most inhibitory synapses have more facilitation in their synapses allowing transmission at higher frequencies than pyramidal neurons (e.g. refs. 38,127,135). If the *in vivo* evidence is correct, then the discharge rates and synaptic dynamics (and in particular their differences with respect to the excitatory system), are crucial parameters of the microcircuit for the dynamic balance of excitation and inhibition. This still does not explain the need for many different types of interneurons.

The rich anatomical, physiological and molecular heterogeneity displayed by neocortical interneurons as well as the diverse types of synapses formed by and onto these inhibitory neurons suggests that each type of inhibitory neuron performs a different functional role in microcircuit operation. The functional role of each interneuron can be considered in terms of a) recruiting inhibition and b) applying inhibition.

In terms of recruiting inhibition, the diversity of glutamatergic synapses, thresholds for spike generation and patterns of spiking responses suggest that each interneuron class is recruited differently[125] and depend on the spatio-temporal structure of pyramidal neuron discharges during excitation. For example, pyramidal neurons utilize facilitating synapses to recruit Martinotti Cells[60] which means that these interneurons would be hardly recruited if the microcircuit is only transiently excited, while Large Basket Cells would be instantly recruited and probably less so during prolonged excitation[38]. Each anatomical-electrophysiological subtype of interneuron is therefore likely to hold it's own conditions for recruitment. With many different types of interneurons the recruitment combination and sequence could be custom configured to match any stimulus structure.

Once the stimulus has configured the recruitment combination and sequence, the recruited interneurons also apply inhibition in strikingly different ways, either inhibiting the tuft dendrites to influence this quite separated compartment (MCs), the distal dendrites (BPCs, BTCs, DBCs, NGCs) to influence dendritic computations, the proximal dendrites and somata (LBCs, NBCs, SBCs) to alter the gain of integration, or the axon initial segment (ChCs) to directly edit the output train of action potentials. The field of influence of each interneuron type also varies with some interneurons distributing highly localized impact to targets within the same layer and column, some others to targets in multiple layers but within the same column, and still others inhibiting targets beyond the 150-200µm radius of a typical column. Synaptic dynamics are also different for each anatomical-electrophysiological subtype of interneuron, indicating that the different neuronal compartments and different layers and columns, will receive markedly different inhibition under different conditions of microcircuit excitation. The synaptic dynamics therefore seem to choreograph the application of inhibition with the computations taking place in different parts of pyramidal neurons, and the computations carried out by pyramidal neurons in different layers and columns.

The diversity of interneuron types and synapses involved in recruiting and applying inhibition may therefore allow the stimulus to custom configure the location, timing and intensity of the inhibition at different points of the computation taking place in neurons and in the microcircuit. If this customized inhibition is combined with the *in vivo* findings suggesting that the excitatory-inhibitory conductances are always balanced, then it would appear that all these parameters must be tuned and aligned relative to each other with extreme care. We have no idea yet how such an alignment of parameters could be achieved with local and/or global learning rules. Nevertheless, these data strongly indicate that the inhibitory system is designed to allow the microcircuit to process information in a near perfectly balanced state under any relevant stimulus condition. We therefore propose that the diversity of interneurons and their circuitry, makes the microcircuit capable of processing each of a vast spectrum of different stimuli with customized specialization. The microcircuit therefore seems to solve a phenomenal computational task of being a "general purpose chip" that can reconfigure at any moment to become a "specialized chip" that is optimal to process the current environmental input.

8. ACKNOWLEDGMENTS

We thank Profs. Gordon Shepherd, Kathleen Rockland, Rafael Yuste, Ed White, Javier DeFelipe, Menachem Segal and Drs. Dirk Jancke and Per Knutsen for critical comments on previous versions of the manuscript. This study was supported by the National Alliance for Autism Research, Boston, USA and an EU grant.

9. REFERENCES

1. Mountcastle V. B. (1998) *Perceptual Neuroscience: the Cerebral Cortex* (Harvard University Press, Cambridge)
2. Kandel E., Schwartz J. H. and Jessel T. M. (2001) *Principles in Neural Science*, 4th ed., Aplleton and Lange, Norwalk
3. Cajal S. R. (1909) *Histology due systeme nerveux de homme et des vertebrates* Maloine, Paris
4. Lorente de No R. (1938) Cerebral cortex: architecture, intracortical connections, motor projections. In: *Physiology of the nervous system*, pp. 291-329, Fulton J. F., ed. (1st. Oxford University Press, NY)
5. Mountcastle V. B. (1978) An organizing principle for cerebral function: the unit module and the distributed system. In *The mindful brain : cortical organization and the group-selective theory of higher brain function*, pp. 7 - 50, Edelman G. M. and Mountcastle V. B. (eds.), MIT Press,
6. Szentagothai J. (1978) The Ferrier Lecture, 1977. The neuron network of the cerebral cortex: a functional interpretation. *Proc R Soc Lond B Biol Sci* **201:** 219-248.
7. Eccles J. C. (1984) The Cerebral neocortex: a theory of its operation. In: *Cerebral cortex: functional properties of cortical cells*, Vol. 2, pp. 1-38, Jones E. G. and Peters A., eds. (Plenum Press, New York)
8. Shepherd G. M. (1988) A basic circuit of cortical organization. In: *Perspectives in memory research*, Gazzaniga, ed. (MIT Press, Cambridge), pp. 93-134
9. Douglas R. and Martin K. A. C. (1998) Neocortex. In: *The synaptic organization of the brain*, pp. 459-511, Shepherd G. M., eds. (Oxford University Press, New York)
10. Mountcastle V. B. (1957) Modality and topographic properties of single neurons of cat's somatic sensory cortex. *J. Neurophysiology* **20:** 408-434.
11. Hubel D. H. and T.N. W. (1962) Receptive fields, binocular interaction and functional architecture in the cat's visual cortex. *J Physiol* **160:** 106-154.
12. Jones E. G., Burton H. and Porter R. (1975) Commissural and cortico-cortical "columns" in the somatic sensory cortex of primates. *Science* **190:** 572-574.
13. Goldman P. S. and Nauta W. J. (1977) Columnar distribution of cortico-cortical fibers in the frontal association, limbic, and motor cortex of the developing rhesus monkey. *Brain Res* **122:** 393-413.
14. Goldman-Rakic P. S. and Schwartz M. L. (1982) Interdigitation of contralateral and ipsilateral columnar projections to frontal association cortex in primates. *Science* **216:** 755-757.
15. Szentagothai J. (1983) The modular architectonic principle of neural centers. *Rev Physiol Biochem Pharmacol* **98:** 11-61.
16. Purves D., Riddle D. R. and LaMantia A. S. (1992) Iterated patterns of brain circuitry (or how the cortex gets its spots). *Trends Neurosci* **15:** 362-368.
17. White E. L. (1989) Cortical Circuits. Synaptic Organization of the Cerebral Cortex Birkhauser, Boston
18. Tanaka K. (1997) Columnar Organization in the Inferotemporal Cortex. In: *Cerebral Cortex : Extrastriate Cortex in Primates*, Vol. 12, pp. 469-498, Rockland K. S. ,Kaas J. H. and Peters A. (eds.), Plenum Press, New York
19. DeFelipe J. and Farinas I. (1992) The pyramidal neuron of the cerebral cortex: morphological and chemical characteristics of the synaptic inputs. *Prog Neurobiol* **39:** 563-607.
20. Jones E. G. (1984) Laminar distribution of cortical efferent cells. In: *Cellular Components of the Cerebral Cortex*, Vol. 1, pp. 521-554, Peters A. and Jones E. G. (eds.), Plenum Press, New York
21. Peters A. and Jones E. G. (1984) *Cellular Componenets of the Cerebral Cortex* Plenum Press, New York
22. DeFelipe J. (1993) Neocortical neuronal diversity: chemical heterogeneity revealed by colocalization studies of classic neurotransmitters, neuropeptides, calcium-binding proteins, and cell surface molecules. *Cereb Cortex* **3:** 273-289.

23. Kawaguchi Y. and Kubota Y. (1997) GABAergic cell subtypes and their synaptic connections in rat frontal cortex. *Cereb Cortex* **7**: 476-486.
24. Cauli B. et al. (1997) Molecular and physiological diversity of cortical nonpyramidal cells. *J Neurosci* **17**: 3894-3906.
25. Toledo-Rodriguez M., Gupta A., Wang Y., Wu C. Z. and Markram H. (2003) Neocortex: basic neuron types. In: *The Handbook of Brain Theory and Neural Networks*. Arbib M. A. (ed.), 2nd ed., MIT Press, Cambridge
26. Peters A. (1987) Synaptic Specificity in the cerebral cortex. In: *Synaptic Function*, pp. 373-397, Edelman G. M. ,Gall W. E. and Cowan J. (eds.), Wiley-Press, New York
27. Somogyi P., Tamas G., Lujan R. and Buhl E. H. (1998) Salient features of synaptic organisation in the cerebral cortex. *Brain Res Brain Res Rev* **26**: 113-135.
28. Gupta A., Wang Y. and Markram H. (2000) Organizing principles for a diversity of GABAergic interneurons and synapses in the neocortex. *Science* **287**: 273-278.
29. Ren J. Q., Aika Y., Heizmann C. W. and Kosaka T. (1992) Quantitative analysis of neurons and glial cells in the rat somatosensory cortex, with special reference to GABAergic neurons and parvalbumin-containing neurons. *Exp Brain Res* **92**: 1-14.
30. Beaulieu C. (1993) Numerical data on neocortical neurons in adult rat, with special reference to the GABA population. *Brain Res* **609**: 284-292.
31. Houser C. R., Vaughn J. E., Hendry S. H. C., Jones E. G. and Peters A. (1984) GABA neurons in the cerebral cortex. In: *Cerebral cortex: functional properties of cortical cells*, Vol. 2, pp. 63-90, Jones E. G. and Peters A. (eds.), Plenum Press, New York
32. DeFelipe J. (2002) Cortical interneurons: from Cajal to 2001. *Prog Brain Res* **136**: 215-238
33. Peters A., Palay S. L. and Webster H. D. (1991) *The fine structure of the nervous system*, 3rd ed., Oxford University Press, New York
34. Peters A. and Jones E. G. (1984) Classification of cortical neurons. In: *Cellular Components of the Cerebral Cortex*, Vol. 1, pp. 107-122, Peters A. and Jones E. G. (eds.), Plenum Press, New York
35. Somogyi P. (1989) Synaptic organisation of GABAergic neurons and GABAa receptors in the lateral geniculate nucleus and visual cortex. In: *Neuronal Mechanisms of Visual Perception*, Proc Retina Res Found Symp, 2, pp. 35-62, Lamm D. K. T. and Gilbert C. D. (eds.), Portfolio Publ., Woodlands, Texas
36. Marin-Padilla M. (1984) Neurons of layer I: a developmental analysis. In: *Cellular Components of the Cerebral Cortex*, Vol. 1Peters A. and Jones E. G. (eds.), Plenum Press, New York
37. DeFelipe J. (1997) Microcircuits in the brain. In: *Biological and artificial computation. lecture notes in computer science*, Vol. 1240, pp. 1-14, Mira J. ,Moreno-Diaz R. and Cabestany J. (eds.), Springer Verlag, Berlin
38. Wang Y., Gupta A., Toledo-Rodriguez M., Wu C. Z. and Markram H. (2002) Anatomical, physiological, molecular and circuit properties of nest basket cells in the developing somatosensory cortex. *Cereb Cortex* **12**: 395-410.
39. Tarczy-Hornoch K., Martin K. A., Jack J. J. and Stratford K. J. (1998) Synaptic interactions between smooth and spiny neurones in layer 4 of cat visual cortex in vitro. *J Physiol* **508**: 351-363.
40. Kisvarday Z. F. (1992) GABAergic networks of basket cells in the visual cortex. *Prog Brain Res* **90**: 385-405.
41. Callaway E. M. (1998) Local circuits in primary visual cortex of the macaque monkey. *Annu Rev Neurosci* **21**: 47-74.
42. Jones E. G. (1975) Varieties and distribution of non-pyramidal cells in the somatic sensory cortex of the squirrel monkey. *J Comp Neurol* **160**: 205-267.
43. Gilbert C. D. (1993) Circuitry, architecture, and functional dynamics of visual cortex. *Cereb Cortex* **3**: 373-386.
44. Keller A. (1993) Intrinsic synaptic organization of the motor cortex. *Cereb Cortex* **3**: 430-441.
45. Lund J. S. (1988) Anatomical organization of macaque monkey striate visual cortex. *Annu Rev Neurosci* **11**: 253-288.
46. O'Leary J. L. (1941) Structure of the area striate of the cat. *J.Comp.Neurol.* **75**: 131-163
47. Valverde F. (1986) Intrinsic neocortical organization: some comparative aspects. *Neuroscience* **18**: 1-23.
48. Feldman M. L. and Peters A. (1978) The forms of non-pyramidal neurons in the visual cortex of the rat. *J Comp Neurol* **179**: 761-793.
49. Martin K. A. C. (1984) Neuronal circuits in cat striate cortex. In: *Cerebral cortex: functional properties of cortical cells*, Vol. 2, pp. 241-284, Jones E. G. and Peters A. (eds.), Plenum Press, New York
50. Peters A. and Fairen A. (1978) Smooth and sparsely-spined stellate cells in the visual cortex of the rat: a study using a combined Golgi-electron microscopic technique. *J Comp Neurol* **181**: 129-171.
51. Jones E. G. and Hendry S. H. C. (1984) Basket cells. In: *Cellular Components of the Cerebral Cortex*, Vol. 1, pp. 309-336, Peters A. and Jones E. G. (eds.), Plenum Press, New York

52. Marin-Padilla M. (1969) Origin of the pericellular baskets of the pyramidal cells of the human motor cortex: a Golgi study. *Brain Res* **14**: 633-646.
53. Marin-Padilla M. (1970) Prenatal and early postnatal ontogenesis of the human motor cortex: a golgi study. II. The basket-pyramidal system. *Brain Res* **23**: 185-191.
54. Marin-Padilla M. and Stibitz G. R. (1974) Three-dimensional reconstruction of the basket cell of the human motor cortex. *Brain Res* **70**: 511-514.
55. Szentagothai J. (1973) Synaptology of the visual cortex. In: *Central Processing of Visual Information*, B. Visual Centers in the Brain, pp. 269-324, Jung R. (ed.), Springer-Verlag, Berlin
56. Kisvarday Z. F., Martin K. A., Whitteridge D. and Somogyi P. (1985) Synaptic connections of intracellularly filled clutch cells: a type of small basket cell in the visual cortex of the cat. *J Comp Neurol* **241**: 111-137.
57. Somogyi P. (1977) A specific 'axo-axonal' interneuron in the visual cortex of the rat. *Brain Res* **136**: 345-350.
58. Somogyi P., Freund T. F. and Cowey A. (1982) The axo-axonic interneuron in the cerebral cortex of the rat, cat and monkey. *Neuroscience* **7**: 2577-2607.
59. Fairen A., DeFelipe J. and Regidor J. (1984) Nonpyramidal Neurons; general account. In: *Cellular Components of the Cerebral Cortex*, Vol. 1Peters A. and Jones E. G. (eds.), m, New York
60. Wu C. Z., Gupta A., Wang Y., Toledo-Rodriguez M., J.Y. L. and Markram H. (2001) Morphological, physiological, molecular and synaptic properties of Martinotti cells in rat somatosensory cortex. *Neural Plast* **8**: 208-208.
61. Peters A. (1984) Bipolar cells. In: *Cellular Components of the Cerebral Cortex*, Vol. 1, pp. 381-408, Peters A. and Jones E. G. (eds.), Plenum Press, New York
62. Peters A. and Harriman K. M. (1988) Enigmatic bipolar cell of rat visual cortex. *J Comp Neurol* **267**: 409-432.
63. Peters A. (1990) The axon terminals of vasoactive intestinal polypeptide (VIP)-containing bipolar cells in rat visual cortex. *J Neurocytol* **19**: 672-685.
64. Somogyi P. and Cowey A. (1981) Combined Golgi and electron microscopic study on the synapses formed by double bouquet cells in the visual cortex of the cat and monkey. *J Comp Neurol* **195**: 547-566.
65. Somogyi P. and Cowey A. (1984) Double bouquet cells. In: *Cellular Components of the Cerebral Cortex*, Vol. 1, pp. 337-360, Peters A. and Jones E. G. (eds.), Plenum Press, New York
66. DeFelipe J., Hendry S. H., Hashikawa T., Molinari M. and Jones E. G. (1990) A microcolumnar structure of monkey cerebral cortex revealed by immunocytochemical studies of double bouquet cell axons. *Neuroscience* **37**: 655-673.
67. Jones E. G. (1984) Neurogliaform or spiderweb cells. In: *Cellular Components of the Cerebral Cortex*, Vol. 1, pp. 409-418, Peters A. and Jones E. G. (eds.), Plenum Press, New York
68. Hestrin S. and Armstrong W. E. (1996) Morphology and physiology of cortical neurons in layer I. *J Neurosci* **16**: 5290-5300.
69. Toemboel T. (1984) Layer VI cells. In: *Cellular Components of the Cerebral Cortex*, Vol. 1, pp. 479-520, Peters A. and Jones E. G. (eds.), Plenum Press, New York
70. McCormick D. A., Connors B. W., Lighthall J. W. and Prince D. A. (1985) Comparative electrophysiology of pyramidal and sparsely spiny stellate neurons of the neocortex. *J Neurophysiol* **54**: 782-806.
71. Connors B. W. and Gutnick M. J. (1990) Intrinsic firing patterns of diverse neocortical neurons. *Trends Neurosci* **13**: 99-104.
72. Thomson A. M., West D. C., Hahn J. and Deuchars J. (1996) Single axon IPSPs elicited in pyramidal cells by three classes of interneurones in slices of rat neocortex. *J Physiol* **496**: 81-102.
73. Thomson A. M. and Deuchars J. (1997) Synaptic interactions in neocortical local circuits: dual intracellular recordings in vitro. *Cereb Cortex* **7**: 510-522.
74. Kawaguchi Y. and Kubota Y. (1996) Physiological and morphological identification of somatostatin- or vasoactive intestinal polypeptide-containing cells among GABAergic cell subtypes in rat frontal cortex. *J Neurosci* **16**: 2701-2715.
75. Kawaguchi Y. and Kubota Y. (1998) Neurochemical features and synaptic connections of large physiologically-identified GABAergic cells in the rat frontal cortex. *Neuroscience* **85**: 677-701.
76. Porter J. T., Cauli B., Staiger J. F., Lambolez B., Rossier J. and Audinat E. (1998) Properties of bipolar VIPergic interneurons and their excitation by pyramidal neurons in the rat neocortex. *Eur J Neurosci* **10**: 3617-3628.
77. Amitai Y. and Connors B. W. (1995) Intrinsic physiology and morphology of single neurons in neocortex. In: *Cerebral cortex: the barrel cortex of rodents*, Vol. 11, pp. 299-331, Jones E. G. and Diamond I. T. (eds.), Plenum Press, New York

78. Gray C. M. and McCormick D. A. (1996) Chattering cells: superficial pyramidal neurons contributing to the generation of synchronous oscillations in the visual cortex. *Science* **274:** 109-113.
79. Kawaguchi Y. and Kubota Y. (1993) Correlation of physiological subgroupings of nonpyramidal cells with parvalbumin- and calbindinD28k-immunoreactive neurons in layer V of rat frontal cortex. *J Neurophysiol* **70:** 387-396.
80. Kawaguchi Y. (1993) Groupings of nonpyramidal and pyramidal cells with specific physiological and morphological characteristics in rat frontal cortex. *J Neurophysiol* **69:** 416-431.
81. Connors B. W., Gutnick M. J. and Prince D. A. (1982) Electrophysiological properties of neocortical neurons in vitro. *J Neurophysiol* **48:** 1302-1320.
82. Cauli B. et al. (2000) Classification of fusiform neocortical interneurons based on unsupervised clustering. *Proc Natl Acad Sci U S A* **97:** 6144-6149.
83. DeFelipe J., Hendry S. H., Jones E. G. and Schmechel D. (1985) Variability in the terminations of GABAergic chandelier cell axons on initial segments of pyramidal cell axons in the monkey sensory-motor cortex. *J Comp Neurol* **231:** 364-384.
84. del Rio M. R. and DeFelipe J. (1995) A light and electron microscopic study of calbindin D-28k immunoreactive double bouquet cells in the human temporal cortex. *Brain Res* **690:** 133-140.
85. Larkum M. E., Zhu J. J. and Sakmann B. (2001) Dendritic mechanisms underlying the coupling of the dendritic with the axonal action potential initiation zone of adult rat layer 5 pyramidal neurons. *J Physiol* **533:** 447-466.
86. Schiller J., Major G., Koester H. J. and Schiller Y. (2000) NMDA spikes in basal dendrites of cortical pyramidal neurons. *Nature* **404:** 285-289
87. Segev I. and London M. (2000) Untangling dendrites with quantitative models. *Science* **290:** 744-750.
88. Hausser M., Spruston N. and Stuart G. J. (2000) Diversity and dynamics of dendritic signaling. *Science* **290:** 739-744.
89. Reyes A., Lujan R., Rozov A., Burnashev N., Somogyi P. and Sakmann B. (1998) Target-cell-specific facilitation and depression in neocortical circuits. *Nat Neurosci* **1:** 279-285.
90. Ahmed B., Anderson J. C., Martin K. A. and Nelson J. C. (1997) Map of the synapses onto layer 4 basket cells of the primary visual cortex of the cat. *J Comp Neurol* **380:** 230-242.
91. Buhl E. H., Tamas G., Szilagyi T., Stricker C., Paulsen O. and Somogyi P. (1997) Effect, number and location of synapses made by single pyramidal cells onto aspiny interneurones of cat visual cortex. *J Physiol* **500:** 689-713.
92. Thomson A. M., West D. C. and Deuchars J. (1995) Properties of single axon excitatory postsynaptic potentials elicited in spiny interneurons by action potentials in pyramidal neurons in slices of rat neocortex. *Neuroscience* **69:** 727-738.
93. Deuchars J. and Thomson A. M. (1995) Innervation of burst firing spiny interneurons by pyramidal cells in deep layers of rat somatomotor cortex: paired intracellular recordings with biocytin filling. *Neuroscience* **69:** 739-755.
94. Krimer L. S. and Goldman-Rakic P. S. (2001) Prefrontal microcircuits: membrane properties and excitatory input of local, medium, and wide arbor interneurons. *J Neurosci* **21:** 3788-3796.
95. Rozov A., Jerecic J., Sakmann B. and Burnashev N. (2001) AMPA receptor channels with long-lasting desensitization in bipolar interneurons contribute to synaptic depression in a novel feedback circuit in layer 2/3 of rat neocortex. *J Neurosci* **21:** 8062-8071.
96. Buhl E. H., Cobb S. R., Halasy K. and Somogyi P. (1995) Properties of unitary IPSPs evoked by anatomically identified basket cells in the rat hippocampus. *Eur J Neurosci* **7:** 1989-2004.
97. Miles R., Toth K., Gulyas A. I., Hajos N. and Freund T. F. (1996) Differences between somatic and dendritic inhibition in the hippocampus. *Neuron* **16:** 815-823.
98. Cobb S. R., Buhl E. H., Halasy K., Paulsen O. and Somogyi P. (1995) Synchronization of neuronal activity in hippocampus by individual GABAergic interneurons. *Nature* **378:** 75-78.
99. Pouille F. and Scanziani M. (2001) Enforcement of temporal fidelity in pyramidal cells by somatic feed-forward inhibition. *Science* **293:** 1159-1163.
100. Larkum M. E., Zhu J. J. and Sakmann B. (1999) A new cellular mechanism for coupling inputs arriving at different cortical layers. *Nature* **398:** 338-341.
101. Segev I. and Rall W. (1998) Excitable dendrites and spines: earlier theoretical insights elucidate recent direct observations. *Trends Neurosci* **21:** 453-460.
102. Segev I. and London M. (1999) A theoretical view of passive and active dendrites. In: *Dendrites* Stuart G. ,Spruston N. and Hausser M. (eds.), Oxford University Press,
103. Williams S. R. and Stuart G. J. (2002) Dependence of EPSP Efficacy on Synapse Location in Neocortical Pyramidal Neurons. *Science* **295:** 1907-1910.
104. Markram H., Lubke J., Frotscher M. and Sakmann B. (1997) Regulation of synaptic efficacy by coincidence of postsynaptic APs and EPSPs. *Science* **275:** 213-215.

105. Magee J. C. and Johnston D. (1997) A synaptically controlled, associative signal for Hebbian plasticity in hippocampal neurons. *Science* **275**: 209-213.
106. Traub R. D. (1995) Model of synchronized population bursts in electrically coupled interneurons containing active dendritic conductances. *J Comput Neurosci* **2**: 283-289.
107. Tamas G., Buhl E. H. and Somogyi P. (1997) Fast IPSPs elicited via multiple synaptic release sites by different types of GABAergic neurone in the cat visual cortex. *J Physiol* **500**: 715-738.
108. Tamas G., Somogyi P. and Buhl E. H. (1998) Differentially interconnected networks of GABAergic interneurons in the visual cortex of the cat. *J Neurosci* **18**: 4255-4270.
109. Koch C. (1999) *Biophysics of computation* Oxford University Press, New York
110. Kisvarday Z. F., Beaulieu C. and Eysel U. T. (1993) Network of GABAergic large basket cells in cat visual cortex (area 18): implication for lateral disinhibition. *J Comp Neurol* **327**: 398-415.
111. Tamas G., Buhl E. H., Lorincz A. and Somogyi P. (2000) Proximally targeted GABAergic synapses and gap junctions synchronize cortical interneurons. *Nat Neurosci* **3**: 366-371.
112. Fukuda T. and Kosaka T. (2000) The dual network of GABAergic interneurons linked by both chemical and electrical synapses: a possible infrastructure of the cerebral cortex. *Neurosci Res* **38**: 123-130.
113. Gibson J. R., Beierlein M. and Connors B. W. (1999) Two networks of electrically coupled inhibitory neurons in neocortex. *Nature* **402**: 75-79.
114. Beierlein M., Gibson J. R. and Connors B. W. (2000) A network of electrically coupled interneurons drives synchronized inhibition in neocortex. *Nat Neurosci* **3**: 904-910.
115. Galarreta M. and Hestrin S. (1999) A network of fast-spiking cells in the neocortex connected by electrical synapses. *Nature* **402**: 72-75.
116. Galarreta M. and Hestrin S. (2001) Electrical synapses between GABA-releasing interneurons. *Nat Rev Neurosci* **2**: 425-433.
117. McBain C. J. and Fisahn A. (2001) Interneurons unbound. *Nat Rev Neurosci* **2**: 11-23.
118. Thomson A. M. (2000) Neurotransmission: Chemical and electrical interneuron coupling. *Curr Biol* **10**: R110-112.
119. Hestrin S. (1993) Different glutamate receptor channels mediate fast excitatory synaptic currents in inhibitory and excitatory cortical neurons. *Neuron* **11**: 1083-1091.
120. Thomson A. M., Deuchars J. and West D. C. (1996) Neocortical local synaptic circuitry revealed with dual intracellular recordings and biocytin-filling. *J Physiol Paris* **90**: 211-215.
121. Thomson A. M., Deuchars J. and West D. C. (1993) Single axon excitatory postsynaptic potentials in neocortical interneurons exhibit pronounced paired pulse facilitation. *Neuroscience* **54**: 347-360.
122. Thomson A. M. (1997) Activity-dependent properties of synaptic transmission at two classes of connections made by rat neocortical pyramidal axons in vitro. *J Physiol* **502**: 131-147.
123. Markram H., Wang Y. and Tsodyks M. (1998) Differential signaling via the same axon of neocortical pyramidal neurons. *Proc Natl Acad Sci U S A* **95**: 5323-5328.
124. Wang Y., Gupta A. and Markram H. (1999) Anatomical and functional differentiation of glutamatergic synaptic innervation in the neocortex. *J Physiol Paris* **93**: 305-317.
125. Gupta A., Wang Y. and Markram H. (2001) Principles of neocortical interneuron recruitment. *Neural Plast* **8**: 176-176
126. Angulo M. C., Rossier J. and Audinat E. (1999) Postsynaptic glutamate receptors and integrative properties of fast-spiking interneurons in the rat neocortex. *J Neurophysiol* **82**: 1295-1302.
127. Galarreta M. and Hestrin S. (1998) Frequency-dependent synaptic depression and the balance of excitation and inhibition in the neocortex. *Nat Neurosci* **1**: 587-594.
128. Rozov A., Burnashev N., Sakmann B. and Neher E. (2001) Transmitter release modulation by intracellular Ca2+ buffers in facilitating and depressing nerve terminals of pyramidal cells in layer 2/3 of the rat neocortex indicates a target cell-specific difference in presynaptic calcium dynamics. *J Physiol* **531**: 807-826.
129. Thomson A. M. and Bannister A. P. (1999) Release-independent depression at pyramidal inputs onto specific cell targets: dual recordings in slices of rat cortex. *J Physiol* **519**: 57-70.
130. Deuchars J. and Thomson A. M. (1995) Single axon fast inhibitory postsynaptic potentials elicited by a sparsely spiny interneuron in rat neocortex. *Neuroscience* **65**: 935-942.
131. Nicoll R. A., Malenka R. C. and Kauer J. A. (1990) Functional comparison of neurotransmitter receptor subtypes in mammalian central nervous system. *Physiol Rev* **70**: 513-565.
132. Thomson A. M. and Destexhe A. (1999) Dual intracellular recordings and computational models of slow inhibitory postsynaptic potentials in rat neocortical and hippocampal slices. *Neuroscience* **92**: 1193-1215.
133. Benardo L. S. (1994) Separate activation of fast and slow inhibitory postsynaptic potentials in rat neocortex in vitro. *J Physiol* **476**: 203-215.

134. Galarreta M. and Hestrin S. (1997) Properties of GABAA receptors underlying inhibitory synaptic currents in neocortical pyramidal neurons. *J Neurosci* **17:** 7220-7227.
135. Varela J. A., Song S., Turrigiano G. G. and Nelson S. B. (1999) Differential depression at excitatory and inhibitory synapses in visual cortex. *J Neurosci* **19:** 4293-4304.

10

FAST-SPIKING CELLS AND THE BALANCE OF EXCITATION AND INHIBITION IN THE NEOCORTEX
The role of short-term synaptic plasticity and electrical synapses

Mario Galarreta and Shaul Hestrin[*]

1. INTRODUCTION

Excitatory glutamatergic neurons largely outnumber inhibitory ones and account for about 80% of the cells in the neocortex. In addition, the axonal collaterals of excitatory neurons make large number of contacts with neighboring neurons. As a result of these extensive recurrent excitatory circuits, the neocortex is a structure particularly prone to epilepsy. Epileptic activity is characterized by both a dramatic increase in the average firing rate and in synchrony across a large population of neurons. It is well known that blockade of inhibition leads to epileptic activity[1], and that drugs (benzodiazepines and barbiturates) that enhance GABA-mediated inhibition are useful in preventing it[2]. Thus, a fundamental role of inhibitory neurons is to stabilize cortical circuits and prevent epileptic activity. However, the mechanisms that adjust the relative level of excitation and inhibition are only poorly understood. In this chapter, we will discuss our recent experimental investigation of the synaptic and network properties of a subtype of inhibitory interneuron, the fast-spiking cell (FS), that may provide powerful anti-epileptic and stabilizing mechanisms. First, we will discuss the observation that prolonged firing induces much less depression at inhibitory synapses made by FS cells than at excitatory synapses among pyramidal neurons. This difference, which is frequency-dependent, could promote stable activity in cortical networks and prevent runaway excitation. In the second part of the chapter, we will discuss how FS cells are selectively interconnected via electrical synapses. The presence of electrical coupling together with a precise spike transmission between pyramidal and FS cells, may allow these neurons to detect synchronous excitation and increase their firing accordingly.

---

[*] M. Galarreta, Dept. of Comparative Medicine, Stanford University School of Medicine, Stanford, CA 94305, USA. S. Hestrin, Dept. of Comparative Medicine and Dept. of Neurology and Neurological Sciences, Stanford University School of Medicine, Stanford, CA 94305, USA.

Excitatory-Inhibitory Balance: Synapses, Circuits, Systems
Edited by Hensch and Fagiolini, Kluwer Academic/Plenum Publishers, 2003

2. SHORT-TERM SYNAPTIC DEPRESSION AND THE BALANCE OF INHIBITION AND EXCITATION

The extensive recurrent excitatory connections among pyramidal cells in the neocortex can produce epileptic activity. Thus, an important role of inhibitory neurons is to balance excitation and regulate cortical activity. However, the effectiveness of inhibition and excitation has to be adjusted to maintain the overall level of activity in the cortex at an operating range. This adjustment has to be dynamic because the overall activity of the cortex covers a wide range. When the activity of excitatory neurons increases, the elevated level of excitation requires an increase of the 'balancing' inhibition. Cortical neurons exhibit an ongoing spiking activity and although the overall activity of cortical circuits varies over time, the firing rates of neocortical neurons does not reach saturating levels, nor does the activity in cortical circuits shut down. Therefore, to produce stable ongoing neocortical spiking, a mechanism must exist that can dynamically adjust the relative effectiveness of inhibition and excitation in response to changes in the global ongoing neuronal activity. Based on our experimental findings we suggest that differences in the short-term synaptic plasticity of excitation vs. inhibition can act as a dynamic mechanism maintaining stability in the cortex.

Short-term synaptic plasticity is a powerful mechanism that adjusts synaptic strength in response to the temporal patterns of synaptic activation[3]. Conventionally, short-term plasticity is studied by testing the postsynaptic responses to a pair or to a brief train of presynaptic stimuli at various time intervals. However, the activity exhibited in the cortex is ongoing and the impact of prolonged activation of presynaptic axons may result in changes of synaptic strength that are not observed under only brief stimulation. Thus, we asked whether there is a differential adjustment of synaptic strength at GABAergic and glutamatergic synaptic connections in response to prolonged stimulation that could play a role in maintaining the balance of inhibition vs. excitation.

Inhibition in the cortex is mainly mediated by a heterogeneous population of GABAergic neurons[4] that may exhibit different forms of short-term plasticity[5]. Important inhibitory control of pyramidal cells is provided by fast-spiking (FS) cells[4], which are a prominent subtype of GABAergic neuron that include basket cells and express parvalbumin[6]. These cells make somatic and proximal dendritic contacts at pyramidal cells and are ideally placed to affect pyramidal cell spiking[7]. Therefore, we decided to focus our study in this type of GABAergic interneuron, and compared the response to prolonged stimulation of inhibitory synapses made by FS cells with that of excitatory synapses made by pyramidal cells.

2.1. FS-Cell Inhibitory Synapses Show Less Frequency-Dependent Depression Than Pyramidal Excitatory Ones

To characterize connections among specific cell types we used paired recordings from cells belonging to identified classes of neocortical neurons. We obtained simultaneous whole-cell recordings from synaptically connected pairs of pyramidal neurons and from pairs consisting of an FS cell and a pyramidal neuron (P). The experiments were performed in layer 5 and in layer 2/3 of the rat somatosensory and visual cortices.

Cells were initially selected according to their appearance using infrared DIC optics[8,9]. To distinguish FS cells from other types of non-FS nonpyramidal cells we examined their pattern of firing[4]. In response to near-threshold current injections, FS cells typically fired one or two initial spikes followed, at variable latency, by a discharge of high-frequency non-adapting action potentials (Fig. 1A). We have found that the FS cells identified in that manner have morphological appearance similar to basket cells and express the calcium-binding protein parvalbumin (Fig. 1B, C)[9,10]. We recorded postsynaptic currents (PSCs) under voltage-clamp while stimulating presynaptic neurons in current-clamp mode. Because we used a chloride-rich intracellular solution, both

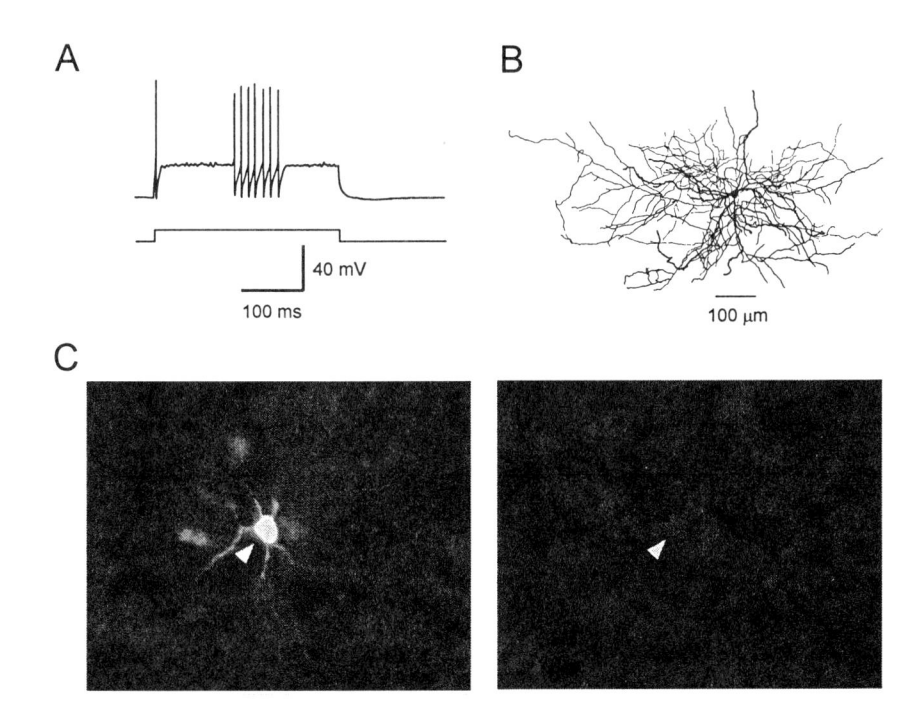

Fig. 1. Characterization of FS cells. A, Typical pattern of firing of a FS cell in response to near-threshold current injection. B, Neurolucida reconstruction of a FS cell filled with biocytin. C, Morphology of a FS cell filled with biocytin (left). Immunoreactivity of the same cell with an antibody specific for parvalbumin

excitatory PSCs (EPSCs) and inhibitory PSCs (IPSCs) were recorded as inward currents (Fig. 2A, B).

Short-term plasticity has been commonly tested by applying two presynaptic stimuli at various time intervals and examining the postsynaptic responses. We found that both P -> P and FS -> P connections exhibited similar degree of paired-pulse depression (50 ms interval, Fig. 2A, B). However, given that neurons *in vivo* are continuously active, we tested the unitary synaptic connections with sustained patterns of activation. Baseline synaptic strength was recorded by stimulating at 0.25 Hz (filled circles Fig. 2C, D). When the stimulation rate was increased; there was a decline of both EPSC and IPSC amplitudes (open circles, Fig. 2C, D). However, whereas the inhibitory synaptic connections were relatively stable after this initial decline, further slowly occurring depression dramatically decreased the amplitude of the EPSCs. This synaptic depression was reversible. After the stimulation rate was reduced to 0.25 Hz, the amplitude of both EPSCs and IPSCs recovered baseline values (Fig. 2C, D).

We defined the steady-state PSC as the average relative amplitudes of the last 200 PSCs (marked "3" in Fig. 2C, D). We found that the average steady-state EPSC, at 20 Hz, of the P -> P connections were dramatically reduced (4.2 $\pm$ 1.4%, Fig. 2C, D) whereas the steady-state IPSC of the connections from fast-spiking to pyramidal neurons were relatively stable (28.9 $\pm$ 2.9%; Fig. 2C, D). On the other hand, the transient

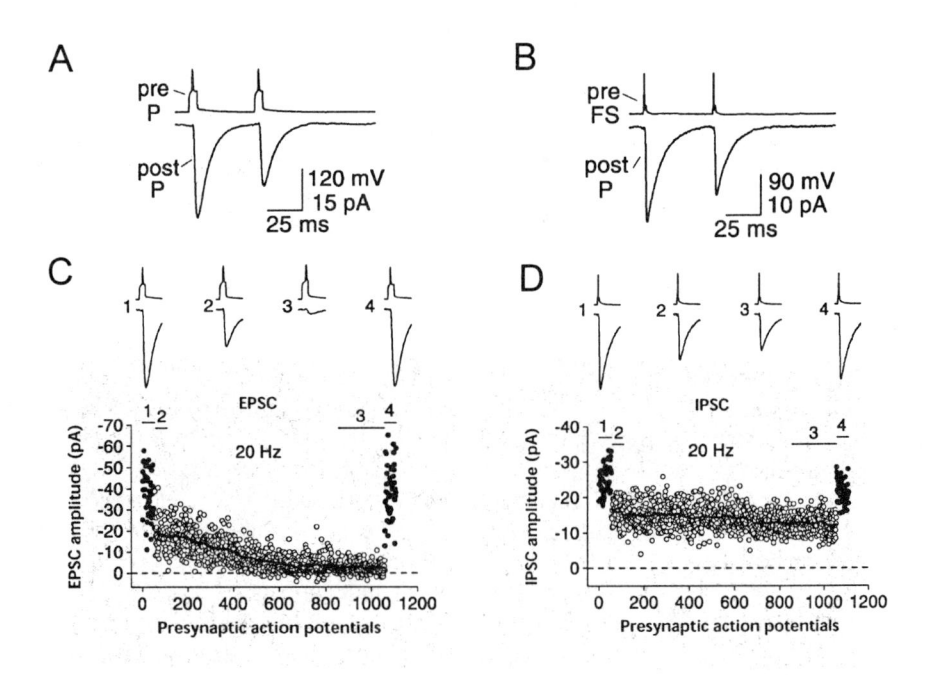

Fig. 2. Inhibitory FS -> P synapses show less depression in response to prolonged high-frequency stimulation than excitatory P-> P ones. A and C, paired recording between two pyramidal neurons. B and D, paired recording between a presynaptic FS cell and a postsynaptic P neuron. Filled circles in C and D represent responses obtained at 0.25 Hz. Open circles represent responses obtained at 20 Hz. With permission from ref. 8.

response at a stimulation frequency of 20 Hz of excitatory and FS inhibitory PSCs showed an initial similar degree of depression (40-50% of baseline value; marked "2" in Fig. 2C, D). Thus, our experiments show that in response to prolonged presynaptic stimulation, IPSCs generated by FS cells onto pyramidal neurons can sustain a much larger synaptic response than the EPSCs generated at P -> P connections.

To investigate the frequency dependence of the synaptic depression induced by sustained synaptic activation, we analyzed the PSCs generated by trains of one thousand action potentials over a range of frequencies (5-40 Hz). Depression of the transient PSC showed a similar frequency dependence at excitatory P -> P and P -> FS and inhibitory FS ->P synaptic connections (Fig. 3A). The transient PSC (average of the first 50 responses) was about 60% of baseline value at a stimulation rate of 5 Hz and about 30% at 40 Hz (Fig. 3A). In contrast, the steady-state EPSCs (from both P -> P and P -> FS synaptic connections) were significantly more depressed than the steady state of inhibition at FS -> P at stimulation frequencies in the range of 10 to 30 Hz (Fig. 3B). These results suggest that excitatory synapses established among pyramidal neurons and at P -> FS connections exhibit stronger depression than inhibitory connections made by FS cells at pyramidal neurons, when relatively high frequencies of activation are sustained.

2. 2. Frequency-Dependent Postsynaptic Impact

Next we studied the impact of excitatory and inhibitory synapses over a range of stimulation frequencies. We compared the synaptic impact, defined as the product of the PSC amplitude and the presynaptic action potential frequency, of both EPSCs and IPSCs. Using this index of synaptic impact, related to the synaptic effect on the average postsynaptic membrane potential, we found that transient increase of stimulation enhanced the synaptic impact at both excitatory and at FS -> P inhibitory connections. Thus, both the excitatory and the inhibitory synapses that we have studied can readily respond to a transient increase in presynaptic activity[8]. Next we studied the steady-state synaptic impact of excitatory and inhibitory synapses of FS cells over a range of frequencies. The plot of the frequency-impact relationship had a positive slope at low frequencies and negative slope at high frequencies (Fig. 3C). Therefore, an increase in the rate of stimulation at low frequencies would enhance synaptic impact. However, an increase in the rate of stimulation at high frequencies would reduce the synaptic impact. Thus, our finding indicates that for each synapse there is an optimal frequency of activation generating maximal postsynaptic impact. At low frequencies, the relative frequency-dependent change of the synaptic impact was similar at excitatory synapses and at inhibitory synapses of FS cells. However, at frequencies above 10 Hz, the relative impact of inhibitory synapses of FS cells increased, whereas the relative impact of excitatory synapses decreased (Fig. 3C). The synaptic impact in response to 20 Hz stimulation relative to that at 5 Hz (Fig. 3D) showed that whereas the impact of the FS -> P inhibitory synapses increased (2.2 fold), that of the excitatory synapses declined (0.32 P -> P and 0.53 P -> FS).

Our main finding is that sustained activation uncovers a much larger depression at excitatory synapses among pyramidal neurons than at the inhibitory connections between fast-spiking and pyramidal neurons. As a result of this difference in frequency-dependent depression, synaptic impact will peak at higher frequencies at FS cell inhibitory synapses than at excitatory synapses, thus promoting stable activity in the cortex.

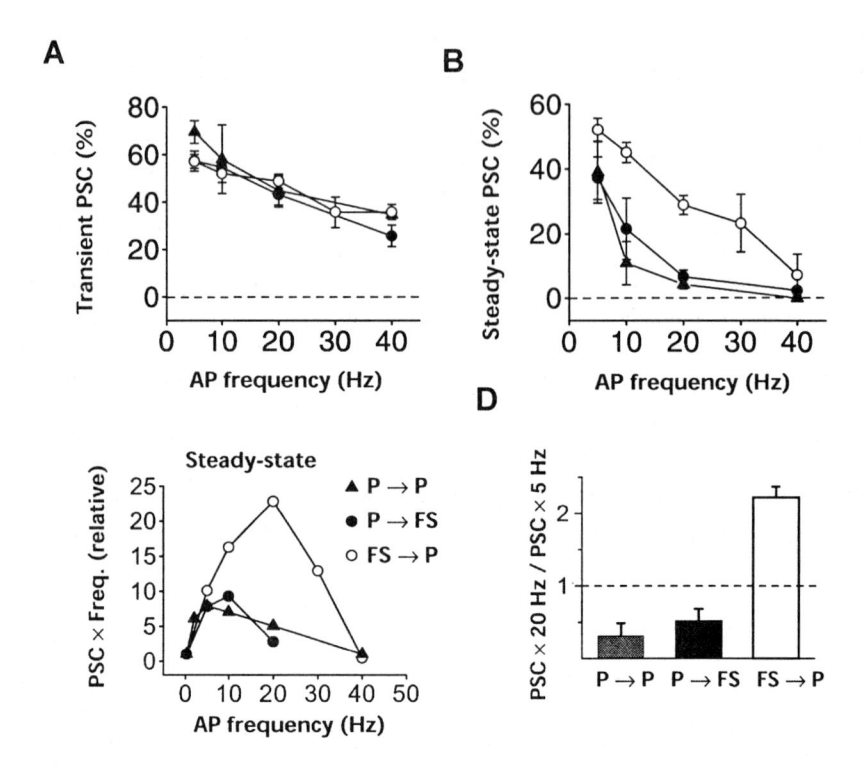

Fig. 3. Frequency dependent depression of inhibitory (FS -> P, open circles) and excitatory (P -> P, filled triangles; and P -> FS, filled circles) unitary connections. A, Transient PSCs are the average response of the first 50 responses. B, Steady-state PSCs are the average of the 800th to the 1000th responses. C, Comparison of the synaptic impact at the steady-state over a range of frequencies. Data obtained from each connection have been normalized to the value obtained at 0.25 Hz. D, Ratio of the steady-state synaptic impact at 20 Hz over that at 5 Hz. With permission from ref. 8.

The differences in the frequency-dependent depression reported here could be one of the mechanisms contributing to the balance of inhibition and excitation and preventing saturation of neuronal firing and epileptic activity. When the average firing rate of cortical neurons is low, the prominent recurrent excitatory connections will further increase neuronal firing. However, prolonged periods of firing at high rates will cause increased inhibition of FS cells relative to excitation and consequently a decrease in firing rate. Taken together, the frequency-dependent depression of fast-spiking and pyramidal neurons can contribute to the stabilization of cortical activity and to the prevention of saturation of neuronal firing rate at the steady state, while allowing a vigorous response to transient signals.

The differences that we found in the frequency-dependent steady-state depression response at FS -> P synapses compared with excitatory synapses has been also reported using extracellular stimulated inhibitory and excitatory synapses in the visual cortex[11], at inhibitory synapses of hippocampal FS cells[12] and at cultured cortical cells[13]. These data suggest that the molecular components of the synaptic machinery underlying short-term plasticity may be different at different classes of excitatory and inhibitory connections.

Indeed recent experiments have revealed that presynaptic proteins linked to short-term plasticity exhibit different pattern of expression at excitatory and inhibitory synapses[13-15]

3. RESPONSE OF NETWORKS OF FAST-SPIKING CELLS TO SYNCHRONOUS EXCITATORY ACTIVITY

During epileptic activity a large number of pyramidal cells fire synchronously[1]. What are the mechanisms that prevent synchronized epileptic activity? The answer to this question remains uncertain, but it is plausible that a mechanism that would increase the firing of inhibitory neurons in response to coherent excitatory activity could prevent the development of full-blown epilepsy. New findings suggest that synaptic interactions of GABAergic interneurons could promote synchrony detection in networks of inhibitory cells. Thus, GABAergic cells could be sensitive to the level of synchronized excitation and increase their firing in response to coherent excitatory inputs. In this section, we will report on these findings and will discuss how networks of GABAergic neurons could be sensitive to synchronized excitation and, thus, may prevent epileptic activity.

3.1. Precise Spike Transmission Between Pyramidal and Fast-Spiking Cells

If FS interneurons detect and respond to synchronized excitatory activity in the neocortex, it is important to determine how sensitive they are to the timing of individual excitatory inputs. To address this question we studied the spike-to-spike transmission (i. e., spike transmission) between pyramidal and GABAergic fast-spiking (FS) cells (Fig. 4; 16). We simultaneously recorded in brain slices from pairs consisting of a presynaptic pyramidal neuron and a postsynaptic FS cell. We focused our study in this type of interneuron because these cells include basket and chandelier cells and exert a powerful inhibitory control over hundreds of local pyramidal cells. In addition, the properties of unitary excitatory connections onto FS cells[8,17], as well as their voltage-dependent conductances[18-20] suggest that these cells might be particularly sensitive to the timing of their excitatory inputs.

Monosynaptic P -> FS cell connections were detected by the generation of short-latency unitary excitatory synaptic potentials (EPSPs) in response to individual pyramidal spikes (mean latency, 0.63 ± 0.05 ms, n=12). To reproduce the ongoing activity that occurs *in vivo*, we injected into the postsynaptic FS cells fluctuating current waveforms resembling synaptic currents that changed from trial to trial (Fig. 4A). These injections produced in an irregular firing of ~ 5-50 Hz at the FS cells. The impact of a presynaptic spike on the postsynaptic firing (i. e., spike transmission) was studied by producing a timed action potential in the pyramidal cell with every trial. Spike transmission was examined by constructing a peristimulus time histogram (PSTH) of the postsynaptic cell. This revealed a sharp increase in the firing frequency of the FS cell after the pyramidal cell presynaptic spike (Fig. 4A). Whenever the FS cell fired in response to the presynaptic action potential, the postsynaptic spike occurred in a very narrow temporal window of only ~ 2 milliseconds. This window was centered at 1.7 ± 0.1 ms (n = 7 pairs) after the peak of the spike in the presynaptic pyramidal cell (Fig. 4B). These experiments indicate, therefore, that individual FS cells can fire with high temporal precision in response to unitary excitatory activation. Sensitivity to the timing of their presynaptic inputs may have important functional consequences and, as we will see below, may allow

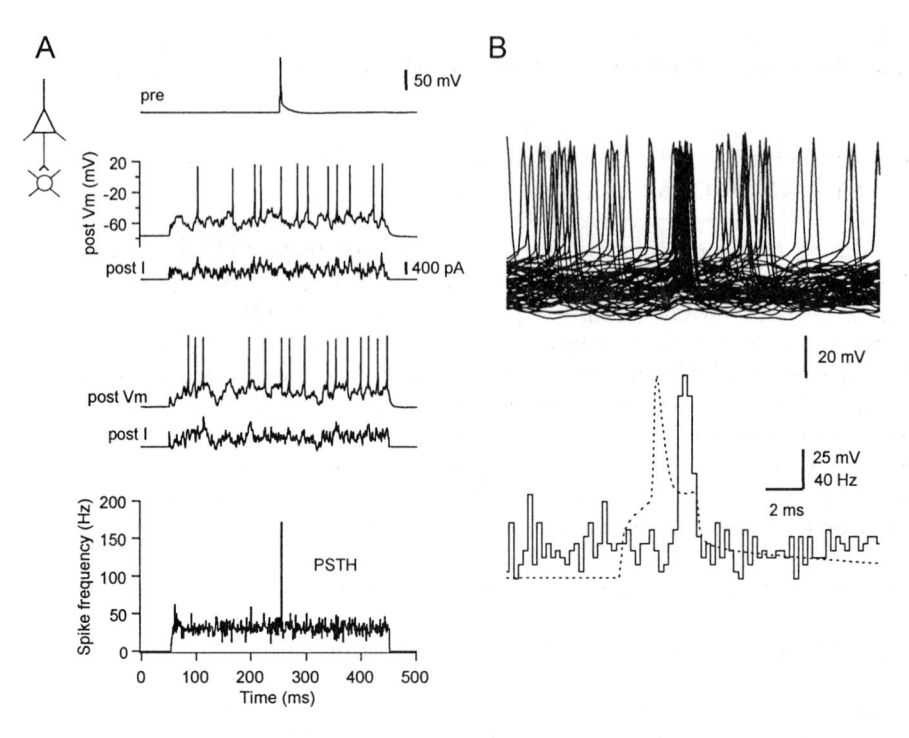

Fig. 4. Precise spike transmission between pyramidal and FS cells. A, Irregular spike activity was generated postsynaptically by injecting waveforms of random fluctuating current resembling synaptic inputs. Two examples are illustrated in the middle traces. A single spike in the presynaptic pyramidal cell (top) produced a sharp peak in the FS cell PSTH (bottom). B, Same experiment as in A. Top, superimposition of 75 consecutive postsynaptic recordings at an expanded time scale. Bottom, temporal relationship between the presynaptic spike (dashed line) and the postsynaptic PSTH. Reprinted with permission from ref. 16. Copyright 2001 American Association for the Advancement of Sciences.

groups of FS cells interconnected by both electrical and chemical synapses to detect synchronous excitatory activity within the neocortex.

Importantly, high temporal precision has also been demonstrated in specific classes of hippocampal interneurons where the EPSP-spike coupling has been examined[21,22].

3. 2. FS Cells Are Interconnected Via Electrical Synapses

How does the firing of a GABAergic FS cell affect the activity of other FS cells? To study the synaptic interactions among neocortical FS cells, we recorded from pairs of electrophysiologically and morphologically identified FS cells in brain slices. We found that in most cases (66 % of pairs) FS cells were electrically coupled[9,23]. Thus, injection of subthreshold current in one of the cells, affected the membrane potential of both the injected and the non-injected neuron (Fig. 5A). Importantly, electrical coupling between pairs of pyramidal cells was not observed (Fig. 5B). In all cases electrical coupling among FS cells was found to be reciprocal, indicating that current could flow bidirectionally through the electrical connections. The anatomical basis of electrical

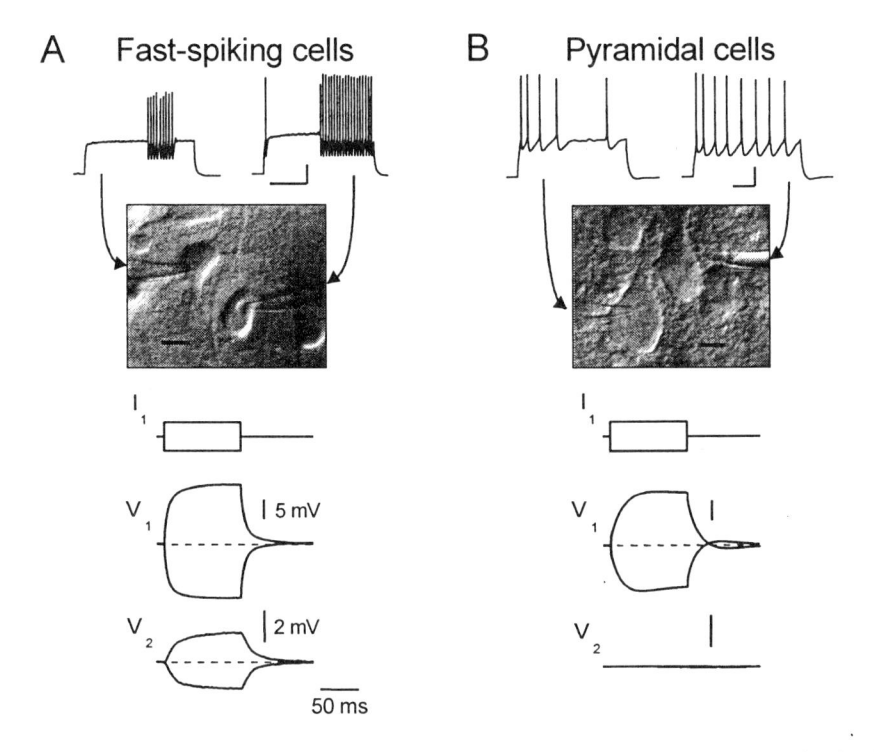

Fig. 5. Specific electrical coupling among FS cells. A, Pattern of spiking (scale bars, 20 mV and 100 ms) and infrared videomicroscopy image of a pair of electrically-coupled FS cells (scale bar, 10 µm). The injection of current in one of the cells affected the membrane voltage of both the injected and the non-injected cell. B, Pairs of neighboring pyramidal cells were not electrically coupled. Scale bar legends in A also apply to B. With permission from ref. 9.

synapses are dendro-dendritic and dendro-somatic gap junctions[24-26]. Gap junctions are made of two hemichannels in close apposition, one in each cell, that establish a direct communication of the cytoplasms of the two connected neurons[27]

It is important to note that FS cells do not seem to establish electrical synapses with other cells including pyramidal neurons, glial cells or other types of nonpyramidal neurons[9,23]. Interestingly, other types of GABAergic interneurons in the neocortex including low-threshold spiking (LTS) cells[23], bipolar cells[28], regular-spiking nonpyramidal (RSNP) cells[29] and late-spiking (LS) cells[30] are also interconnected via electrical synapses with cells belonging to the same type. Thus, electrical coupling does not occur randomly among GABAergic interneurons, but it is cell specific, connecting inhibitory cells belonging to the same functional type for review see ref. 31. In addition, functional electrical synapses among FS cells have been also demonstrated using neocortical slices from adult animals[10,32].

Spikes in FS cells are fast and are followed by a prominent afterhyperpolarization. Consequently, the signal transmitted to the postsynaptic cell through an electrical synapse in response to a presynaptic spike is also biphasic (Fig. 6A, bottom traces). It consists of a brief depolarization (i. e. *spikelet*) reflecting the spike itself, and a slower hyperpolarizing component reflecting the afterhyperpolarization. Thus, a spike in an FS cell can transiently excite and then inhibit other electrically-coupled FS cells.

In addition, FS cells are also commonly connected via GABA-mediated chemical synapses. In this case, the generation of a presynaptic action potential results in a hyperpolarizing postsynaptic potential (IPSP) (Fig. 6A, top traces). When both types of connections coexist, the spikelet is followed by a hyperpolarization mediated by both the GABAergic IPSP and the electrical transmission of the AHP (Fig. 6B).

3. 3. Networks of FS cells are sensitive to the level of excitatory synchrony

We propose that the combination of these synaptic properties: temporal precision of spike transmission from excitatory cells together with the interconnection via electrical

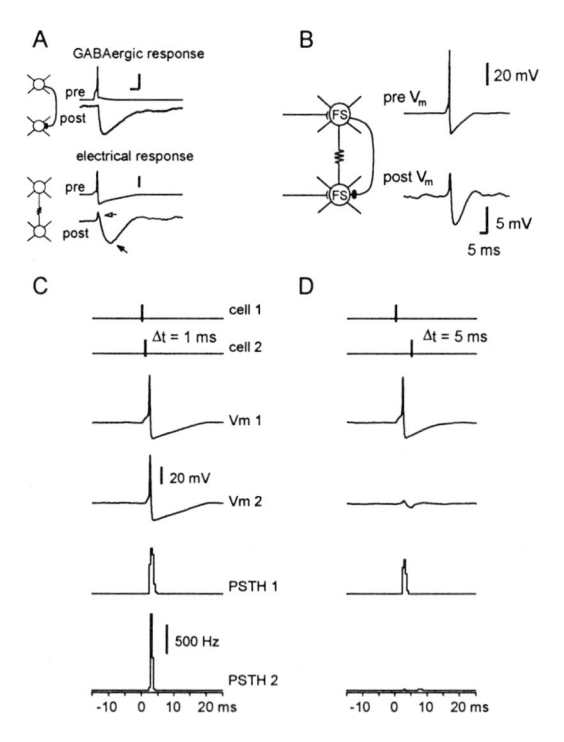

Fig. 6. Pairs of FS connected by both electrical and chemical synapses can detect synchronous excitatory activity. A, (Top) Example of a $GABA_A$-mediated IPSP in a pair of FS cells chemically connected. (Bottom) Postsynaptic response produced by a presynaptic spike in a pair of FS cells electrically-coupled. B, Recording from a pair of FS cells connected by both electrical and GABAergic synapses. C, Two simulated EPSPs separated by 1 ms produced synchronous firing of both FS cells. D, If two sim-EPSPs were separated by 5 ms, the firing of the postsynaptic cell was dramatically reduced. With permission from ref. 16.

and GABAergic synapses, may allow networks of FS cells to be sensitive to the level of synchronous firing of excitatory cells.

Nearly synchronous excitatory inputs (1 ms apart) at different FS cells within the network will be aided by the mutual depolarizing phase of the electrical response (*spikelet*), and synchronous firing of the FS cells will be promoted (Fig. 6C). On the other hand, nonsynchronous inputs (5 ms apart) will be relatively inefficient, because FS cells receiving late inputs would be influenced by the hyperpolarizing response from neurons receiving early inputs and their firing diminished (Fig. 6D).

Based on these results, we propose a model in which a network of FS cells embedded in the neocortex would be sensitive to the level of synchronous activity of excitatory axons (Fig. 7). If this network receives a number of nonsynchronous excitatory inputs, the firing of the FS cells will not be promoted because FS cells firing earlier will inhibit FS cells receiving the later inputs. As a result of this mutual inhibition, the number of FS cells firing will be diminished. In contrast, if the same number of excitatory inputs arrives synchronously, the firing of FS cells in the network will be promoted through their mutual excitation via the electrical connections. Consequently, more FS cells will simultaneously fire. We suggest that the properties of their synaptic connections allow networks of FS cells to selectively increase their firing in response to synchronous excitatory activity, a mechanism that could prevent the generation of epileptic seizures.

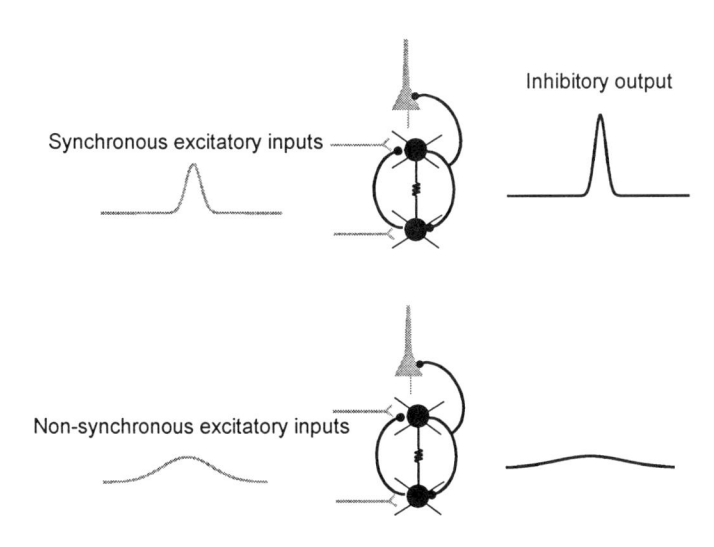

Fig. 7. Networks of FS cells would increase their firing in response to synchronous excitatory excitation. Top, when a network of FS cell receives synchronous excitatory excitation, FS cell firing will be promoted by their electrical connections. Bottom, if the same number of excitatory inputs arrives asynchronously, the firing of some FS cell will inhibit other one via the chemical and the electrical synapses and the global firing of the network will be diminished.

4. CONCLUDING REMARKS

The properties of the synapses that FS cells make and receive suggest that this class of GABAergic interneurons could be particularly well suited for dynamically controlling cortical activity. Networks of FS cells connected via electrical and chemical synapses could detect the synchronous excitation that characterizes the epileptic activity, and increase their firing accordingly. In addition, FS cell synapses show less depression in response to sustained high-frequency activity than pyramidal excitatory synapses. Thus, the relative strength of FS cell inhibition would increase in response to a global increase of cortical activity.

5. REFERENCES

1. Dichter, M. A., and Ayala G. F. (1987) Cellular mechanisms of epilepsy: a status report, *Science* **237**:157.
2. Bazil, C. W., and Pedley T. A. (1998) Advances in the medical treatment of epilepsy, *Annu. Rev. Med.* **49**:135.
3. Zucker, R. S., and Regehr W. G. (2002) Short-term synaptic plasticity, *Annu. Rev. Physiol.* **64**:355.
4. Kawaguchi, Y., and Kubota Y. (1997) GABAergic cell subtypes and their synaptic connections in rat frontal cortex, *Cereb. Cortex* **7**:476.
5. Gupta, A., Wang Y., and Markram H. (2000) Organizing principles for a diversity of GABAergic interneurons and synapses in the neocortex, *Science* **287**:273.
6. Kawaguchi, Y., and Kubota Y. (1993) Correlation of physiological subgroupings of nonpyramidal cells with parvalbumin- and calbindinD28k-immunoreactive neurons in layer V of rat frontal cortex, *J. Neurophysiol.* **70**:387.
7. Tamás, G., Buhl E. H., and Somogyi P. (1997) Fast IPSPs elicited via multiple synaptic release sites by different types of GABAergic neurone in the cat visual cortex, *J. Physiol.* **500**:715.
8. Galarreta, M., and Hestrin S. (1998) Frequency-dependent synaptic depression and the balance of excitation and inhibition in the neocortex, *Nat. Neurosci.* **1**:587.
9. Galarreta, M., and Hestrin S. (1999) A network of fast-spiking cells in the neocortex connected by electrical synapses, *Nature* **402**:72.
10. Galarreta, M., and Hestrin S. (2002) Electrical and chemical synapses among parvalbumin fast-spiking GABAergic interneurons in adult mouse neocortex, *Proc. Natl. Acad. Sci. U. S. A.* **99**:12438.
11. Varela, J. A., Song S., Turrigiano G. G., and Nelson S. B. (1999) Differential depression at excitatory and inhibitory synapses in visual cortex, *J. Neurosci.* **19**:4293.
12. Kraushaar, U., and Jonas P. (2000) Efficacy and stability of quantal GABA release at a hippocampal interneuron-principal neuron synapse, *J. Neurosci.* **20**:5594.
13. Luthi, A., Di Paolo G., Cremona O., Daniell L., De Camilli P., and McCormick D. A. (2001) Synaptojanin 1 contributes to maintaining the stability of GABAergic transmission in primary cultures of cortical neurons, *J. Neurosci.* **21**:9101.
14. Schoch, S., Castillo P. E., Jo T., Mukherjee K., Geppert M., Wang Y., Schmitz F., Malenka R. C., and Sudhof T. C. (2002) RIM1alpha forms a protein scaffold for regulating neurotransmitter release at the active zone, *Nature* **415**:321.
15. Ho, A., Morishita W., Hammer R. E., Malenka R. C., and Sudhof T. C. (2003) A role for Mints in transmitter release: Mint 1 knockout mice exhibit impaired GABAergic synaptic transmission, *Proc. Natl. Acad. Sci. U. S. A.* **100**:1409.
16. Galarreta, M., and Hestrin S. (2001b) Spike transmission and synchrony detection in networks of GABAergic interneurons, *Science* **292**:2295.
17. Geiger, J. R., Lubke J., Roth A., Frotscher M., and Jonas P. (1997) Submillisecond AMPA receptor-mediated signaling at a principal neuron- interneuron synapse, *Neuron* **18**:1009.
18. Martina, M., and Jonas P. (1997) Functional differences in Na+ channel gating between fast-spiking interneurones and principal neurones of rat hippocampus, *J. Physiol.* **505 (Pt 3)**:593.

19. Martina, M., Schultz J. H., Ehmke H., Monyer H., and Jonas P. (1998) Functional and molecular differences between voltage-gated K+ channels of fast-spiking interneurons and pyramidal neurons of rat hippocampus, *J. Neurosci.* **18**:8111.
20. Chow, A., Erisir A., Farb C., Nadal M. S., Ozaita A., Lau D., Welker E., and Rudy B. (1999) K(+) channel expression distinguishes subpopulations of parvalbumin- and somatostatin-containing neocortical interneurons, *J. Neurosci.* **19**:9332.
21. Fricker, D., and Miles R. (2000) EPSP amplification and the precision of spike timing in hippocampal neurons, *Neuron* **28**:559.
22. Maccaferri, G., and Dingledine R. (2002) Control of feedforward dendritic inhibition by NMDA receptor-dependent spike timing in hippocampal interneurons, *J. Neurosci.* **22**:5462.
23. Gibson, J. R., Beierlein M., and Connors B. W. (1999) Two networks of electrically coupled inhibitory neurons in neocortex, *Nature* **402**:75.
24. Sloper, J. J., and Powell T. P. (1978) Gap junctions between dendrites and somata of neurons in the primate sensori-motor cortex, *Proc. R. Soc. Lond. B. Biol. Sci.* **203**:39.
25. Katsumaru, H., Kosaka T., Heizmann C. W., and Hama K. (1988) Gap junctions on GABAergic neurons containing the calcium-binding protein parvalbumin in the rat hippocampus (CA1 region), *Exp Brain Res* **72**:363.
26. Tamás, G., Buhl E. H., Lorincz A., and Somogyi P. (2000) Proximally targeted GABAergic synapses and gap junctions synchronize cortical interneurons, *Nat. Neurosci.* **3**:366.
27. Bennett, M. V., Barrio L. C., Bargiello T. A., Spray D. C., Hertzberg E., and Saez J. C., 1991, Gap junctions: new tools, new answers, new questions, *Neuron* **6**:305.
28. Venance, L., Rozov A., Blatow M., Burnashev N., Feldmeyer D., and Monyer H. (2000) Connexin expression in electrically coupled postnatal rat brain neurons, *Proc. Natl. Acad. Sci. U. S. A.* **97**:10260.
29. Szabadics, J., Lorincz A., and Tamás G. (2001) Beta and gamma frequency synchronization by dendritic gabaergic synapses and gap junctions in a network of cortical interneurons, *J. Neurosci.* **21**:5824.
30. Chu, Z., Galarreta M., and Hestrin S. (2003) Synaptic interactions of late-spiking neocortical neurons in layer 1, *J Neurosci* **23**:96.
31. Galarreta, M., and Hestrin S. (2001a) Electrical synapses between GABA-releasing interneurons, *Nat. Rev. Neurosci.* **2**:425.
32. Meyer, A. H., Katona I., Blatow M., Rozov A., and Monyer H. (2002) In vivo labeling of parvalbumin-positive interneurons and analysis of electrical coupling in identified neurons, *J. Neurosci.* **22**:7055.

11

HOMEOSTATIC REGULATION OF EXCITATORY-INHIBITORY BALANCE

Gina Turrigiano[*]

1. INTRODUCTION

Activity plays an important role in refining synaptic connectivity during development, but the precise mechanisms that underlie this activity-dependent refinement are still unknown. There is considerable evidence that correlation-based, or Hebbian, plasticity mechanisms akin to long-term potentiation and depression (LTP and LTD) play an important role in synaptic refinement[1-3]. However, it is not clear that Hebbian plasticity is sufficient for understanding activity-dependent development because the dramatic changes in synaptic strength produced by this kind of plasticity tend to destabilize the activity of neural circuits[4-6]. Additionally, LTD and LTP do not automatically generate competition between inputs[5], but such competition is clearly an important aspect of sensory system plasticity[1]. It is therefore likely that LTP and LTD-like phenomena co-exist with other cellular plasticity mechanisms that provide competition and stability during activity-dependent development.

In addition to the destabilizing effects of Hebbian plasticity, in many developing circuits synapse number can change dramatically through both addition and loss of synaptic contacts. For example, during the first few weeks of development in rodent visual cortex neurons go from receiving almost no synaptic input, to receiving thousands of excitatory and inhibitory synaptic connections. For highly recurrent cortical networks maintaining stability in the face of this synaptic rearrangement is an especially difficult problem. There are extensive positive feedback connections between excitatory pyramidal neurons both within and between cortical layers, which are kept in check by feedback and feedforward inhibition mediated by complex networks of inhibitory interneurons. Even small reductions in this inhibition can result in epileptiform activity[7,8], disrupt sensory responses in primary visual cortex[9,10], and profoundly alter experience-dependent plasticity[11-14]. This suggests that excitation and inhibition must be delicately

---

[*] G. Turrigiano. Department of Biology and Volen National Center for Complex Systems, Brandeis University, Waltham, MA 02493 USA

balanced both to keep cortical networks functional, and to allow normal experience-dependent refinement to occur. This raises the question of how this balance can be maintained in complex recurrent circuits when the number and strength of inhibitory and excitatory connections are changing dramatically. An idea that is beginning to emerge from recent work is that a number of homeostatic plasticity mechanisms cooperate to dynamically adjust the excitatory-inhibitory (E-I) balance to keep network activity within some functional range.

2. HOMEOSTATIC REGULATION OF FIRING RATES

In both invertebrate central pattern generators and vertebrate spinal networks, pharmacological blockade of rhythmic activity engages compensatory mechanisms that cause activity to resume after a period of hours to days[15-17]. In cortical and hippocampal networks a similar homeostatic regulation of firing rates has been demonstrated. Chronically reducing inhibition in cortical networks initially raises firing rates, but over a period of many hours firing rates return to control levels[18]. The converse experiment, in which activity is inhibited by overexpression of a potassium channel in individual hippocampal neurons, initially lowers firing rates, but again firing rates recover over a time course of many hours despite the continued expression of the K channel[19]. These experiments suggest that neurons and networks have some "set-point" of activity that is dynamically maintained through adjustments in the strengths of synaptic connections (Figure 1).

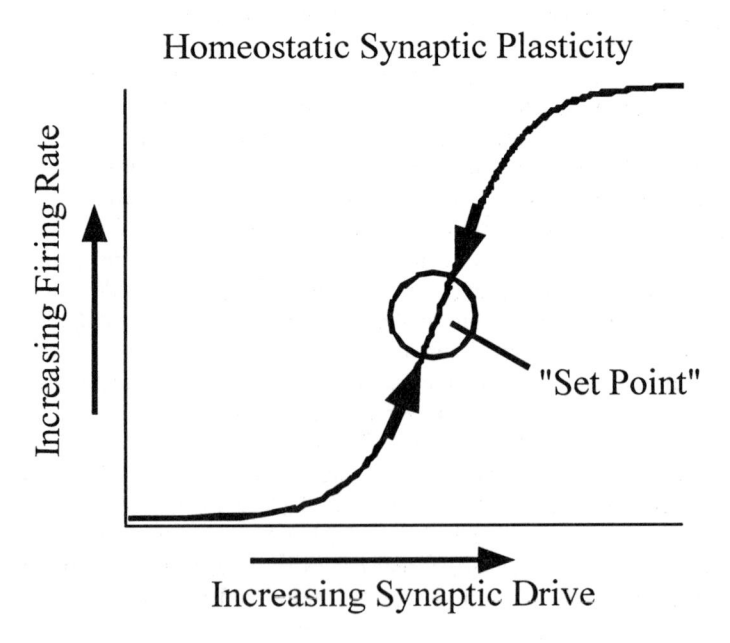

Fig. 1. Homeostatic synaptic plasticity adjusts total synaptic strength to keep firing rates within some "set point". If synaptic drive increases too much and firing rates rise, all excitatory synaptic strengths are decreased until firing rates fall again. Conversely, if firing rates fall too low, excitatory synaptic strengths are increased until firing rates are once more within the target range.

3. HOMEOSTATIC CHANGES IN EXCITATORY SYNAPTIC CURRENTS

One mechanism through which cortical networks adapt to altered drive is by scaling the strength of all excitatory connections onto excitatory pyramidal neurons up or down as a function of how active they are[6, 20]. This process has been studied extensively in networks of dissociated central neurons, where activity can be pharmacologically manipulated for long periods of time. Increased activity scales down excitatory synaptic strengths onto pyramidal neurons, while decreased activity scales synaptic strengths up. This synaptic scaling occurs largely through changes in the accumulation of glutamate receptors in the postsynaptic membrane, which increases or decreases the amplitude of excitatory miniature postsynaptic currents (mEPSCS)[18, 21-23], and thus modifies the amplitude of spike-mediated transmission[18]. Experiments in cortical cultures have shown no effect of lowered activity on mEPSC frequency or on the number of functional excitatory synapses[18, 24]. In contrast, some experiments in hippocampal cultures have found changes in mEPSC frequency and in the uptake of lipophilic dyes such as FM1-43 by presynaptic terminals[25, 19]. This suggests that, in addition to postsynaptic changes in receptor accumulation, homeostatic plasticity in hippocampal neurons may involve additional alterations in transmitter release or in the number of release sites.

Synaptic scaling differs in a number of important ways from Hebbian, synapse-specific forms of synaptic plasticity such as LTP[2] and LTD[26]. First, it appears to be independent of NMDA receptor activation (except that NMDA receptor block may itself alter activity levels and thus indirectly influence homeostatic synaptic plasticity). Second, it is quite slow, requiring hours or days of altered activity to modify synaptic strengths. This is probably important for allowing neurons to regulate average activity over a long time scale, without responding to moment-to-moment fluctuations in firing rates. Finally, the data are consistent with a model in which all of a neuron's excitatory synapses are scaled up or down proportionally, by the same multiplicative factor[18, 20]. This multiplicative scaling of synaptic strengths has the right characteristics to preserve relative differences between inputs (such as those produced by LTP or LTD), while allowing a neuron to adjust the total amount of synaptic excitation it receives. Synaptic scaling may also generate a form of synaptic competition, because when inputs are strengthened through Hebbian mechanisms the entire distribution of synaptic weights will be scaled down, effectively reducing the strength of all other inputs[27].

4. HOMEOSTATIC CHANGES IN INHIBITION

An important aspect of synaptic scaling in cultured cortical networks is that the direction of change of a synapse depends on the identity of both the presynaptic neuron (pyramid or interneuron) and the postsynaptic neuron[18, 28, 24]. For example, excitatory connections onto pyramidal neurons are regulated differently from those onto interneurons[28]. In addition, inhibitory synapses onto pyramidal neurons are regulated in the opposite direction from excitatory synapses[24]. The overall pattern of changes in synaptic strengths produced by altered activity is consistent with a role for synaptic scaling in stabilizing the activity of these highly recurrent networks.

Visual experience has long been known to regulate inhibition in primary visual cortex of rodents and primates. Visual deprivation or inhibition of retinal activity with

TTX can decrease immunoreactivity for both GABA[30-32] and for $GABA_A$ receptors[33], and can decrease the amount of evoked inhibitory currents onto layer 2/3 pyramidal neurons[34]. A similar phenomenon has been demonstrated by us and others in dissociated cortical and hippocampal networks, where activity-blockade reversibly decreases GABA immunoreactivity[35-37], and reduces the amount of functional inhibition received by pyramidal neurons[37]. These studies raise the possibility that inhibitory synaptic strengths, like excitatory synaptic strengths, can be scaled up or down by long-lasting changes in activity.

To directly test this idea we recorded miniature inhibitory postsynaptic currents (mIPSCs) from cultured pyramidal neurons after blocking all spiking activity with TTX for two days. We found that activity-blockade reduced inhibition through both a decrease in quantal amplitude, and a 50% reduction in the number of functional inhibitory synapses[24]. Interestingly, the number of presynaptic terminals that were co-localized with excitatory synaptic markers was not affected by activity blockade. Additionally, the total number of presynaptic contacts did not decrease, indicating that there was a selective decrease in the proportion of presynaptic terminals that were colocalized with $GABA_AR$[24], (Wierenga, Ibata, and Turrigiano, unpublished data). This suggests that inhibitory presynaptic terminals are still present, but are no longer functional due to loss of postsynaptic receptor clusters. Consistent with these observations, the frequency of mEPSCs (which should be proportional to the number of excitatory contacts) did not change following activity blockade[18, 23], while the frequency of mIPSCs was decreased by about 50%[24]. Inhibition and excitation are therefore regulated in fundamentally

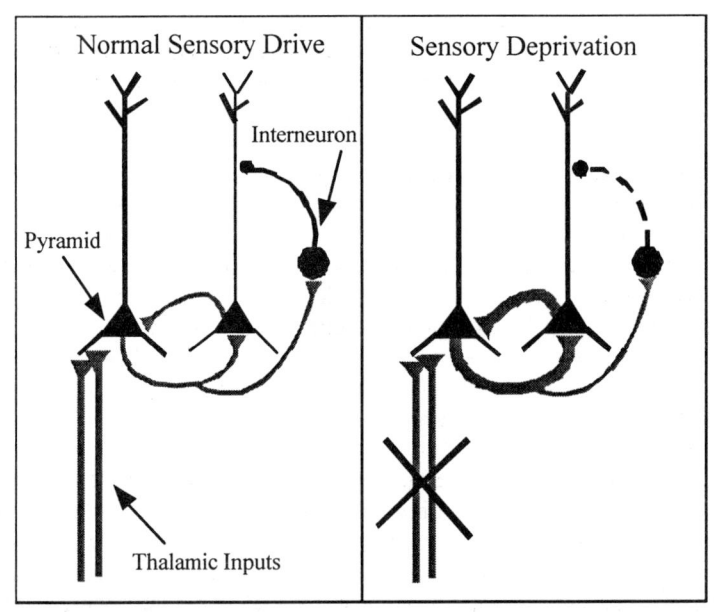

Fig. 2. Effects of sensory deprivation on the E-I balance of the cortical microcircuit. Left: visual cortex normally receives external visual drive, relayed through thalamic afferents to cortex. Right: when this drive is reduced by sensory deprivation, there are dramatic changes in intracortical circuitry. Excitatory recurrent synapses between pyramids are increased in strength, while inhibitory drive to pyramidal neurons is reduced. This compensates for reduced sensory drive by increasing the excitability of the intracortical circuitry and raising spontaneous firing rates.

different ways by activity blockade. While both excitation and inhibition are regulated at the level of changes in postsynaptic receptor accumulation, inhibition is additionally regulated through profound changes in the number of functional inhibitory synapses.

Most experiments on the role of activity in regulating cortical inhibition have not distinguished between different classes of interneuron, or examined synaptic connections between interneurons[30-32, 37, 24, 34]. Cortical interneurons are extremely heterogeneous in their pattern of connectivity, peptide content, morphology, and firing properties, and are likely to play very different roles within the cortical microcircuit. A global reduction in inhibition would therefore have very different consequences for cortical function than would selective changes in particular connections. Cultured cortical neurons derived from postnatal rodents appear to maintain many of the properties of neurons *in vivo*, including morphology, firing properties, and transmitter expression[37-39, 18].

However, a major limitation of cultured cortical networks is that the laminar specificity of different neuronal connections is not preserved, nor is it clear that functional differences between different classes of interneuron are maintained. A detailed analysis of how inhibition between interneurons is regulated by activity has thus awaited an *in vivo* paradigm for inducing homeostatic cortical plasticity. This paradigm now exists (see the following section), so future work is likely to focus on this important issue.

These data from *in vitro* cortical networks suggest that homeostatic synaptic plasticity rules function to independently adjust excitatory and inhibitory feedback loops within recurrent cortical networks so that activity is preserved despite changes in drive. When activity falls too low (because, for example, sensory drive is reduced), excitation between pyramidal neurons is boosted, while feedback inhibition is reduced (Figure 2). This should have the net effect of raising the firing rates of pyramidal neurons. Conversely, when activity rises too high, excitation between pyramids is reduced, while excitation onto interneurons and inhibitory synapses back onto pyramids are increased, thus boosting feedback inhibition. This should have the net effect of lowering the activity of pyramidal neurons. An important question is whether a similar dynamic regulation of the balance of excitation and inhibition occurs *in vivo*.

5. *IN VIVO* REGULATION OF THE EXCITATORY-INHIBITORY BALANCE

The work on synaptic scaling in culture raises the legitimate question of whether this form of plasticity is also present in intact cortical networks during activity-dependent development. In recent experiments we have begun to address this by using classic manipulations of visual experience to alter cortical activity *in vivo*, and then use slice recordings to measure the resulting changes in intracortical connectivity. As a first step we asked how excitatory synaptic strengths change during development as synapse number rises. An analysis of mEPSCs from principal neurons in slices from rodent V1 demonstrated an inverse relationship between mEPSC amplitude and frequency during the third and fourth weeks of postnatal development, indicating that as excitatory synapse number rises (thus increasing mEPSC frequency) synaptic strength is reduced. The developmental decrease in mEPSC amplitude was prevented by dark-rearing[40], as predicted if the reduction in synaptic strength was driven by increased visual drive to layer 4 pyramids as synapse number increased.

To test more thoroughly the idea that mEPSC amplitude is scaled up or down as a function of cortical activity, we used a 2-day monocular deprivation (MD) paradigm to reduce activity levels in rodent monocular primary visual cortex. This method is known to selectively reduce the expression of a number of markers of activity in the deprived region of cortex[41-43]. Because much of rodent primary visual cortex is driven exclusively by one eye, MD effectively deprives one hemisphere of visual drive, while leaving the other unperturbed. After two days of MD we cut slices of primary visual cortex and compared the quantal amplitude of excitatory synaptic connections from the deprived and control hemispheres. We found evidence for a global rescaling of synapses identical to that observed in culture[40], and that operated in a laminar- and age-dependent manner, as do other forms of activity-dependent plasticity observed *in vivo*[44-46]. These results demonstrate that the quantal amplitude of excitatory currents in primary visual cortex can be globally scaled up or down as a function of altered sensory experience, and suggest that homeostatic synaptic scaling could play an important role in the activity-dependent refinement of central circuits.

Many forms of developmental plasticity exhibit pronounced critical periods during which sensory experience is able to alter circuit properties, but outside of which sensory experience has little or no effect. To investigate whether the effects of MD on mEPSCs was developmentally regulated, rats were subjected to MD beginning at either P14 (just before eye opening), or at P21, and then mEPSCs were measured 2 days later. We found that in layer 4 MD scaled up mEPSC amplitude when begun at P14, but had no effect when begun at P21. In contrast, mEPSCs onto layer 2/3 pyramids were unaffected by MD begun at P14, but were significantly increased by MD begun at P21. These results demonstrate that homeostatic plasticity of excitatory synaptic strengths exhibits critical periods that close early in layer 4 but persist in layer 2/3. This is strongly reminiscent of the critical periods for ocular dominance (OD) plasticity in binocular visual cortex[44,47,48] and for whisker deprivation plasticity in somatosensory cortex[45,49]. In both of these forms of experience-dependent plasticity the critical period in layer 4 closes after the first few weeks of life, whereas plasticity in layer 2/3 extends significantly later in development.

Because the critical period for excitatory synaptic scaling in layer 4 closes before the classical critical period for OD plasticity in binocular rodent visual cortex, scaling at this synapse is unlikely to contribute to the synaptic rearrangements that underlie OD plasticity. On the other hand, scaling in layer 2/3 (and also in layer 5, unpublished data) are present during the OD critical period, and so could play an important role in OD plasticity. Several pieces of evidence suggest that OD shifts in layer 2/3 result from changes in intracortical connectivity, rather than simply reflect changes in thalamocortical connectivity in layer 4. First, as described above, OD shifts in layer 2/3 can be induced developmentally even after shifts in layer 4 can no longer be induced. Second, OD shifts occur more rapidly in layer 2/3 than they do in layer 4 (ref. 50). This suggests that the changes in layer 2/3 precede and possibly instruct the changes that occur in layer 4. Finally, OD shifts are larger in layer 2/3 and V than in layer IV[51] Changes in intracortical circuitry are therefore important in inducing OD plasticity.

The effects of MD on visual system development and plasticity have been interpreted almost solely in terms of alterations in Hebbian plasticity or in competition between inputs from the two eyes[1,52]. Our results (as well as the long-standing literature on changes in inhibition, see above) indicate that these manipulations also produce dramatic increases in the strengths of intracortical synapses. In preliminary MD studies we have found that slices from the deprived hemisphere have higher levels of spontaneous activity than slices from the control hemisphere, and that the net charge through excitatory synapses increases, while net charge through inhibitory synapses decreases (Maffei, Nelson, and Turrigiano, unpublished data). Taken together, these data suggest that *in vivo* as *in vitro*, MD shifts the balance between excitation and inhibition within the intracortical circuitry to favor excitation (Figure 2).

Changes in the E-I balance could have dramatic effects on Hebbian plasticity, because the amount of inhibition influences the ease with which LTP can be induced. In addition, this balance can influence the opening and closing of critical periods for competitive plasticity[53,54,11,13]. To date, studies of homeostatic plasticity in visual cortex have been confined to monocular cortex, where there is no competition between thalamic inputs from the two eyes. If visual experience also regulates the E-I balance of intracortical circuitry within binocular cortex, as seems likely, then such shifts could profoundly influencing the segregation of thalamic afferents following MD. The drop in drive from one eye is initially predicted to increase the excitability of the intracortical circuitry. This in turn should make it easier for the non-deprived eye to drive cortical neurons, undergo potentiation, and induce competitive loss of the inactive inputs. The exact manner in which Hebbian, competitive, and homeostatic plasticity will interact in the wake of MD remains to be determined. Nonetheless, it seems clear that interpreting the effects of visual deprivation on sensory response properties and plasticity will require a detailed understanding of how intracortical circuitry is affected.

6. REFERENCES

1. Shatz, C.J., (1990) Impulse activity and the patterning of connections during CNS development. *Neuron* **5**:745–756.
2. Malenka, R.C., and Nicoll, R.A., (1993) NMDA-receptor-dependent synaptic plasticity: multiple forms and mechanisms. *Trends Neurosci.* **16**,12,:521–527.
3. Bear, M.F., and Malenka, R.C., (1994) Synaptic plasticity: LTP and LTD. *Curr. Opin. Neurobiol.* **4**:389–399.
4. Abraham, W. C., and Bear, M.F., (1996) Metaplasticity: the plasticity of synaptic plasticity. *Trends Neurosci.* **19**:126–130.
5. Miller, K.D.,(1996) Synaptic economics: competition and cooperation in synaptic plasticity. *Neuron* **17**:371–374.
6. Turrigiano, G.G., (1999) Homeostatic plasticity in neuronal networks: the more things change, the more they stay the same. *Trends Neurosci.* **22**:221–227.
7. Kreigstein, A.R., Suppes, T., and Prince, D.A., (1987). Cellular and synaptic physiology and epileptogenesis of developing rat neocortical neurons in vitro. *Brain Res. Dev. Brain Res.* **34**:161–171.
8. Chagnac-Amitai, Y., and Conners, B.W., (1989) Synchronized excitation and inhibition driven by intrinsically bursting neurons in neocortex. *J. Neurophysiol.* **62**:1149–1162.
9. Sillito, A.M., (1975) The contribution of inhibitory mechanisms to the receptive field properties of neurones in the striate cortex of the cat. *J. Physiol.* **250**:305–329.
10. Nelson, S.B., (1991) Temporal interactions in the cat visual system. III. Pharmacological studies of cortical suppression suggest a presynaptic mechanism. *J. Neurosci.* **11**:369–380.
11. Kirkwood, A., and Bear, M.F., (1994) Hebbian synapses in visual cortex. *J. Neurosci.* **14**:1634–1645.
12. Kirkwood, A., Lee, H.K., and Bear, M.F., (1995) Co-regulation of long-term potentiation and experience-dependent synaptic plasticity in visual cortex by age and experience. *Nature* **375**:328–331.
13. Hensch, T.K., Fagiolini, M., Mataga, N., Stryker, M.P., Baekkeskov, S., and Kash, S.F., (1998) Local GABA circuit control of experience-dependent plasticity in the developing visual cortex. *Science* **282**:1504–1508.
14. Huang, Z.J., Kirkwood, A., Pizzorusso, T., Porciatti, V., Morales, B., Bear, M.F., Maffei, L., and Tonegawa, S., (1999) BDNF regulates the maturation of inhibition and the critical period of plasticity in mouse visual cortex. *Cell* **17**:739–755.
15. Chub, N., and O'Donovan, M.J., (1998) Blockade and recovery of spontaneous rhythmic activity after application of neurotransmitter antagonists to spinal networks of the chick embryo. *J. Neurosci.* **18**:294–306.

16. Thoby-Brisson, M., and Simmers, J., (1998) Neuromodulatory inputs maintain expression of a lobster motor pattern-generating network in a modulation-dependent state: evidence from long-term decentralization in vitro. *J. Neurosci.* **18**:2212–225
17. Golowasch, J., Casey, M., Abbott, L.F., and Marder, E., (1999) Network stability from activity-dependent regulation of neuronal conductances. *Neural Comput.* **11**:1079–1096.
18. Turrigiano, G.G., Leslie, K.R., Desai, N.S., Rutherford, L.C., and Nelson, S.B., (1998) Activity-dependent scaling of quantal amplitude in neocortical neurons. *Nature* **391**:892–896.
19. Burrone J, O'Byrne M., and Murthy V.N., (2003) Multiple forms of synaptic plasticity triggered by selective suppression of activity in individual neurons. *Nature* **420**:414-418
20. Turrigiano, G.G., and Nelson, S.B., (2000) Hebb and homeostasis in neuronal plasticity. *Curr. Opin. Neurobiol.* **10**:358–364.
21. O'Brien, R.J., Kamboj, S., Ehlers, M.D., Rosen, K.R., Fischbach G.D., and Huganir, R.L., (1998) Activity-dependent modulation of synaptic AMPA receptor accumulation. *Neuron* **21**:1067–1078.
22. Lissin, D.V., Gomperts, S.N., Carroll, R.C., Christine, C.W., Kalman, D., Kitamura, M., Hardy, S., Nicoll, R.A., Malenka, R.C., and von Zastrow, M. (1998) Activity differentially regulates the surface expression of synaptic AMPA and NMDA glutamate receptors. *Proc. Natl. Acad. Sci. USA* **95** : 7097-7102.
23. Watt, A., van Rossum, M., MacLeod, K., Nelson, S.B., and Turrigiano, G.G., (2000) Activity co-regulates quantal AMPA and NMDA currents at neocortical synapses. Neuron **26**:659–670.
24. Kilman, V., van rossum, M.C.W., and Turrigiano, G.G., (2002) Activity deprivation reduces mIPSC amplitude by decreasing the number of postsynaptic GABAa receptors clustered at neocortical synapses. *J. Neurosci.*, **22**:1328-1337
25. Murthy, V.N., Schikorski, T., Stevens, C.F., and Zhu, Y., (2001) Inactivity produces increases in neurotransmitter release and synapse size. *Neuron* **32**:673–682.
26. Linden, D.J., and Connor, J.A., (1995) Long-term synaptic depression. *Annu. Rev. Neurosci.* **18**:319–357.
27. van Rossum MCW., Bi G.Q., and Turrigiano G.G., (2000) Stable hebbian learning from spike timing-dependent plasticity. J. Neurosci. **20**:8812-8821
28. Rutherford, L.C., Nelson, S.B., Turrigiano, G.G. (1998) BDNF has opposite effects on the quantal amplitude of pyramidal interneuron excitatory synapes. *Neuron* **21** : 3, 521-530.
29. Hendry S.H.C., and Jones, E.G., (1986). Reduction in number of immunostained GABAergic neurons in deprived-eye dominance columns of monkey area 17. *Nature* **320**:750–753.
30. Hendry, S.H.C., and Jones, E.G., (1988) Activity-dependent regulation of GABA expression in the visual cortex of adult monkeys. *Neuron* **1**:701–712.
31. Benevento, L.A., Bakkum, B.W., and Cohen, R.S., (1995) Gamma-aminobutyric acid and somatostatin immunoreactivity in the visual cortex of normal and dark-reared rats. *Brain Res.* **689**:172–182.
32. Hendry S.H.C., Huntsman, M-M., Viñuela, A., Möhler, G., de Blas, A.L., and Jones, E.G., (1994) GABA$_A$ receptor subunit immunoreactivity in primate visual cortex: distribution in macaques and humans and regulation by visual input in adulthood. *J. Neurosci.* **14**:2383–2401.
33. Morales, B., Choi, S.Y., Kirkwood, A. (2002) Darkrearing alters the development of GABAergic transmission in visual cortex. *J. Neurosci.* **22** : 18, 8084-8090.
34. Marty, S., Berninger, B., Carroll, P., and Thoenen, H., (1996a,) GABAergic stimulation regulates the phenotype of hippocampal interneurons through the regulation of brain-derived neurotrophic factor. *Neuron* **16**: 565–570.
35. Marty, S., Carroll, P., Cellerino A., Castrén E., Staiger V., Thoenen, H., and Lindholm, D., (1996b). Brain-derived neurotrophic factor promotes the differentiation of various hippocampal nonpyramidal neurons, including Cajal-Retzius cells, in organotypic slice cultures. *J. Neurosci.* **16**:675–687.
36. Rutherford, C.L, Dewan, A., Lauer, H.M., and Turrigiano, G.G., (1997) BDNF mediates the activity-dependent regulation of inhibition in neocortical cultures. *J. Neurosci.* **17**:4527–4535.
37. Huettner, J.E., and Baughman, R.W., (1986). Primary cultures of identified neurons from the visual cortex of postnatal rats. *J. Neurosci.* **6**:3044–3060.
38. Jones, K.A., and Baughman, R.W., (1991) Both NMDA and non-NMDA subtypes of glutamate receptors are concentrated at synapses on cerebral cortical neuron sin culture. *Neuron* **7**:593–603.
39. Desai, N.S., Cudmore, R., Nelson, S.B., and Turrigiano, G.G., (2002) Critical Periods for experience-dependent synaptic scaling in visual cortex. *Nature Neurosci.* **5**:783-789
40. Worley, P.F., Christy, B.A., Nakabeppu, Y., Bhat, R.V., Cole, A.J., and Baraban, J.M., (1991) Constitutive expression of zif268 in neocortex is regulated by synaptic activity. *Proc. Natl. Acad. Sci. U. S. A.* **88**:5106–5110.
41. Caleo, M., Lodovichi, C., Maffei, L. (1999a) Effects of nerve growth factor on visual cortical plasticity require afferent electrical activity. *Eur. J. Neurosci.* **11**: 8, 2979-2984

42. Caleo, M., Lodovichi, C., Pizzorusso, T., and Maffei, L., (1999b) Expression of the transcription factor Zif268 in the visual cortex of monocularly-deprived rats: effects of nerve growth factor. *Neuroscience* **91**:1017–1026.
43. Daw, N.W., Fox, K., Sato, H., and Czepita, D., (1992) Critical period for monocular deprivation in the cat visual cortex. *J. Neurophysiol.* **67**:197–202.
44. Fox, K., (1992) A critical period for experience-dependent synaptic plasticity in rat barrel cortex. *J. Neurosci.* **12**:1826–1838.
45. Stern, E.A., Maravall, M., and Svoboda, K., (2001) Rapid development and plasticity of layer 2/3 maps in rat barrel cortex in vivo. *Neuron* **31**:305–315.
46. Hubel, D.H., and Wiesel, T.N., (1970) The period of susceptibility to the physiological effects of unilateral eye closure in kittens. *J. Physiol.* **206**:419–436.
47. Fagiolini, M., Pizzorusso, T., Berardi, N., Domenici, L., Maffei, L. (1994) Functional postnatal development of the rat primary visual cortex and the role of visual experience : Dark rearing and monocular deprivation. *Vision Res.* **34**(6), 709-720.
48. Diamond, M.E., Armstrong-James, M., and Ebner, F.F., (1993) Experience-dependent plasticity in adult rat barrel cortex. *Proc. Natl. Acad. Sci. U. S. A.* **90**:2082-2086.
49. Trachtenberg, J., Trepel, C., and Stryker, M.P., (2000) Rapid extragranular plasticity in the absence of thalamocortical plasticity in the developing primary visual cortex. *Science* **287**:2029–2032.
50. Gordon, J.A., and Stryker, M.P., (1996) Experience-dependent plasticity of binocular responses in the primary visual cortex of the mouse. *J. Neurosci.* **16**:3274–3286.
51. Katz, L.C., and Shatz, C.J., (1996) Synaptic activity and the construction of cortical circuits. *Science* **274**:1133–1138.
52. Reiter, H.O., and Stryker, M.P., (1988) Neural plasticity without postsynaptic action potentials: less-active inputs become dominant when kitten visual cortical cells are pharmacologically inhibited. *Proc. Natl. Acad. Sci. U. S. A.* **85**:3623–3627.
53. Hata Y, Tsumoto T, Stryker MP, (1999) Selective pruning of more active afferents when cat visual cortex is pharmacologicallyt inhibited. *Neuron* **22**:375-381

12

ADULT NEUROGENESIS CONTROLS EXCITATORY-INHIBITORY BALANCE IN THE OLFACTORY BULB

Pierre-Marie Lledo, Armen Saghatelyan, and Gilles Gheusi[*]

1. INTRODUCTION

In the cortex, networks of inhibitory interneurons play a crucial role in the modulation of the electrical activity patterns of the principal neurons known as projecting (glutamatergic) neurons. Inhibitory interneurons containing ☐-aminobutyric acid (GABA), with distinct connectivities and neurochemical features, carry out specific functions within cortical networks. They govern, for instance, the activity of the profusely interconnected ensembles of projecting neurons, and are responsible for the precise timing of individual principal cell discharges in relation to the emergent behavior of the entire cell assembly. Today, one of the central issue in developmental neurobiology is the need to characterize how both projecting neurons and the various interneurons migrate along precise pathways to find the correct sites for their final differentiation and integration. This highly temporally and spatially orchestrated migration is essential for proper brain development and function.

For a long time, morphological studies described only a radial mode of migration of postmitotic neurons from the ventricular zone[1], which produces cortical projection neurons that incorporate into the developing layers of the cerebral cortex. This was a comfortable notion implying that the neocortex is built according to a carefully crafted blueprint with only local, outward migration to create the cortical columns. This notion was however upset by studies which provided evidence for distinctly non-radial routes of cellular

---

[*] P-M. Lledo, A. Saghatelyan, G. Gheusi. Pasteur Institute, Laboratory of Perception and Memory, Centre National de la Recherche Scientifique, URA 2182, 25 rue du Dr. Roux, 75724 Paris, France.

migration. Hence, lineage studies, among others, clearly demonstrated that clones of pyramidal and non-pyramidal neurons have distinct patterns of organization[2]. Subsequent studies clearly showed that clones of radially arranged neurons contain glutamate, the neurochemical signature of pyramidal neurons, whereas tangentially dispersed cells contain GABA[3]. As a result, two germinative zones were identified in the developing brain. Since they provide new neurons that follow distinct modes of migration, these neurogenic areas have been broadly subdivided into the dorsal and the ventral forebrain (Fig. 1).

Postmitotic cells migrate radially from the dorsal ventricular zone to differentiate into excitatory projection neurons that integrate the different layers of the cerebral cortex (black arrows in Fig. 1). In contrast, the lateral (LGE) and medial (MGE) ganglionic eminences of the ventral forebrain represent the best-established sources of migrating neurons that travel tangentially (blue arrows in Fig. 1). Early on during development, cells originating from the LGE migrate ventrally and anteriorly to give rise to medium spiny neurons of the striatum and to interneurons of the olfactory bulb. Cells deriving from the MGE migrate dorsally and spread across most of the dorsal forebrain to give rise to inhibitory interneurons that integrate into the developing neocortex. This ventral to dorsal migration has been defined more precisely in genetic experiments using tissue from mice lacking the transcription factors *Dlx* and *Nkx*[4]. With such tissue, it has been possible to distinguish two populations of GABAergic interneurons which migrate towards the cortex. An early population of cells migrates dorsally from the MGE to the cortex, and a later population migrates along this trajectory from the LGE. The importance of this ventral to dorsal mode of migration is illustrated by the rather amazing fact that GABAergic neurons represent about 20 % of the total neuronal population in the cortex. More importantly, these observations point to an unexpected degree of complexity, with a mixture of

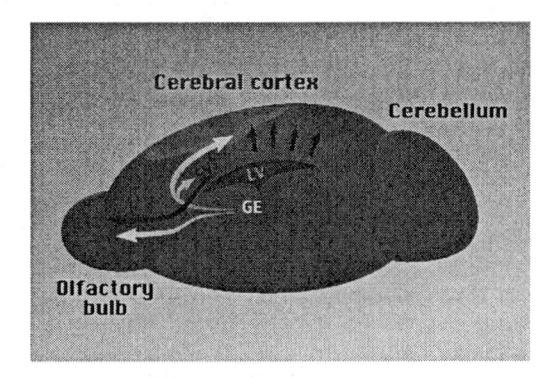

Fig. 1. Radial and tangential migrations of postmitotic neurons generated in the developing and adult brain. Black arrows indicate radial migration of cortical projection neurons from the lateral ventricle (LV) to the developing layers of the cerebral cortex. Blue arrows show tangential migration of neuroblasts from the ganglionic eminence (GE) to the neocortex, the olfactory bulb and the subventricular zone (SVZ) of the lateral ventricle, during embryogenesis. The red arrow illustrates cell migration from the SVZ to the olfactory bulb, during adulthood.

radial and tangential cell trafficking required to build a normally functioning brain where excitatory and inhibitory actions are successfully balanced.

Not only does the LGE send interneurons dorsally to the overlying cortex, but it also provides the subventricular zone (SVZ) of the postnatal lateral ventricle with proliferating neural stem cells (indicated by the small blue arrow in Fig. 1). In this zone, they constitute a germinal region that persists into adulthood. The SVZ, once thought to provide the overlying cortex with glia late in development, is now also considered to contribute a steady supply of new interneurons for the adult olfactory bulb[5] (red arrow in Fig. 1). Indeed, young neurons born in the postnatal SVZ travel to the olfactory bulb where they differentiate into bulbar interneurons named granule cells and periglomerular neurons (Fig. 2). To reach their target, newborn neurons need to travel tangentially in the rostral migratory stream (RMS) (Fig. 2A). It is noteworthy that they continue to proliferate while migrating towards the olfactory bulb. Cells within this stream undergo an unusual mode of migration, as a chain of neurons unsheathed by a protective layer of glial cells. This cell population has given rise to great interest since it continues to generate new neurons after birth, well into adulthood, providing both a supply of new GABAergic interneurons for the olfactory bulb and a population of partially committed stem cells that can be differentiated into interneurons when transplanted into the septum, thalamus, and midbrain of the embryonic brain. Below we discuss some of the recent findings that support functional implications for producing newborn neurons into an adult brain. Due to space constraints, this review focuses exclusively on the olfactory system of vertebrates, and particularly on that of mammals.

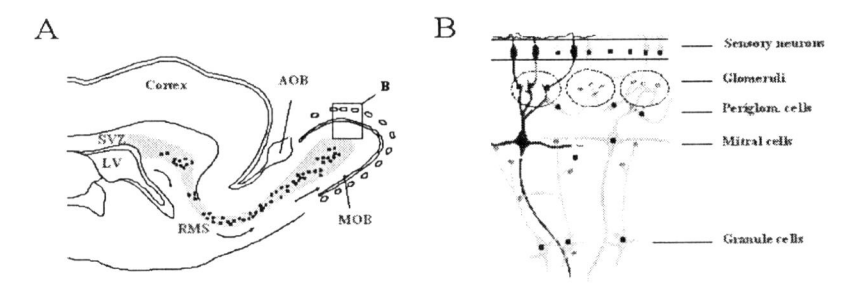

Fig. 2. Tangential migration of neuroblasts toward the olfactory bulb (A) and synaptic organization of the main olfactory bulb (B). Note that neurogenesis refers here to the replacement of bulbar local interneurons. AOB, Accessory olfactory bulb; LV, Lateral ventricle; MOB, Main olfactory bulb; RMS, rostral migratory stream; SVZ, subventricular zone.

2. SYNAPTIC ORGANIZATION OF THE MAMMALIAN OLFACTORY BULB

The olfactory bulb is the first central relay of the olfactory system that receives inputs from sensory neurons located in the olfactory epithelium (Fig.

2B). This information is then encoded and refined prior to its transmission to the olfactory cortex. It is worthwhile to state that there is no strict spatial relationship between the arrangement of the projections of excitatory olfactory neurons in the olfactory bulb and the regions of mucosa from which they originate, whereas in other sensory systems the afferent inputs are organized in a rather precise topographical mode. Similarly, much evidence indicates that bulbar outputs do not have point-to-point topographical projections to their target structures[6], which are characteristic of other sensory systems.

Nevertheless, olfactory stimuli evoke spatially organized patterns of neuronal activity in the vertebrate olfactory bulb. While other parameters are likely to be as important, spatial activity patterns have been hypothesized for a long time to play an important role in olfactory information encoding[7]. Several studies have provided experimental evidence in favor of this hypothesis and have explored the relationship between spatial patterns of olfactory bulb activity and features of olfactory stimuli. The initial mapping of olfactory stimuli across the spatial dimension of the olfactory bulb arises from the precise convergence of receptor neurons expressing the same odorant receptor onto only a few olfactory bulb glomeruli in a stereotyped location. This spatial map of receptor neuron activation may be shaped by synaptic interactions within glomeruli, between neurons in adjacent glomeruli, or between neurons in subglomerular layers, as the olfactory input is transferred to higher-order neurons in the bulb.

Two potential sites of odor processing are distinguished according to the topographical organization of the olfactory bulb circuit. The first one resides in the glomeruli where local interneurons shape excitatory inputs coming from sensory neurons (Fig. 2B). The second one lies in the external plexiform layer of the olfactory bulb where reciprocal dendro-dendritic synapses, between dendritic spines of local interneurons and the dendrite of principal neurons (*i.e.* mitral or tufted cells), are heavily distributed (Fig. 2B). These principal neurons, in addition to their primary dendrites which are confined to a single glomerulus (at least in mammals), extend secondary dendrites over long distances (up to 1,000 µm) laterally within the external plexiform layer. Since secondary dendrites have large projection fields and extensive reciprocal connections with interneurons, each local neuron may contact the dendrites of numerous output neurons. This suggests that dendro-dendritic connections mediate lateral inhibition between projecting neurons that innervate different glomeruli. Consequently, it has been demonstrated that bulbar projecting neurons connected to different glomerular units, and which respond to a wide range of related odor molecules, also receive inhibitory inputs from neighboring glomerular units via lateral inhibition at dendro-dendritic connections[8]. In other words, weak responses to a given odorant in one glomerular unit may be suppressed by strong responses from other units. Thus, by sharpening the receptive fields of individual projecting neurons and weakening the activity of neighboring ones, lateral inhibition mediated by local interneurons enhances the quantity and quality of information transferred to higher cortical areas.

Interestingly, theoretical models have suggested that inhibitory inputs to projecting neurons are also implicated in the temporal control of odor processing by synchronizing the activity patterns of principal cells[9]. Lord Adrian, who studied the olfactory bulb of the hedgehog more than 50 years ago[10], reported an oscillatory activity of the local field potential (LFP). Today, spontaneous or odor-induced LFP oscillations are believed to be a universal feature of olfactory processing systems. It is thought that the dendro-dendritic reciprocal synapse between principal neurons and local interneurons is responsible for the oscillatory behavior reported in the olfactory bulb (Fig. 3). The emerging view is that interaction of both subthreshold and spiking activity of principal neurons with GABAergic synaptic inputs are critical both for generating and maintaining gamma rhythms. Such oscillations could play an important role in odor processing: the relative timing of active modules or cells between each other and relative to an oscillatory baseline drive could be used to code odor identity. In this respect, experimental findings in insects or limax strongly indicate that these oscillations play a key role in odor discrimination and that GABAergic inputs to principal neurons are crucial for network synchrony and odor perception[7, 11].

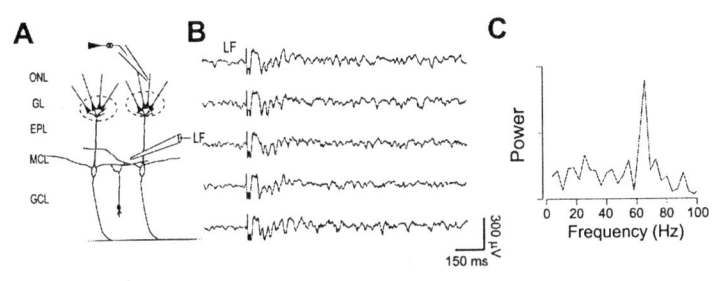

Fig. 3. Olfactory nerve stimulation triggers local field potential (LFP) oscillations. (A) Experimental scheme, *arrowhead*: stimulating electrodes, *LF*: Local field recording electrode placed in the mitral cell layer. *ONL*, olfactory nerve layer; *GL*, glomerular layer; *EPL*, external plexiform layer; *MCL*, mitral cell body layer; *GCL*, granule cell layer. (B) Individual traces illustrating induced LFP oscillations. (C) Fast Fourier Transformation (FFT) of LFP signals measured just after the stimulation.

3. PROCESSING IN THE OLFACTORY BULB REQUIRES A PROPER BALANCE BETWEEN EXCITATION AND INHIBITION

In order to understand the mechanism by which the central nervous system (CNS) processes odor information, one needs to take into account the rationality of the synaptic organization at the first central relay. The intrabulbar circuit includes two classes of interneurons that participate actively to this process: periglomerular and granule cells (Fig. 2B). Briefly, a principal neuron of the olfactory bulb (*i.e.* mitral or tufted cell) receives glomerular excitatory synaptic inputs via the distal tuft of a primary (or apical) dendrite extending vertically from the soma. At this level, the dendrites of most of the periglomerular cells (GABAergic or dopaminergic interneurons) are restricted to one glomerulus and impinge onto olfactory nerve terminals and/or primary dendrites. The secondary dendrites of mitral cells that extend across the external plexiform layer form reciprocal dendro-dendritic synapses with the dendrites of the largest population of bulbar interneurons: the granule cells. Therefore, synaptic transmission between dendrites represents the major neuronal interaction in the olfactory bulb. It consists of a balance between inhibitory and excitatory signals (Fig. 4). At dendro-dendritic synapses between mitral and granule cells, inhibition is provided by a reciprocal circuit that forms the basis of a reliable, spatially localized, recurrent inhibition of mitral cells. Glutamate released by lateral dendrites of mitral cells excites the dendrites of granule cells which, in turn, release GABA onto mitral cells. This feedback inhibition has been proposed to be crucial for the complex dynamics of olfactory network responses[12].

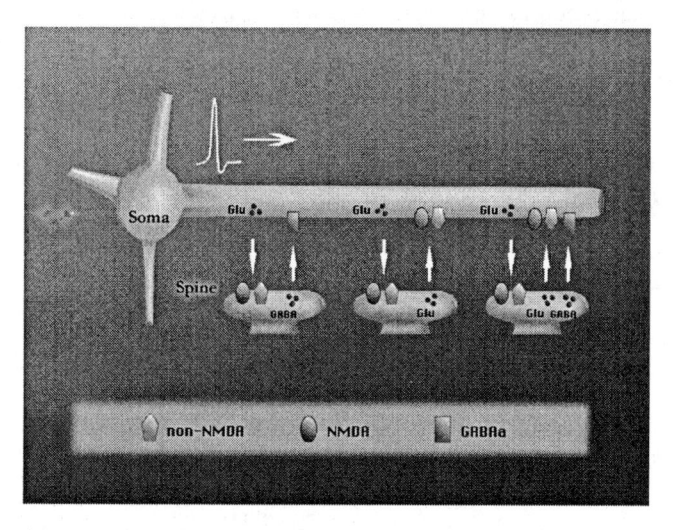

Fig. 4. Model representing different reciprocal synapses mediated by the same or different subtypes of granule cells. The term 'reciprocal' implies that a mitral cell releases glutamate that activates both non-NMDA and NMDA receptors which in turn trigger neurotransmitter release by granule cells at the same synapse (arrows).

It has been reported that the lateral dendrites of mitral cells receive, in addition to inhibitory inputs from granule cells, large excitatory inputs when either inhibition is antagonized or magnesium is removed from the external medium[13, 14]. Using electronic microscopy, excitatory synapses of mitral cells have been localized exclusively on the apical dendritic tufts that receive primary sensory afferents. The origin of the excitatory inputs to the lateral dendrites of mitral cells thus remained unknown, until recently. Employing a combination of whole-cell recording of the patch-clamp technique with immunogold staining, we have observed that mitral cells receive glutamate from local interneurons located in the granule cell layer. Importantly, these interneurons were previously thought to exert only an inhibitory action. Such a mixture of feedback inhibition and excitation from local interneurons (Fig. 4) may provide an effective mechanism of olfactory information processing by generating rhythms that play a key role in temporal and spatial coding[15].

Two lessons can thus be learned from the examination of the bulbar synaptic organization. First, odor information coding strategies are much more dynamic than previously thought. Second, even paleocortical structures may play an integral part in higher cognitive processing rather than simply act as relay stations whose sole functions would be contrast enhancement and the improvement of signal-to-noise ratio. As above described, bulbar neurons that relay activity from the odor receptors to the brain interact extensively with each other, both directly and through a network of coupled inhibitory and excitatory interneurons. Finally, since local bulbar interneurons are so important in shaping network activity, adjusting their number by a continuous renewal during the animal's lifetime may fine-tune the network synchrony and therefore olfactory perception and memory[16].

4. THE OLFACTORY BULB: A MODEL FOR STUDYING ADULT NEUROGENESIS

There are at least two germinative zones in the adult mammalian brain: one in the SVZ in the forebrain[4] and one in the dentate gyrus of the hippocampus[18]. Within these regions, the neural precursor cells are considered to be stem cells since they proliferate and give rise to several different cell types. Progeny from the SVZ give rise to neural precursors that reach the olfactory bulb where they turn radially to invade both the granule and periglomerular cell layers[16].

Although this ongoing neurogenesis and migration have been extensively documented, their function remains unknown. We postulated that if newborn interneurons generated throughout life were indeed necessary for bulbar function, then modifications of the processes of migration, leading to changes in the recruitment of newborn interneurons, would affect olfactory processing. We tested this hypothesis by quantifying the level of migration and the size of the interneuron population in NCAM-mutant mice. We found that the reduction in the number of newborn granule cells (of about 40 %) in adult mutants is accompanied with impaired odor discrimination[19]. This led us to propose a specific role for the newborn interneurons in downstream coding of olfactory information. According to our theory, a critical level of inhibition, mediated by the activation of $GABA_A$ receptors localized on the secondary dendrites of bulbar principal neurons, is crucial for olfactory processing. Furthermore, since the olfactory bulb is also involved in consolidation processes associated with long-term odor memory, we propose that a modulation of the number of GABAergic interneurons by continual neurogenesis regulates olfactory perception and learning.

To shed light on this issue, we investigated whether a change in the number of newborn bulbar interneurons alters olfactory memory. To increase bulbar activity, and therefore the number of newly formed interneurons, adult mice were subjected to an odor-enriched environment. We found that such enriched-mice retain learned olfactory information for longer periods of time than controls[20]. In particular, animals raised in an enriched environment were able to recognize familiar odors in a more durable and stronger way (*i.e.*, they expressed a high resistance to retroactive interference) as compared to animals raised in standard conditions. This increase in the rate of neurogenesis was specific to the olfactory bulb since the second neurogenic zone of the adult CNS, the dentate gyrus of the hippocampus, remained unaltered by the enrichment. Although the potential consequences of an enriched environment on synaptic efficacy and/or synchronization of mitral cells were not specifically explored, the results of our study support the existence of a link between the size of the population of newborn interneurons and olfactory memory.

5. MATURATION AND FUNCTIONAL INTEGRATION OF NEWBORN INTERNEURONS

As above described, bulbar interneurons play an essential role in shaping the olfactory information which reaches the olfactory cortex. Since adult-generated cells born in the SVZ differentiate exclusively into bulbar interneurons, they probably contribute to essential aspects of olfactory processing. Although modeling and experimental studies have already provided arguments in favor of a role for newborn granule cells in olfactory discrimination, a functional role remains to be characterized. For instance, whereas much work has been devoted to elucidating the mechanisms of birth and migration of new SVZ neuroblasts, no information is available concerning the temporal sequence of electrophysiological changes that accompany their maturation. Such information is essential to appreciate how newly formed neurons become functionally integrated into adult neural networks.

To address this question, we characterized the electrophysiological properties of newborn cells during their migration and differentiation, using a replication-defective retrovirus that carries GFP. Following different survival times after viral injection into the SVZ, GFP$^+$ labeled neurons were recorded from acute olfactory bulb slices (Fig. 5). We characterized, for the first time, the unique sequence that migrating neuroblasts follow to become mature interneurons in the adult living brain. We found that not only do SVZ precursor-derived neurons integrate the adult neuronal circuit, but also that they receive synaptic inputs during their radial migration in the olfactory bulb. Interestingly, GABAergic and glutamatergic synapses impinging onto newborn interneurons were established sequentially, GABAergic synapses being established first. In contrast to the sequential acquisition of synapses during embryogenesis, we found NMDA responses in neurons that were unable to fire and had started to receive GABA$_A$ and AMPA inputs.

The rules that govern the incorporation of adult-generated neurons into mature neuronal circuits may thus differ from those previously described in the developing brain. Furthermore, this study demonstrates that, within a month, newborn neurons express synaptic and firing properties similar to those found in older interneurons. The correlated maturation of intrinsic electrical properties with synaptic activity of newborn cells may influence both maturation and integration processes and therefore represents a mechanism by which neuronal activity may regulate neurogenesis.

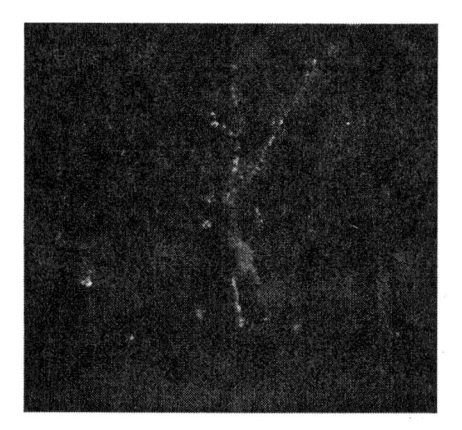

Fig. 5. A replication-defective retrovirus expressing GFP-GAP43 was injected in the ventricle to label SVZ precursors of adult mice. Several days later, GFP$^+$ cells are seen in acute OB slices stained with the neuronal marker NeuN. The GFP-GAP43 fusion targets the transduced protein in the plasma membrane, staining the entire surface of newborn interneurons.

6. ADULT NEUROGENESIS: A NEURAL BASIS FOR EXPERIENCE-INDUCED PLASTICITY

Similarly to the development of the immature brain, neurogenesis that takes place in the adult brain includes cell division, differentiation, migration, synapse formation and programmed cell death. However, the maturation of newly generated neurons in the adult brain does not recapitulate that of embryogenesis, partly because of the high neuronal activity of the adult mature circuitry. In central sensory pathways, this activity is elicited by external stimuli that lead to experience-dependent anatomical changes in connectivity during late development and in adulthood. The study of the maturation of synaptic transmission at newly formed synapses may provide important insight into how experience modifies already functioning neural circuits. In this respect, the late expression of functional NMDA receptors and the maturation sequence of glutamatergic synapses that we found in newborn neurons indicate the existence of unique developmental sequences specific to experience-dependent maturation of neural circuits.

The hypothesis according to which olfactory nerve-olfactory bulb connections are initially highly precise, rather than initially crude and only gradually refined, has been supported by molecular studies of the formation of glomeruli[21]. These studies indicate that the initial projections laid down during the establishment phase are remarkably precise. Although some controversy remains, considerable evidence indicates that the initial patterning in these segregated systems does not depend upon correlated activity patterns, whereas maintenance of the segregated state requires activity[22].

Olfactory sensory neurons which exhibit the same type of odorant receptor possess axons that converge onto few distinct glomeruli in the olfactory bulb. At first sight, this seems to be an ideal organization whereby correlation-based sorting may be involved in segregating axons into discrete glomeruli, since all the axons bearing the same receptor would presumably show highly correlated activity. However, several genetic manipulations have conclusively shown that neither spontaneous nor odorant-induced activity is required for the initial specification of glomeruli[23,24].

Although map formation does not require activity, there is considerable evidence that maintenance of such a map requires sensory inputs[25]. Hence, the mechanisms underlying the development of visual cortical columns, barrels in somatosensory cortex and glomeruli in the olfactory bulb share the same overall features: a precise, rapid establishment of initial connections that is relatively immune to manipulations of sensory activity, followed by a subsequent period of sensitivity to manipulations of activity level at the sensory periphery.

Due to continuous neurogenesis in adult animals, the olfactory system provides a unique model to study the role of sensory-driven activity during different life stages. In the olfactory bulb, the role of environmental-induced activity can be explored not only during the neonatal period, when refinement and stabilization of developing structure takes place, but also in adulthood, when a more delicate regulation occurs. Many studies have demonstrated that modified levels of activity do not affect proliferation and tangential migrations of neuroblasts[26,27]. In contrast, elevated neuronal death in the olfactory bulb has been documented in odor-deprived animals while odor enrichment has been found to increase the number of newly generated granule cells in adults[27-29]. This indicates that sensory-driven activity may play an important role in the survival of progenitors and/or in their guidance from the rostral migratory stream to the olfactory bulb. It would lead to great changes in the number of newborn interneurons and therefore alter the inhibitory-excitatory balance. This, in turn, would modify odor processing by affecting the synchronizing activity of relay neurons.

However, when functional consequences were explored following modulation of neurogenesis in young and adult animals, opposite results were found. Challenging the level of sensory activity in adults alters their olfactory memory[20], whereas sensory-deprivation performed at early stages does not affect olfactory performance. Thus, changing the level of sensory inputs may trigger homeostatic compensatory mechanisms only in newborn animals. During development, sensory activity might be important for refinement and maturation of anatomical and physiological connections established through an activity-independent manner, thus playing an *instructive* role. In contrast, sensory activity during adulthood might play an *informative* role for newborn interneurons arriving to the olfactory bulb, allowing them to reach a proper site in the mature neuronal network. Supporting this hypothesis are observations showing that almost half of adult-generated neurons die shortly after reaching the olfactory bulb[30].

Thus, it is conceivable that an alteration in the inhibitory-excitatory balance during development, as compared to during adulthood, might have a much more profound effect on animal survival, and thus might lead to evolution of some compensatory mechanisms that will operate at that stage of life helping to preserve normal functioning of olfactory system. Further research is needed in order to determine whether specific compensatory mechanisms operate exclusively during development. Alternatively, the inhibitory-excitatory balance may be modified by sensory activity in a different way from one life stage to another. Following odor deprivation, several mechanisms that are affected differently in the neonatal *versus* adult brain have been reported. Probably the most striking difference which has been found is a 40 % reduction in the size of the olfactory bulb following olfactory deprivation performed at early stages but not during adulthood.

The effect of sensory activity in shaping the olfactory system during the neonatal period resembles the role of environmentally induced stimuli in the establishment of appropriate patterns of neuronal circuit in other sensory systems. In the latter, experience-induced activity is particularly influential during the sensitive period of development called "critical period", and maturation of GABAergic synaptic transmission during this period is believed to be one of the essential factors contributing to functional remodeling[31]. Interestingly, the duration of the critical period is correlated with the duration of the expression of extracellular molecular cues that surround soma and dendrites of a subset of inhibitory interneurons[32].

Altogether, these results suggest that inhibitory interneurons may not only play a major role in sculpting the architecture of the developing sensory systems, but may also regulate the duration of the critical period during which functional remodeling take place.

7. FUNCTIONAL CONTRIBUTION OF NEWBORN INTERNEURONS TO INFORMATION PROCESSING

Perhaps the most intriguing question regarding adult neurogenesis is whether newly generated neurons have important functions, especially in learning. In songbirds, there are peaks of neuronal loss and replacement that correlate with seasonal periods of song instability and restabilization, respectively. Similarly, in mammals, previous observations in the hippocampus suggest that new neurons bring unique properties into adult circuits. For instance, new neurons respond preferentially to the modulation mediated by hormones as well as to spatial learning and exercise. Similarly, we found that increasing the number of newborn interneurons in the olfactory bulb improves olfactory memory whereas decreasing neurogenesis alters odor discrimination. The multiple stages in the maturation of newborn neurons may serve as various substrates for plasticity. Those newly formed interneurons could provide the olfactory bulb with a large repository of plastic cells to better respond to environmental changes and to different life experiences. Interestingly, one of the forms of plasticity may reside in the neurotransmitter type released by newborn bulbar interneurons. We reported that neurons located in the granule cell layer, considered as a homogeneous population of GABAergic local interneurons, might also release glutamate onto mitral cell dendrites. It is possible that new bulbar granule cells use both neurotransmitter systems or that they release glutamate only transiently, at early stages of their maturation.

Although recent estimates suggest that the quantity of neurons produced in adulthood is much greater than previously thought, the rate of neuronal production in adulthood remains lower than during development. Supposing that adult-generated neurons bring similar functional properties to those brought by neurons generated in early life, functional consequences of adult neurogenesis then would be negligible[33]. In contrast, if adult-generated neurons were to have unique properties that increase their impact relative to more mature neurons, then their continuous integration into the functional circuitry would have tremendous effects. Studies in the adult hippocampus and olfactory bulb support the second hypothesis. Young granule cells in the adult dentate gyrus appear to exhibit robust plasticity that, in contrast to mature granule cells, cannot be inhibited by GABA[34]. These newborn neurons may respond preferentially to the modulation induced by stress hormones, while learning increases their number. Running increases both the number of new dentate gyrus cells and performance on a hippocampal-dependent task, whereas a decrease in the number of new granule neurons is correlated with altered spatial memory.

Our recent results, which show that an increase in the number of newborn interneurons improves olfactory memory, demonstrate that functions thought to belong exclusively to hippocampal newborn neurons can be extended to other brain areas. Since new neurons are structurally plastic, they are highly susceptible to changes in their environment and to different life experiences. This suggests that adult neurogenesis is important since it results in a continual influx of neurons that are, at least temporarily, immature with unique physiological properties. It is however noteworthy that the functional benefit from adult neurogenesis cannot be acute since it takes several weeks to generate a new functionally integrated interneuron. We have shown that newborn cells extend neurites within a couple of weeks after cell birth, and it is obvious that the new connections cannot benefit the particular functional event that triggered neurogenesis since it will be long over when the new neurons are in place. Thus, recruitment of bulbar neurons has to be considered as a long-term adjustment of bulbar circuitry to an experienced level of higher complexity governed by olfaction.

In addition to long-term changes, there is now evidence that olfactory learning also involves short-term synaptic plasticity. For instance, the synaptic strength at reciprocal dendro-dendritic synapses may be subjected to modulation. Hence, it has been reported that extracellular bulbar concentrations of glutamate and GABA may change following learning of odors. These aspects have been well documented in different biologically relevant models that we here describe briefly.

Female mice form a memory of their mating male's pheromones, a phenomenon known as the Bruce effect. This memory prevents the newly mated female from sustaining her pregnancy when exposed to pheromones of other males[35]. Such a memory is formed during mating and requires a period of about 3 to 4 hours of exposure to the male's pheromones. Lesions of the vomeronasal organ (VNO) prevent the pregnancy-block effect, demonstrating that the accessory olfactory system mediates the estrus-inducing effects of male pheromones[36]. It has recently been shown that, contrary to the sensory neurons located in the main olfactory epithelium, vomeronasal sensory neurons appear to be highly selective detectors[37]. Although the functions it possesses are different, the accessory olfactory bulb (AOB) shares the same synaptic organization as the main olfactory bulb (MOB). Mitral cells from the AOB receive pheromonal inputs and transmit the pregnancy-block signal centrally. They express glutamate and in turn receive GABAergic inhibitory feedback from granule cells at reciprocal synapses.

With a microdialysis approach, Brennan and colleagues[38] explored the level changes of a range of neurotransmitters occurring when female mice form a memory of their stud male. The main changes are a significant increase in noradrenaline release during mating and a decrease in the glutamate/GABA ratio following memory formation in response to the mating male's pheromones. The increase in GABA release illustrates a higher inhibitory tone at the reciprocal synapses. The authors interpreted such a mechanism as a way of reducing the excitation from mitral cells, which mediate the input from the mating male's pheromones, in order to prevent a pregnancy block. In contrast, exposure to a different male would activate a different subpopulation of mitral cells, not previously submitted to an increased inhibitory control, thus eliciting central pregnancy blocking mechanisms.

These processes do not occur only during formation of a memory of the mating male's pheromones, at the AOB. Similar events have been found in the MOB of ewes when they form a specific ability to recognize their own lamb and reject alien ones that attempt to suckle[39]. Ewes develop a selective bond to their own lambs within 2 hours following parturition[40]. Bulbectomy or lesion of the olfactory epithelium, but not section of the vomeronasal nerve, lead to a complete loss of selectivity in postpartum ewes. This demonstrates the key role played by the MOB in this type of learning[41]. After parturition, in response to odors from her own lamb but not to odors from alien lambs, an increase in glutamate and GABA release occurs in the ewe's MOB[42]. Once the ewe has established a selective bond with its lamb, the ratio of glutamate to GABA is decreased in response to the lamb's odor, suggesting, once again, the existence of changes in the gain of reciprocal synapses between mitral and granule cells.

These two examples illustrate changes in excitatory-inhibitory balance in the olfactory bulb occurring during the establishment of specific social preferences. In addition, similar changes have recently been shown to occur in circumstances involving olfactory learning in a broader context. For instance, Brennan and colleagues[43] used a conditioning procedure in which adult mice were trained with odor added to clean wood shavings through which they dug to obtain a buried piece of sugar. Following conditioning, the presentation of the odor previously associated with the reward induced an increase in both glutamate and GABA levels in the MOB and a decrease in the balance existing between excitatory and inhibitory transmitters. In contrast, no change was reported when animals were exposed to a non-conditioning odor, indicating that the increase inhibition only occurs for relevant odors. These findings emphasize the significance of changes in the excitatory-inhibitory ratio that occur in the olfactory bulb following odor learning. Finally, it is noteworthy that the neurochemistry of inhibition enhancement is reminiscent of the up-regulation of the permanent supply of bulbar interneurons that correlates with a stronger memory in odor-enriched mice.

8. CONCLUDING REMARKS

The sense of smell has played essential roles during mammalian evolution. Odor representation is dynamic and highly complex, perhaps requiring a unique mechanism of plasticity. Neurogenesis, migration and replacement of new neurons in the olfactory bulb are likely to play a part in this adaptive mechanism. It takes about two weeks for a new SVZ neuron to become part of the existing circuit in the olfactory bulb and during this time it undergoes a series of unique maturational stages, as described above. Each of these maturational stages could contribute to plasticity. The presence of young neurons at different stages of differentiation in the active olfactory bulb may allow for adjustments in the integration of these cells (*e.g.*, how they grow their dendrites, establish synapses, use neurotransmitters, or modify their pattern of gene expression). Since it takes time for new neurons to mature and become synaptically integrated, adult neurogenesis may contribute to slow, long-term adjustments of the olfactory bulb circuitry, rather than to fast and acute plastic changes. However, other mechanisms may also take advantage of new neurons. More than half of the new interneurons in the olfactory bulb die within a month after having reached their mature state. Our previous results confirm that newborn granule cells are synaptically connected during this period of fast cell death. Selective elimination of neurons may allow for a rapid modification of the circuitry. The ability to mold the integration of new neurons and to continually eliminate older cells without depleting the neuronal population may bring into neuronal networks a degree of circuit adaptation unmatched by synaptic plasticity alone.

9. ACKNOWLEDGMENTS
This work was supported by the Pasteur Institute, The Annette Gruner-Schlumberger Foundation, the CNRS, and a grant from the French Ministry of

Research and Education (ACI Biologie du Développement et Physiologie Intégrative, 2000).

10. REFERENCES

1. Hatten, M.E. (1999) Central neurons system neuronal migration. *Ann. Rev. Neurosci.* **22**, 511-539.
2. Kornack D.R. and Rakic, P. (1995) Radial and horizontal deployment of clonally related cells in the primate neocortex: relationship to distinct mitotic lineages. *Neuron* **15**, 311-321.
3. Letinic, K., Zoncu, R. and Rakic, P. (2002) Origin of GABAergic nerurons in the human neocortex. *Nature* **417**, 645-649.
4. Anderson, S. A., Eisenstat, D. D., Shi, L. and Rubenstein, J. L. (1997) Interneuron migration from basal forebrain to neocortex: dependence on Dlx genes, *Science* **278**: 474-476.
5. Luskin, M.B. (1993) Restricted proliferation and migration of postnatally generated neurons derived from the forebrain subventricular zone. *Neuron* **11**, 173-189.
6. Zou, Z., Horowitz, L.F., Montmayeur, J.P., Snapper, S. and Buck L.B. (2001) Genetic tracing reveals a stereotyped sensory map in the olfactory cortex. *Nature* **414**, 173-179.
7. Laurent, G. (2002) Olfactory network dynamics and the coding of multidimensional signals. *Nat. Rev. Neurosci.* **3**, 884-895.
8. Mori, K., Nagao, H. and Yoshihara, Y. (1999) The olfactory bulb: coding and processing of odor molecule information, *Science* **286**: 711-715.
9. Rall, W. and Shepherd, G.M. (1968) Theoretical reconstruction of field potentials and dendrodendritic synaptic interactions in olfactory bulb. *J. Neurophysiol.* **31**, 884-915.
10. Adrian, E.D. (1950) The Electrical activity of the mammalian olfactory bulb. *Electroencephalogr. Clin. Neurophysiol.* **2**, 377-388.
11. Gelperin, A., Kleinfeld, D., Denk, W. and Cooke, I.R. (1996) Oscillations and gaseous oxides in invertebrate olfaction. *J. Neurobiol.* **30**, 110-122.
12. Xiong, W. and Chen, W.R. (2002) Dynamic gating of spike propagation in the mitral cell lateral dendrites. *Neuron* **34**, 115-126.
13. Isaacson, J.S. (1999) Glutamate spillover mediates excitatory transmission in the rat olfactory bulb. *Neuron* **23**, 377-384.
14. Aroniadou-Anderjaska, V., Ennis, M. and Shipley, M.T. (1999) Dendrodendritic recurrent excitation in mitral cells of the rat olfactory bulb. *J. Neurophysiol.* **82**, 489-494.
15. Didier, A., Carleton, A., Bjaalie, J.G., Vincent, J.D., Ottersen, O.P., Storm-Mathisen, J. and Lledo, P.-M. (2001) A dendrodendritic reciprocal synapse provides a recurrent excitatory connection in the olfactory bulb, *Proc. Natl. Acad. Sci. USA* **98**: 6441-6446.
16. Alvarez-Buylla, A. and Garcia-Verdugo, M. (2002) Neurogenesis in adult subventicular zone. *J. Neurosci.* **22**, 629-634.
17. Temple, S. and Alvarez-Buylla, A. (1999) Stem cells in the adult mammalian central nervous system, *Curr. Opin. Neurobiol.* **9**: 135-141.
18. Gage, F.H. (2002) Neurogenesis in the adult brain, *J. Neurosci.* **22**: 612-613.
19. Gheusi, G, Cremer, H., McLean, H., Chazal, G., Vincent, J.D. and Lledo, P.-M.(2000) Importance of newly generated neurons in the adult olfactory bulb for odor discrimination, *Proc. Natl. Acad. Sci. USA* **97**: 1823-1828.
20. Rochefort, C., Gheusi, G., Vincent, J.D. and Lledo, P.-M. (2002) Enriched odor exposure increases the number of newborn neurons in the adult olfactory bulb and improves odor memory, *J. Neurosci.* **22**: 2679-2689.
21. Mombaerts, P. (2001) How smell develops. *Nat. Neurosci.* **4** Suppl: 1192-1198.
22. Zhao, H. and Reed. R.R. (2001) X inactivation of the OCNC1 channel gene reveals a role for activity-dependent competition in the olfactory system. *Cell.* **104**, 651-660.
23. Bulfone, A. et al. (1998) An olfactory sensory map develops in the absensce of normal projection neurons or GABAergic interneurons. *Neuron* **21**, 1273-1282.
24. Lin, D.M. et al. (2000) Formation of precise connections in the olfactory bulb occurs in the absence of odorant-evoked neuronal activity. *Neuron* **26**, 69-80.
25. Cline, H.T. (2001) Dendritic arbor development and synaptogenesis, *Curr. Opin. Neurobiol.*, **11**:118-126.

26. Kirschenbaum, B., Doetsch, F., Lois, C. and Alvarez-Buylla, A. (1999) Adult subventricular zone neuronal precursors continue to proliferate and migrate in the absence of the olfactory bulb. *J. Neurosci.* **19**, 2171-2180.
27. Frazier-Cierpial, L. and Brunjes, P.L. (1989) Early postnatal cellular proliferation and survival in the olfactory bulb and rostral migratory stream of normal and unilaterally oder-deprived rats. *J. Comp. Neurol.* **289**, 481-492.
28. Najbauer, J. and Leon, M. (1995) Olfactory experience modulated apoptosis in the developing olfactory bulb. *Brain. Res.* **674**, 245-251.
29. Fiske, B.K. and Brunjes, P.C. (2001) NMDA receptor regulation of cell death in the rat olfactory bulb. *J. Nerobiol.* **47**, 223-232.
30. Petreanu, L and Alvarez-Buylla A. (2002) Maturation and death of adult-born olfactory bulb granule neurons: role of olfaction, *J. Neurosci.* **22**: 6106-6113.
31. Fagiolini, M. and Hensch, T.K. (2000) Inhibitory threshold for critical-period activation in primary visual cortex, *Nature* **404**: 183-186.
32. Pizzorusso, T., Medini, P., Berardi, N., Chierzi, S., Fawcett, J.W. and Maffei, L. (2002) Reactivation of ocular dominance plasticity in the adult visual cortex, *Science* **298**: 1248-1251.
33. Nottebohm, F. (2002) Why are some neurons replaced in adult brain? *J. Neurosci.* **22**: 624-628.
34. Snyder, J.S., Kee, N. and Wojtowicz, J.M. (2001) Effects of adult neurogenesis on synaptic plasticity in the rat dentate gyrus, *J. Neurophysiol.* **85**: 2423-2431.
35. Bruce, H. M.(1959) An exteroceptive block to pregnancy in the mouse, *Nature* **184**: 105.
36. Lloyd-Thomas, A. and Keverne, E. B. (1982) Role of the brain and accessory olfactory system in the block to pregnancy in mice, *Neuroscience* **7**: 907-913.
37. Leinders-Zufall, T., Lane, A. P., Puche, A. C., Ma, W., Novotny, M., Shippley, M. T. and Zufall, F. (2000) Ultrasensitive pheromone detection by mammalian vomeronasal neurons, *Nature* **405**: 792-796.
38. Brennan, P. A., Kendrick, K. M. and Keverne, E. B. (1995) Neurotransmitter release in the accessory olfactory bulb during and after the formation of an olfactory memory in mice, *Neuroscience* **69**: 1075-1086.
39. Poindron, P. and Levy, F. (1990) Physiological, sensory and experiential determinants of maternal behaviour in sheep, In Mammalian Parenting, pp. 133-156. Eds N.A. Krasnegor and R.S. Bridges. Oxford University Press: New York.
40. Poindron, P. and Le Neindre, P. (1980) Endocrine and sensory regulation of maternal behavior in the ewe, *Adv. Stud. Behav.* **11**: 75-119.
41. Baldwin, B. A. and Shillito, E. E. (1974) The effects of ablation of the olfactory bulbs on parturition and maternal behaviour in soay sheep, *Anim. Behav.* **22**: 220-223.
42. Kendrick, K. M., Levy, F. and Keverne, E. B. (1992) Changes in the sensory processing of olfactory signals induced by birth in sheep, *Science* **256**: 833-836.
43. Brennan, P.A., Schellinck, H.M., de la Riva, C., Kendrick, K.M. and Keverne, E.B. (1998) Changes in neurotransmitter release in the main olfactory bulb following an olfactory conditioning procedure in mice. *Neuroscience*, **87**(3): 583-590.

Excitatory-Inhibitory Balance:

Systems

13

GABA_A RECEPTOR SUBTYPES: MEMORY FUNCTION AND NEUROLOGICAL DISORDERS

Jean-Marc Fritschy, Florence Crestani, Uwe Rudolph and Hanns Möhler[*]

1. INTRODUCTION

GABA is the main inhibitory neurotransmitter in the vertebrate CNS and it modulates every aspect of brain function. On the molecular level, the action of GABA is mediated by ionotropic $(GABA_A)$[1] and metabotropic $(GABA_B)$[2] receptors, which are ubiquitously expressed in developing and adult brain. The significance of $GABA_A$ receptor-mediated neurotransmission is underscored by the multiple neurological and psychiatric diseases for which alterations in the GABAergic system have been postulated, including epilepsy[3-5], Huntington's disease[6], Angelman syndrome[7], autism[8,9], anxiety disorders[10], bipolar disorders, schizophrenia[11-13], and alcohol use disorders[14]. In particular, $GABA_A$ receptors represent a major site of action for clinically important drugs, including benzodiazepines and barbiturates, some general anesthetics such as halothane, etomidate and propofol, as well as ethanol[15-18].

The purpose of this chapter is to review the significance of the molecular heterogeneity of $GABA_A$ receptors for the organization of neuronal circuits, the role of $GABA_A$ receptors in associative learning and memory, and the contribution of $GABA_A$ receptors to the pathophysiology of major neurological and neuropsychiatric disorders.

2. GABA_A RECEPTOR SUBTYPES

2.1. Identification and neuron-specific expression of GABA_A receptor subtypes

$GABA_A$ receptors belong to the superfamily of ligand-gated ion channels[19,20]. Along with glycine receptors, they mediate fast inhibitory neurotransmission in vertebrate CNS

---

[*] J.M. Fritschy[1], F. Crestani[1], U. Rudolph[1], H. Möhler[1,2] Institute of Pharmacology and Toxicology, Univ. of Zurich; [2]Dept. of Applied Biosciences, Swiss Federal Institute of Technology (ETH), 8057 Zurich, Switzerland.

Excitatory-Inhibitory Balance: Synapses, Circuits, Systems
Edited by Hensch and Fagiolini, Kluwer Academic/Plenum Publishers, 2003

by gating Cl⁻ ions through an integral membrane channel. $GABA_A$ receptors form multimeric complexes assembled from a family of at least 21 constituent subunits (α1-6, β1-4, γ1-4, δ, ρ1-3, θ, π)[1, 21]. The molecular heterogeneity of $GABA_A$ receptors is far greater than that of any other ligand-gated ion channel, a fact that has largely hampered their functional analysis.

With regard to the structure and stoichiometry of native $GABA_A$ receptors, the evidence currently available indicates that they form pentameric complexes containing $2\alpha/2\beta/1\gamma$ subunit variants[22-25]. The majority of $GABA_A$ receptors contain a single type of α and β subunit variant, with the $\alpha1\beta2\gamma2$ combination representing the largest population of $GABA_A$ receptors, followed by $\alpha2\beta3\gamma2$ and $\alpha3\beta3\gamma2$. Receptors containing the $\alpha4$, $\alpha5$, or $\alpha6$ subunit, as well as the $\beta1$, $\gamma1$, $\gamma3$, δ, π and θ subunit, form minor receptor populations. The ρ subunits are primarily expressed in the retina and correspond to $GABA_C$ receptors, which are sometimes classified separately[26].

Pharmacologically, several ligands allow differentiation between $GABA_A$ receptor subtypes (Table 1). The α1-, α2-, α3-, and α5-$GABA_A$ receptors represent diazepam-sensitive receptors, whereas α4- and α6-$GABA_A$ receptors are insensitive to diazepam[27, 28]. The former are distinguished further by their affinity to the imidazopyridine zolpidem ($\alpha1>\alpha2=\alpha3>>\alpha5$) and various β-carbolines ($\alpha1>\alpha2=\alpha3$)[15, 29]. Beyond this classical pharmacological distinction, neurosteroids, which are positive allosteric modulators of recombinant and native $GABA_A$ receptors[30], are most potent on receptors containing the δ subunit[31, 32]. Furthermore, novel ligands are being introduced, which show little selectivity based on affinity but possess intrinsic activity only on specific $GABA_A$ receptor subtypes[33-36].

Functionally, distinct subunit-specific properties have been identified in both recombinant and native receptors, which include ligand affinity, kinetic properties, and desensitization kinetics. In particular, the type of α subunit determines the kinetics of receptor deactivation[37-39] and the presence of the δ subunit results in markedly increased agonist affinity and apparent lack of desensitization[32, 40, 41]. The functional relevance of these differences for network properties has been analyzed extensively in mutant mice lacking the δ subunit[42-44].

Morphologically, the distribution of $GABA_A$ receptor subtypes differentiated by their α subunit variants exhibit a striking region- and neuron-specific distribution pattern that is largely conserved across species (Table 1)[45-48]. Importantly, these findings underscored that the functional relevance of a given $GABA_A$ receptor subtype depends on the neuronal circuits to which it belongs and is not necessarily correlated with its relative abundance. For example, α3-$GABA_A$ receptors, which represent only about 10% of all $GABA_A$ receptors in brain, are the main subtype expressed by monoaminergic and basal forebrain cholinergic neurons[49], suggesting a modulatory role in various brain functions as diverse as motor function and cognition. The α3-$GABA_A$ receptors are also the sole receptor subtype expressed by neurons of the thalamic reticular nucleus, thereby potentially modulating the entire thalamo-cortical network (Table 1)[50]. Conversely, morphological studies revealed an unsuspected abundance of $GABA_A$ receptor subtypes in apparently simple circuits. For instance, cerebellar granule cells receive their GABAergic innervation from a single cell type, the Golgi type II cells. Nevertheless, granule cells express probably more than a dozen of $GABA_A$ receptor subtypes[51], with largely overlapping subcellular distribution. The role of these individual receptor subtypes in granule cells has remained largely elusive.

2.2. Circuit-specific segregation of GABA$_A$ receptor subtypes

The combination of pharmacological and morphological studies reviewed in the previous section provides strong evidence that GABA$_A$ receptor subtypes are functionally specialized and allocated to defined neuronal circuits (Table 1). Although the functional relevance at the system's level of having, for instance, an α1- instead of an α2-GABA$_A$ receptor in a given type of neuron, remains speculative, the specificity of subtype expression is underscored by the remarkable selectivity of action of diazepam in knock-in mutant mice carrying "custom-made" diazepam-insensitive GABA$_A$ receptor subtypes[52]. Thus, abolition of diazepam binding on α2-GABA$_A$ receptors *in vivo* by exchanging the conserved histidine 101 residue with an arginine residue results in a selective loss of the anxiolytic action of diazepam. A similar histidine/arginine point mutation in the α1-, α3- or α5-GABA$_A$ receptors does not affect this particular drug effect[53]. Although the α2 subunit has a widespread distribution in numerous brain areas, these findings indicate that the α2-GABA$_A$ receptors are strategically located in circuits mediating diazepam anxiolysis. The motor depressant effects of diazepam and zolpidem, but not those of pentobarbital or the neurosteroid 3α-hydroxy-5β-pregnan-20-one, are abolished in mice carrying α1^{H101R}-GABA$_A$ receptors[54-56]. Likewise, diazepam, in contrast to the muscarinic receptor antagonist scopolamine, does not induce anterograde amnesia in α1^{H101R} mice[54].

Finally, the anticonvulsant action of diazepam is strongly reduced in α1^{H101R} mice, whereas its myorelaxant action is retained. α1-GABA$_A$ receptors appear therefore to be a selective molecular target mediating three of the major actions of diazepam, sedation, anterograde amnesia, and seizure protection[54]. This selectivity of drug action was extended to both positive and negative allosteric modulators of the benzodiazepine-binding site. Indeed, a recent study reveals a switch in the *in vivo* action of Ro 15-4513 from inverse agonism in wild type mice to agonism in α1^{H101R}-mice, resulting in awake immobility and anticonvulsant action[57]. This indicates that the α1-GABA$_A$ receptors mediate specific behavioral output produced by benzodiazepine site ligands with regard to motor activity and seizure protection, irrespective of the direction of the intrinsic activity of these ligands.

The cellular corollary of the specificity of diazepam action is the synapse-specific distribution of GABA$_A$ receptor subtypes, in particular in neurons expressing multiple GABA$_A$ receptor subtypes, such as hippocampal pyramidal neurons (Table 1). In these cells, a high level of α1, α2, and α5 subunit expression has been reported, along with β1-3 and γ2 subunit[47, 48], suggesting that they express at least three main GABA$_A$ receptor subtypes. α1-GABA$_A$ receptors are located postsynaptically in a majority of somatodendritic synapses and to a lesser extent in the axon initial segment; in contrast, α2-GABA$_A$ receptors are particularly abundant in the axon initial segment and are only few in somatodendritic synapses[58,59]. Finally, α5-GABA$_A$ receptors have an extrasynaptic localization, being distributed throughout the somatodendritic compartment of hippocampal pyramidal cells without being aggregated at postsynaptic sites[60, 61]. Their functional role is discussed in Section 3.

Strikingly, in the soma of hippocampal pyramidal cells, α1- and α2-GABA$_A$ receptor subtypes are segregated to distinct synapses formed by two separate populations of basket cells[62, 63]. These interneurons, which selectively innervate the somatic region of pyramidal cells, are subdivided into two groups based on differential expression of the

Table 1. GABA$_A$ receptor subtypes*

Composition	Pharmacological Characteristics[a]	Neuronal Localisation[b]	
$\alpha_1\beta_2\gamma_2$	Major subtype (60 % of all GABA$_A$ receptors). Mediates the sedative, amnestic and – to a large extent – the anticonvulsant action of benzodiazepine site agonists. High affinity for classical benzodiazepines, zolpidem and the antagonist flumazenil.	Synaptic and extrasynaptic	Cerebral cortex[71, 72] Hippocampus, dentate gyrus (interneurons and principal cells)[58, 59, 73] Pallidum[74], Striatum (interneurons)[75] Thalamic relay nuclei[76] Olfactory bulb (mitral cells and interneurons)[77] Cerebellum (Purkinje cells, stellate, basket cells and granule cells)[72, 74, 78-80], deep cerebellar nuclei[72]
$\alpha_2\beta_3\gamma_2$	Minor subtype (15-20 %). Mediates anxiolytic action of benzodiazepine site agonists. High affinity for classical benzodiazepine agonists and the antagonist flumazenil. Intermediate affinity for zolpidem.	Synaptic	Cerebral cortex[59] Hippocampus, dentate gyrus (principal cells mainly on the axon initial segment)[58, 59, 72] Olfactory bulb (granule cells)[72] Striatum (spiny stellate cells)[59] Inferior olivary neurons (dendrites)[39]
$\alpha_3\beta_n\gamma_2$	Minor subtype (10-15 %). High affinity for classical benzodiazepine agonists and the antagonist flumazenil. Intermediate affinity for zolpidem	Synaptic Extrasynaptic	Cerebral cortex (principal cells in particular in layers V and VI; some axon initial segments)[59] Hippocampus (some hilar cells)[60] Olfactory bulb (tufted cells)[77] Thalamic reticular nucleus[59] Cerebellum (Golgi cells)[72] Medullary reticular formation[59] Inferior olivary neurons[39]
$\alpha_4\beta_n\gamma/\alpha_4\beta_n\delta$	Less than 5 % of all receptors. Insensitive to classical benzodiazepine agonists and zolpidem.	Extrasynaptic	Dentate gyrus (granule cells)[81]
$\alpha_5\beta_{1/3}\gamma_2$	Less than 5 % of all receptors; High affinity for classical benzodiazepine agonists and the antagonist flumazenil. Very low affinity for zolpidem.	Synaptic Extrasynaptic	Spinal trigeminal nucleus, superior olivary neurons (Fritschy, unpublished) Cerebral cortex (Fritschy, unpublished) Hippocampus (pyramidal cells)[61] Olfactory bulb (granule cells)[61]
$\alpha_6\beta_{2,3}\gamma_2$ $\alpha_6\beta_n\delta$	Less than 5 % of all receptors. Insensitive to classical benzodiazepine agonists and zolpidem. Minor population. Lacks benzodiazepine site.	Synaptic Extrasynaptic	Cerebellum (granule cells)[78-80] Cerebellum (granule cells)[72, 78-80]
ρ	Homomeric receptors: Insensitive to bicuculline, barbiturates, baclofen and all benzodiazepine site ligands. Also termed GABAc receptor.	Synaptic	Retina[82]

a) reviewed in refs. 1, 52, 83-85; the term classical benzodiazepines refers to diazepam and structurally related agonists in clinical use.
b) Synaptic localization is based mainly on colocalization with gephyrin[72] or on ultrastructural evidence and refers to the respective type of α subunit variant.
*modified from ref. 17.

neurochemical markers parvalbumin and cholecystokinin[64]. Parvalbumin-positive basket cells form GABAergic synapses containing the α1 subunit[62], whereas cholecystokinin-positive basket cells form GABAergic synapses containing the α2 subunit[63]. It is of note that CB1-type of cannabinoid receptors are localized presynaptically, selectively on the terminals of CCK-positive basket cells in hippocampus and amygdala[65, 66]. These receptors have been shown to mediate depolarization-induced suppression of inhibition[67], a form of short-term GABAergic plasticity[68-70]. Therefore, endogenous cannabinoids potentially modulate hippocampal circuits containing α2-GABA$_A$ receptors, while they will have no effect on circuits containing α1-GABA$_A$ receptors. These observations underscore the existence of parallel, functionally specialized neuronal circuits within the same brain structure. Such segregation of GABA$_A$ receptor subtypes is not unique to the hippocampus, but is likely to be found in many other brain regions, such as the amygdala, which exhibits striking nucleus-specific differences in GABA$_A$ receptor subunit expression[45, 48]. Mutant mice carrying diazepam-insensitive GABA$_A$ receptor subtypes therefore represent a powerful tool to unravel neuronal networks that control brain functions that are influenced by benzodiazepine site ligands.

3. GABA$_A$ RECEPTORS AND ASSOCIATIVE LEARNING AND MEMORY

Several forms of network oscillations have been described in the hippocampus based on frequency, waveform, amplitude, and correlation to behavior[86-94]. The dynamics of neural networks are largely shaped by the activity pattern of interneurons, most of which are GABAergic[62, 64, 92, 95-97]. The activity of these interneurons is thought to set the spatio-temporal conditions for synaptic plasticity and hippocampus-dependent learning[92, 96-100].

On the molecular level, spatial learning and memory were initially linked to changes in the function of hippocampal NMDA receptors. Infusion of AP5 into the brain caused a deficit in spatial learning performance[101-103]. When NMDA receptors were targeted exclusively in the CA1 subregion of the hippocampus (NR1 subunit CA1-knockout mice) the representation of space was impaired, as shown by a deficit in the coordination of place cell firing[104] and a deficit in developing a navigational strategy in the water maze task[105]. Nevertheless the mutant animals continued to perform normally in non-spatial learning tasks[105]. Similarly, NMDA receptors in CA1 pyramidal cells were found to be crucial for the formation of memories that associate events across time, as shown by trace fear conditioning. In fear conditioning experiments, when the conditioned and unconditioned stimuli (CS, US) were separated by a 30s trace, the NR1 subunit CA1-knockout mice failed to memorize the association but were indistinguishable from normal animals when the US co-terminated with the CS (delay conditioning)[106]. Using a similar regional restricted knockout strategy, NMDA receptors in the CA3 region were shown to be essential for associative memory recall for pattern completion[107]. Conversely, when the NR2B subunit of the NMDA receptor was overexpressed in the forebrain of transgenic mice, the animals exhibited better learning and memory in various behavioral tasks[108]. Thus, activity-dependent modifications of synapses on hippocampal principal cells mediated by NMDA receptors play an essential role in the acquisition of spatial and temporal memory.

3.1. Inhibition of hippocampal pyramidal cells

The excitatory input to hippocampal pyramidal cells is closely balanced by a diversity of neuronal inhibition arising from interneurons, which are strikingly different in domain specificity, morphology, and response properties[62, 64, 92, 96, 97, 109, 110]. The knowledge of the spatial-temporal activity pattern in distinct classes of interneurons is essential to an understanding of the cellular mechanisms of learning and memory. On the molecular level, the inhibitory GABAergic regulation of hippocampal principal cells is mediated by various structurally diverse $GABA_A$ receptors expressed in a domain-specific manner (see section 2.2). Among these receptors, the α5 subtype displays a privileged site of expression, being located closely to the excitatory synapses at the dendritic spines of the principal cells. Indeed, as shown in primary hippocampal cultures, the α5-$GABA_A$ receptors are located extrasynaptically on spines and on dendrite shafts of pyramidal cells[60]. The α5-$GABA_A$ receptors were therefore considered to be able to regulate the transduction of the signal arising at excitatory synapses and, by doing so, would operate as control elements in learning and memory processes in their own right.

3.2. Extrasynaptic α5-$GABA_A$ receptors

To analyze the functional relevance of the α5-$GABA_A$ receptors, which are mainly expressed in the hippocampus (pyramidal cells), mutant mice were generated in which the benzodiazepine site was rendered diazepam insensitive by a (H105R) point mutation in the α5 subunit gene[61]. In these mice, the complement of α5-$GABA_A$ receptors was reduced exclusively in the hippocampus by 30-40% due to a molecular mechanism that is not understood. The remaining α5-$GABA_A$ receptors showed a distribution corresponding to that in wild type mice. There was no indication for adaptive changes of other $GABA_A$ receptors expressed in the same cells[61]. Behaviorally, the partial deficit of hippocampal α5-$GABA_A$ receptors resulted in an enhanced conditioned fear response in a trace fear conditioning task, in which a 1-sec trace interval was interposed between the CS and the US. However, in a delay fear conditioning task, the α5 (H105R) mice showed performance similar to that of wild type animals. These results were supported by the analysis of another mouse line in which the α5-$GABA_A$ receptors were deleted in the entire brain by targeting the α5 subunit gene[111, 112]. These mice showed a significantly improved spatial learning performance in the water maze task, but not in a two-way active avoidance paradigm[111]. In addition, the absence of α5-$GABA_A$ receptors was associated with decreased amplitude of hippocampal IPSCs and enhanced paired-pulse facilitation of field EPSP amplitudes. These data strongly suggest that α5-$GABA_A$ receptors play a crucial role in certain forms of learning by regulating hippocampal synaptic transmission.

A deficit in α5-$GABA_A$ receptor activation can conceivably be mimicked pharmacologically. Partial inverse agonists acting at the benzodiazepine site of α5-$GABA_A$ receptors would be expected to facilitate hippocampus-dependent learning and memory. Such agents may provide a novel avenue for the treatment of memory disorders.

3.3. Balance between hippocampal NMDA-receptors and α5-GABA$_A$ receptors

It is striking that the behavioral consequences of an impairment of α5-GABA$_A$ receptors are opposite to that of a NMDA receptor deficit: while the NR1 subunit-CA1 knockout mice show impaired spatial memory and trace fear conditioning, the mice with a deficit in α5-GABA$_A$ receptors display an improvement in trace fear conditioning. Thus, it appears that these two receptor systems may play antagonistic roles in controlling signal transduction of the hippocampal principal cells. This interaction is presumably facilitated by the close subcellular apposition of these two receptor systems.

Extrasynaptic GABA$_A$ receptors were long considered to mediate only tonic inhibition[81, 113, 114]. However, the results outlined above point to a pronounced influence of the extrasynaptic α5-GABA$_A$ receptor on hippocampal function. The spatio-temporal activation pattern of these receptors may be governed by synchronous discharges of interneurons. Rhythmic neuronal activities can provide the associative firing needed to trigger changes in synaptic strength[109, 115-118]. The simultaneous firing of interneurons is known to occur[119, 120] and GABA can spill over to adjacent extrasynaptic receptors[81, 121]. In this way, extrasynaptic α5-GABA$_A$ receptors in hippocampal pyramidal cells may contribute to the modulation of dendritic excitability and of the efficacy of excitatory inputs. The α5-GABA$_A$ receptor thereby participates in synaptic plasticity processes regulating the association of spatial and temporal cues.

4. ALTERATIONS IN GABA$_A$ RECEPTOR EXPRESSION AND FUNCTION IN NEUROLOGICAL AND PSYCHIATRIC DISORDERS

Recent evidence is reviewed about long-term regulation of GABA$_A$ receptor expression and localization in various disease states. This overview includes experimental studies as well as clinical investigations to underscore the relevance of these findings for the pathophysiology of neurological and psychiatric disorders.

4.1. Regulation of GABA$_A$ receptor expression in Huntington's disease

Degeneration of GABAergic neurons is one of the hallmarks of Huntington's disease. In the caudate nucleus and putamen, it is accompanied by a profound reduction in benzodiazepine binding sites and GABA$_A$ receptor subunit-immunoreactivity, corresponding to the loss of neurons[6, 122]. However, in the globus pallidus, a major target of striatal neurons, GABA$_A$ receptors are increased, suggesting compensatory up-regulation in the remaining synapses. Experimentally, the quinolinic acid-induced lesion of the striatum, which mimics the pattern of neuronal degeneration of Huntington's disease, results in an increased GABA$_A$ receptor β2/3 subunit-immunoreactivity in the substantia nigra pars reticulata[123]. A detailed electron microscopic analysis, using post-embedding techniques, revealed a selective increase of GABA$_A$ receptor labeling in symmetric synapses, but not of AMPA receptors in asymmetric synapses, in lesioned animals[124]. Most strikingly, the increased expression of GABA$_A$ receptors is long-lasting (> 15 months) but is induced very rapidly, being detectable by autoradiography within 2 hours following intrastriatal quinolinic acid injection[123]. The signals involved in this rapid induction have not been investigated.

4.2. GABA$_A$ receptors in epilepsy

Genetic evidence that GABA$_A$ receptors are involved in human idiopathic epilepsy has been provided for several mutations in the γ2 subunit gene and one mutation in the α1 subunit gene. The γ2^{K289M} point mutation in the extracellular loop between TM2 and TM3 of the γ2 subunit was reported in a family with generalized epilepsy with febrile seizures[125]. In recombinant α1β2γ2 receptors expressed in *Xenopus* oocytes, this mutation reduces the amplitude of GABA-induced currents. However, potentiation by diazepam is not affected. Another mutation, γ2^{R43Q}, has been associated with childhood absence epilepsy and febrile seizures[126, 127]. In recombinant expression systems, the γ2^{R43Q} mutation results in an altered pharmacological profile and an increased desensitization of the mutated receptors. The latter suggests that decreased inhibitory transmission underlies hyperexcitability of thalamo-cortical networks[128]. A single nucleotide exchange at the splice donor site of intron 6 of the γ2 subunit, most likely resulting in a non-functional allele, has also been reported in a family with childhood absence epilepsy and febrile convulsions[129], whereas another truncation mutation, γ2^{Q351X}, was shown to prevent surface expression of the corresponding recombinant receptors tagged with GFP[130]. Finally, a loss-of-function mutation of α1-GABA$_A$ receptors (α1^{A322D}) was detected in a family with an autosomal dominant form of juvenile myoclonic epilepsy[131]. It was not established whether this mutation affects GABA$_A$ receptor function or surface expression. It should also be noted that a polymorphism in the GABA$_B$ receptor R1 subunit (G1465A) has been associated with drug-resistant temporal lobe epilepsy[132]. Although the functional consequences of these mutations can be readily investigated in recombinant expression systems, their relevance for seizure development, and the possible contribution of additional genes, remain poorly understood.

Alterations in GABA$_A$ receptor distribution have been studied extensively in human mesial temporal lobe epilepsy with hippocampal sclerosis and in various rodent models of this disorder[133]. In human, the profound neuronal loss in CA1, which is one of the characteristic features of hippocampal sclerosis, is accompanied by a marked decrease in benzodiazepine binding sites[134, 135]. Alterations have also been reported in white matter of the ipsilateral temporal lobe and in the superior and medial frontal gyrus[136]. Autoradiography of resected tissue with ^{125}I-iomazenil revealed a reduction of binding sites exceeding neuronal loss, notably in the CA1 area[137]. In contrast, an immunohistochemical analysis with antibodies against the α1, α2, α3, β2,3, and γ2 subunit revealed a complex pattern of changes, characterized above all by an increased staining intensity on surviving neurons, and by subtype-specific changes in subcellular distribution of GABA$_A$ receptors in epileptic tissue[138]. Among the most pronounced and consistent changes was the increased α1 and α2 subunit-IR in the soma and apical dendrites of dentate gyrus granule cells and an apparent translocation of the α3 subunit-immunoreactivity from the somatic region to distal dendrites in CA2 pyramidal cells[138]. These alterations in GABA$_A$ receptor expression and cellular distribution suggest extensive remodeling of neuronal circuits in the affected hippocampal formation. It is not known whether these changes are causally related to epileptogenesis and neuronal death, or whether they represent protective mechanisms in response to seizures.

An important observation, based on electrophysiological recordings in slices of tissue resected from patients with intractable mesial temporal lobe epilepsy, was the presence of spontaneously active neurons in the subiculum[139], a region which relays the output of the hippocampus to the cerebral cortex. The spontaneous discharges of subicular neurons were shown to correspond to interictal spikes-and-waves measured electroencephalographically prior to surgery. The rhythmic discharges of subicular neurons were mediated by $GABA_A$ receptors and were due to a depolarizing action of GABA, presumably caused by altered function or expression of chloride-extrusion pumps, such as KCC2[139]. The presence of an epileptic focus outside of the sclerotic part of the hippocampus explains the persistence of recurrent seizures in spite of extensive neuronal death in CA1 and CA3. In addition, these results suggest that molecular alterations affecting the function of $GABA_A$ receptors give rise to interictal discharges in mesial temporal lobe epilepsy.

In experimental temporal lobe epilepsy, changes in $GABA_A$ receptor subunit expression have been analyzed in several animal models, with largely convergent results. The main observation was overall increased expression of $GABA_A$ receptors in dentate gyrus granule cells, with changes in pharmacological properties suggesting aberrant expression of $GABA_A$ receptors in these cells. Notably, an increase in $\alpha 3$, $\alpha 4$, and δ subunit expression has been reported in rats experiencing chronic recurrent seizures following intraperitoneal injection of the muscarinic receptor agonist, pilocarpine or the glutamate receptor agonist, kainic acid[140-142].

However, chronic recurrent seizures induced by intraperitoneal injection of kainic acid or pilocarpine do not mimic the complex partial seizures experienced by most patients with temporal lobe epilepsy. Also the pattern of neuronal loss is significantly different from that reported in neuropathological studies of hippocampal sclerosis. These limitations are partially overcome in a mouse model of temporal lobe epilepsy, in which spontaneous recurrent partial seizures are induced following unilateral injection of kainic acid into the dorsal hippocampus[143, 144]. In these mice, dentate gyrus granule cells become hypertrophic and undergo a prominent dispersion. A marked increase in $\alpha 1$, $\alpha 2$, $\alpha 5$, and $\gamma 2$ subunit-immunoreactivity was observed in epileptic dentate gyrus[145], which corresponded to an increase in the size and density of postsynaptic $GABA_A$ receptor clusters[146]. These findings are strongly suggestive of the formation of novel GABAergic synapses, possibly reflecting sprouting of GABAergic axons in the epileptic dentate gyrus. The formation of aberrant GABAergic connections might thus also be considered as a contributing factor of temporal lobe epilepsy.

4.3. Synapse-specific alterations of $GABA_A$ receptors in schizophrenia

Alterations in several biochemical and anatomical markers of GABAergic transmission have been reported in schizophrenic patients, including changes in GAD expression, muscimol binding, and number of interneurons[11-13, 147]. The regions affected include the hippocampus, anterior cingulate cortex, prefrontal cortex, and medial temporal cortex. The specificity of these alterations for schizophrenia remains debated since similar changes might also occur in bipolar disorder and major depressive disorder[148-151]. A selective alteration of GABAergic input from chandelier cells onto the axon-initial segment of pyramidal cells has been demonstrated in areas 9 and 46 of prefrontal cortex in sections from schizophrenic patients labeled with antibodies to GAT1[152]. This decrease was not seen in age-matched non-schizophrenic patients, and was independent of anti-psychotic medication at the time of death. Since the number and size of parvalbumin-positive neurons, which include chandelier neurons, was not affected, these results were taken as evidence for an alteration in GAT1 expression and not for a decrease in the number of axon terminals. In the same cohort, α2-GABA$_A$ receptor-immunoreactivity in the axon initial segment was enhanced, again selectively in schizophrenic patients independently of the anti-psychotic medication[153]. Collectively, these findings support the view that altered GABAergic control of pyramidal cell output contributes to cognitive dysfunction in schizophrenia[148].

5. CONCLUDING REMARKS

There are multiple modes of GABA$_A$ receptor-mediated neurotransmission, including phasic inhibition produced by distinct GABA$_A$ receptor subtypes, localized in defined neuronal circuits operating in parallel within a given brain structure, as shown for hippocampal pyramidal cells, and tonic inhibition, mediated by functionally- and pharmacologically-specialized extrasynaptic receptors, as shown for the α5-GABA$_A$ receptors, which play a crucial role in learning and memory processes by regulating hippocampal excitatory synaptic transmission. From a pharmacological point-of-view, these findings provide a rationale to target specific GABA$_A$ receptor subtypes. Thus, new anxiolytic drugs devoid of sedative action and novel memory enhancing drugs, acting by reducing the function of α5-GABA$_A$ receptors, appear feasible.

Direct evidence for alterations in GABAergic neuron morphology and in distribution of GABA$_A$ receptors is available for the most prevalent neurological and psychiatric disorders, most likely reflecting changes in GABAergic circuits, including axonal sprouting and formation of novel synapses. While the analysis of causative mechanisms is impossible in human studies, some of these changes can be reproduced experimentally in animal models, and their molecular and cellular bases analyzed *in vitro*. The ultimate goals of these studies are to further our understanding of brain function and to provide effective treatments or relief of symptoms for neurological and psychiatric disorders. Given the clinical importance of GABA$_A$ receptor pharmacology, the elucidation of the role of GABA$_A$ receptor subtypes in specific neuronal circuits will represent a major advance towards these goals.

6. ACKNOWLEDGEMENTS

Our work was supported by the Swiss National Science Foundation.

7. REFERENCES

1. Barnard, E.A., Skolnick, P., Olsen, R.W., Mohler, H., Sieghart, W., Biggio, G., Braestrup, C., Bateson, A.N. and Langer, S.Z., *Pharmacol. Rev.* **50**, 291 (1998).
2. Bowery, N.G., Bettler, B., Froestl, W., Gllagher, J.P., Marshall, F., Raiteri, M., Bonner, T.I. and Enna, S.J., *Pharmacol. Rev.* **54**, 247 (2002).
3. Duncan, J.S., *Rev. Neurol.* **155**, 482 (1999).
4. Olsen, R.W., DeLorey, T.M., Gordey, M. and Kang, M.H., *Adv. Neurol.* **79**, 499 (1999).
5. Coulter, D.A., *Int. Rev. Neurobiol.* **45**, 237 (2001).
6. Kunig, G., Leenders, K.L., Sanchez-Pernaute, R., Antonini, A., Vontobel, P., Verhagen, A. and Gunther, I., *Ann. Neurol.* **47**, 644 (2000).
7. DeLorey, T.M., Handforth, A., Anagnostaras, S.G., Homanics, G.E., Minassian, B.A., Asatourian, A., Fanselow, M.S., Delgado-Escueta, A., Ellison, G.D. and Olsen, R.W., *J. Neurosci.* **18**, 8505 (1998).
8. Fatemi, S.H., Halt, A.R., Stary, J.M., Kanodia, R., Schulz, S.C. and Realmuto, G.R., *Biol. Psychiatry* **52**, 805 (2002).
9. Buxbaum, J.D., Silverman, J.M., Smith, C.J., Greenberg, D.A., Kilifarski, M., Reichert, J., Cook, E.H., Fang, Y., Song, C.Y. and Vitale, R., *Mol. Psychiatry* **7**, 311 (2002).
10. Malizia, A.L., *J. Psychopharmacol.* **13**, 372 (1999).
11. Nutt, D.J. and Malizia, A.L., *Brit. J. Psychiat.* **179**, 390 (2001).
12. Blum, B.P. and Mann, J.J., *Int. J. Neuropsychopharm.* **5**, 159 (2002).
13. Lewis, D.A., *Brain Res. Rev.* **31**, 270 (2000).
14. Morrow, A.L., VanDoren, M.J., Penland, S.N. and Matthews, D.B., *Brain Res. Rev.* **37**, 98 (2001).
15. Sieghart, W., *Pharmacol. Rev.* **47**, 181 (1995).
16. Dilger, J.P., *Br. J. Anaest.* **89**, 41 (2002).
17. Mohler, H., Fritschy, J.M. and Rudolph, U., *J. Pharm. Exp. Ther.* **300**, 2 (2002).
18. Grobin, A.C., Matthews, D.B., Devaud, L.L. and Morrow, A.L., *Psychopharmacology.* **139**, 2 (1998).
19. Unwin, N., *Neuron* **10 Suppl.**, 31 (1993).
20. Barnard, E.A., in: *Pharmacology of GABA and glycine neurotransmission, Handbook Expt. Pharmacol., vol. 150*, edited by H. Mohler, (Springer-Verlag, Berlin, 2001) p. 79.
21. Whiting, P.J., *Neurochem. Int.* **34**, 387 (1999).
22. Knight, A.R., Stephenson, F.A., Tallman, J.F. and Ramabahdran, T.V., *Recept. Channels* **7**, 213 (2000).
23. Baumann, S.W., Baur, R. and Sigel, E., *J. Biol. Chem.* **276**, 36275 (2001).
24. Klausberger, T., Sarto, I., Ehya, N., Fuchs, K., Furtmuller, R., Mayer, B., Huck, S. and Sieghart, W., *J. Neurosci.* **21**, 9124 (2001).
25. Farrar, S.J., Whiting, P.J., Bonnert, T.P. and McKernan, R.M., *J. Biol. Chem.* **274**, 10100 (1999).
26. Bormann, J., *Trends Pharmacol. Sci.* **21**, 16 (2000).
27. Benson, J.A., Low, K., Keist, R., Mohler, H. and Rudolph, U., *FEBS Lett.* **431**, 400 (1998).
28. Wingrove, P.B., Safo, P., Wheat, L., Thompson, S.A., Wafford, K.A. and Whiting, P.J., *Eur. J. Pharmacol.* **437**, 31 (2002).
29. Mohler, H., Fritschy, J.M., Luscher, B., Rudolph, U., Benson, J. and Benke, D., in: *Ion channels, vol. 4*, edited by T. Narahashi, (Plenum Press, New York, 1996) p. 89.
30. Lambert, J.J., Belelli, D., Harney, S.C., Peters, J.A. and Frenguelli, B.G., *Brain Res. Rev.* **37**, 68 (2001).
31. Wohlfarth, K.M., Bianchi, M.T. and Macdonald, R.L., *J. Neurosci.* **22**, 1541 (2002).
32. Adkins, C.E., Pillai, G.V., Kerby, J., Bonnert, T.P., Haldon, C., McKernan, R.M., Gonzalez, J.E., Oades, K., Whiting, P.J. and Simpson, P.B., *J. Biol. Chem.* **276**, 38934 (2001).
33. Sigel, E. and Dodd, R.H., *Drugs Future* **26**, 1191 (2001).
34. Collins, I., Moyes, C., Davey, W.B., Rowley, M., Bromidge, F.A., Quirk, K., Atack, J.R., McKernan, R.M., Thompson, S.A., Wafford, K., Dawson, G.R., Pike, A., Sohal, B., Tsou, N.N., Ball, R.G. and Castro, J.L., *J. Med. Chem.* **45**, 1887 (2002).
35. Selleri, S., Bruni, F., Costagli, C., Costanzo, A., Guerrini, G., Ciciani, G., Gratteri, P., Bonaccini, C., Malmberg Aiello, P., Besnard, F., Renard, S., Costa, B. and Martini, C., *J. Med. Chem.* **46**, 310 (2003).
36. Griebel, G., Perrault, G., Simiand, J., Cohen, C., Granger, P., Decobert, M., Françon, D., Avenet, P., Depoortere, H., Tan, S., Oblin, A., Schoemaker, H., Evanno, Y., Sevrin, M., George, P. and Scatton, B., *J. Pharmacol. Exp. Ther.* **298**, 753 (2001).
37. Verdoorn, T.A., Draguhn, A., Ymer, S., Seeburg, P.H. and Sakmann, B., *Neuron* **4**, 919 (1990).
38. Hutcheon, B., Morley, P. and Poulter, M.O., *J. Physiol.* **522**, 3 (2000).
39. Devor, A., Fritschy, J.M. and Yarom, Y., *J. Neurophysiol.* **85**, 1686 (2001).
40. Burgard, E.C., Tietz, E.I., Neelands, T.R. and Macdonald, R.L., *Mol. Pharmacol.* **50**, 119 (1996).
41. Fisher, J.L. and Macdonald, R.L., *J. Physiol.* **505**, 283 (1997).

42. Mihalek, R.M., Bowers, B.J., Wehner, J.M., Kralic, J.E., VanDoren, M.J., Morrow, A.L. and Homanics, G.E., *Alcohol. Clin. Exp. Res.* **25**, 1708 (2001).
43. Spigelman, I., Li, Z., Banerjee, P.K., Mihalek, R.M., Homanics, G.E. and Olsen, R.W., *Epilepsia* **43**, 3 (2002).
44. Porcello, D.M., Huntsman, M.M., Mihalek, R.M., Homanics, G.E. and Huguenard, J.R., *J. Neurophysiology* **89**, 1378 (2003).
45. Schwarzer, C., Berresheim, U., Pirker, S., Wieselthaler, A., Fuchs, K., Sieghart, W. and Sperk, G., *J. Comp. Neurol.* **433**, 526 (2001).
46. Waldvogel, H.J., Fritschy, J.M., Mohler, H. and Faull, R.L.M., *J. Comp. Neurol.* **397**, 297 (1998).
47. Pirker, S., Schwarzer, C., Wieselthaler, A., Sieghart, W. and Sperk, G., *Neurosci.* **101**, 815 (2000).
48. Fritschy, J.M. and Mohler, H., *J. Comp. Neurol.* **359**, 154 (1995).
49. Gao, B., Fritschy, J.M., Benke, D. and Mohler, H., *Neurosci.* **54**, 881 (1993).
50. Huntsman, M.M., Porcello, D.M., Homanics, G.E., DeLorey, T.M. and Huguenard, J.R., *Science* **283**, 541 (1999).
51. Sieghart, W., Fuchs, K., Tretter, V., Ebert, V., Jechlinger, M., Hoger, H. and Adamiker, D., *Neurochem. Int.* **34**, 379 (1999).
52. Rudolph, U., Crestani, F. and Mohler, H., *Trends Pharmacol. Sci.* **22**, 188 (2001).
53. Löw, K., Crestani, F., Keist, R., Benke, D., Brünig, I., Benson, J.A., Fritschy, J.M., Rulicke, T., Bluethmann, H., Mohler, H. and Rudolph, U., *Science* **290**, 131 (2000).
54. Rudolph, U., Crestani, F., Benke, D., Brünig, I., Benson, J., Fritschy, J.M., Martin, J.R., Bluethmann, H. and Mohler, H., *Nature* **401**, 796 (1999).
55. McKernan, R.M., Rosahl, T.W., Reynolds, D.S., Sur, C., Wafford, K.A., Atack, J.R., Farrar, S., Myers, J., Cook, G., Ferris, P., Garrett, L., Bristow, L., Marshall, G., Macaulay, A., Brown, N., Howell, O., Moore, K.W., Carling, R.W., Street, L.J., Castro, J.L., Ragan, C.I., Dawson, G.R. and Whiting, P.J., *Nature Neurosci.* **3**, 587 (2000).
56. Crestani, F., Martin, J.R., Mohler, H. and Rudolph, U., *Brit. J. Pharmacol.* **131**, 1251 (2000).
57. Crestani, F., Assandri, R., Täuber, M., Martin, J.R. and Rudolph, U., *Neuropharmacol.* **43**, 679 (2002).
58. Nusser, Z., Sieghart, W., Benke, D., Fritschy, J.M. and Somogyi, P., *Proc. Natl. Acad. Sci. USA* **93**, 11939 (1996).
59. Fritschy, J.M., Weinmann, O., Wenzel, A. and Benke, D., *J. Comp. Neurol.* **390**, 194 (1998).
60. Brünig, I., Scotti, E., Sidler, C. and Fritschy, J.M., *J.Comp. Neurol.* **443**, 43 (2002).
61. Crestani, F., Keist, R., Fritschy, J.M., Benke, D., Vogt, K., Prut, L., Bluethmann, H., Mohler, H. and Rudolph, U., *Proc. Natl. Acad. Sci. USA* **99**, 8980 (2002).
62. Klausberger, T., Roberts, J.D. and Somogyi, P., *J. Neurosci.* **22**, 2513 (2002).
63. Nyiri, G., Freund, T.F. and Somogyi, P., *Eur. J. Neurosci.* **13**, 428 (2001).
64. Freund, T.F. and Buzsaki, G., *Hippocampus* **6**, 345 (1996).
65. Katona, I., Rancz, E.A., Acsady, L., Ledent, C., Mackie, K., Hajos, N. and Freund, T.F., *J. Neurosci.* **21**, 9506 (2001).
66. Katona, I., Sperlágh, B., Sík, A., Káfalvi, A., Vizi, E.S., Mackie, K. and Freund, T.F., *J. Neurosci.* **19**, 4544 (1999).
67. Wilson, R.I., Kunos, G. and Nicoll, R.A., *Neuron* **31**, 453 (2001).
68. Llano, I., Leresche, N. and Marty, A., *Neuron* **6**, 565 (1991).
69. Pitler, T.A. and Alger, B.E., *J. Neurosci.* **12**, 4122 (1992).
70. Wilson, R.I. and Nicoll, R.A., *Science* **296**, 678 (2002).
71. Somogyi, P., Takagi, H., Richards, J.G. and Mohler, H., *J. Neurosci.* **9**, 2197 (1989).
72. Sassoè-Pognetto, M., Panzanelli, P., Sieghart, W. and Fritschy, J.M., *J. Comp. Neurol.* **420**, 481 (2000).
73. Nusser, Z., Roberts, J.D.B., Baude, A., Richards, J.G., Sieghart, W. and Somogyi, P., *Eur. J. Neurosci.* **7**, 630 (1995).
74. Somogyi, P., Fritschy, J.M., Benke, D., Roberts, J.D.B. and Sieghart, W., *Neuropharmacology* **35**, 1425 (1996).
75. Waldvogel, H.J., Kubota, Y., Trevallyan, S.C., Kawaguchi, Y., Fritschy, J.M., Mohler, H. and Faull, R.L., *Neurosci.* **80**, 775 (1997).
76. Somogyi, P., in: *Neural mechanisms of visual perception*, edited by D.K.T. Lam and C.D. Gilbert, (Portfolio Publishing Co, Houston, 1989) p. 35.
77. Giustetto, M., Kirsch, J., Fritschy, J.M., Cantino, D. and Sassoè-Pognetto, M., *J. Comp. Neurol.* **395**, 231 (1998).
78. Nusser, Z., Sieghart, W., Stephenson, F.A. and Somogyi, P., *J. Neurosci.* **16**, 103 (1996).
79. Nusser, Z., Sieghart, W. and Somogyi, P., *J. Neurosci.* **18**, 1693 (1998).
80. Nusser, Z., Ahmad, Z., Tretter, V., Fuchs, K., Wisden, W., Sieghart, W. and Somogyi, P., *Eur. J. Neurosci.* **11**, 1685 (1999).

81. Mody, I., *Neurochem. Res.* **26**, 907 (2001).
82. Koulen, P., Brandstatter, J.H., Enz, R., Bormann, J. and Wassle, H., *Eur. J. Neurosci.* **10**, 115 (1998).
83. Macdonald, R.L. and Olsen, R.W., *Annu. Rev. Neurosci.* **17**, 569 (1994).
84. Mohler, H., in: *Pharmacology of GABA and Glycine Neurotransmission, Handbook Expt. Pharmacol., vol. 150*, edited by H. Mohler, (Springer-Verlag, Berlin Heidelberg, 2000) p. 101.
85. Whiting, P.J., Bonnert, T.P., McKernan, R.M., Farrar, S., Le Bourdelles, B., Heavens, R.P., Smith, D.W., Hewson, L., Rigby, M.R., Sirinathsinghji, D.J., Thompson, S.A. and Wafford, K.A., *Ann. NY Acad. Sci.* **868**, 645 (1999).
86. Brun, V.H., Otnaess, M.K., Molden, S., Steffenach, H.A., Witter, M.P., Moser, M.B. and Moser, E.I., *Science* **296**, 2243 (2002).
87. Engel, A.K., Fries, P. and Singer, W., *Nature Rev. Neurosci.* **2**, 704 (2001).
88. Harris, K.D., Henze, D.A., Hirase, H., Leinekugel, X., Dragoi, G., Czurko, A. and Buzsaki, G., *Nature* **417**, 738 (2002).
89. Metha, M.R., Lee, A.K. and Wilson, M.A., *Nature* **417**, 741 (2002).
90. O'Keefe, J. and Nadel, L., The hippocampus as a cognitive map, Clarendon Press, Oxford (1978).
91. O'Keefe, J. and Recce, M.L., *Hippocampus* **3**, 317 (1993).
92. Paulsen, O. and Moser, E.I., *Trends Neurosci.* **21**, 273 (1998).
93. Skaggs, W.E., McNaughton, B.L., Wilson, M.A. and Barnes, C.A., *Hippocampus* **6**, 149 (1996).
94. Traub, R.D., Draguhn, A., Whittington, M.A., Baldeweg, T., Bibbig, A., Buhl, E.H. and Schmitz, D., *Rev. Neurosci.* **13**, 1 (2002).
95. Buzsaki, G. and Chrobak, J.J., *Curr. Opin. Neurobiol.* **5**, 504 (1995).
96. Miles, R., *Science* **287**, 244 (2000).
97. Klausberger, T., Magill, P.J., Marton, L.F., Roberts, J.D.B., Cobden, P.M., Buzsaki, G. and Somogyi, P., *Nature* **421**, 844 (2003).
98. Burgess, N., Maguire, E.A. and O'Keefe, J., *Neuron* **35**, 625 (2002).
99. Csicsvari, J., Hirase, H., Czurko, A., Mamiya, A. and Buzsaki, G., *J. Neurosci.* **19**, 274 (1999).
100. Maccaferri, G., Roberts, J.D.B., Szucs, P., Cottingham, C.A. and Somogyi, P., *J. Physiol.* **524**, 91 (2000).
101. Morris, R.G.M., *J. Neurosci.* **9**, 3040 (1989).
102. Morris, R.G.M., Anderson, A., Lynch, G.S. and Baudry, M., *Nature* **319**, 774 (1986).
103. Davis, S., Butcher, S.P. and Morris, R.G.M., *J. Neurosci.* **12**, 21 (1992).
104. McHugh, T.J., Blum, K., Tsien, J.Z., Tonegawa, S. and Wilson, M.A., *Cell* **87**, 1339 (1996).
105. Tsien, J.Z., Huerta, P.T. and Tonegawa, S., *Cell* **87**, 1327 (1996).
106. Huerta, P.T., Sun, L.D., Wilson, M.A. and Tonegawa, S., *Neuron* **25**, 473 (2000).
107. Nakazawa, K., Quirk, M.C., Chitwood, R.A., Watanabe, M., Yeckel, M.F., Sun, L.D., Kato, A., Carr, C.A., Johnston, D., Wilson, M.A. and Tonegawa, S., *Science* **297**, 211 (2002).
108. Tang, Y.P., Shimizu, E., Dube, G.R., Rampon, C., Kerchner, G.A., Zhuo, M., Liu, G. and Tsien, J.Z., *Nature* **401**, 63 (1999).
109. Gupta, A., Wang, Y. and Markram, H., *Science* **287**, 273 (2000).
110. Fricker, D. and Miles, R., *Neuron* **32**, 771 (2001).
111. Collinson, N., Kuenzi, F.M., Jarolimek, W., Maubach, K.A., Cothliff, R., Sur, C., Smith, A.D., Otu, F.M., Howell, O., Atack, J.R., McKernan, R.M., Seabrook, G.R., Dawson, G.R., Whiting, P.J. and Rosahl, T.W., *J. Neurosci.* **22**, 5572 (2002).
112. Whiting, P.J., *Drug Discov. Today* **in press** (2003).
113. Alger, B.E. and LeBeau, F.E.N., in: *Pharmacology of GABA and glycine neurotransmission. Handbook Expt. Pharmacol. vol. 150*, edited by H. Möhler, (Springer, New York, 2001) p. 3.
114. Nusser, Z. and Mody, I., *J. Neurophysiol.* **87**, 2624 (2002).
115. Cobb, S.R., Buhl, E.H., Halasy, K., Paulsen, O. and Somogyi, P., *Nature* **378**, 75 (1995).
116. Whittington, M.A., Traub, R.D. and Jefferys, J.G.R., *Nature* **373**, 612 (1995).
117. Wallenstein, G., Eichenbaum, H. and Hasselmo, M., *Trends in Neurosci* **21**, 317 (1998).
118. Tamas, G., Buhl, E.H., Lorincz, A. and Somogyi, P., *Nature Neurosci.* **3**, 366 (2000).
119. Galarretta, M. and Hestrin, S., *Nature* **402**, 72 (1999).
120. Gibson, J.R., Beierlein, M. and Connors, B.W., *Nature* **402**, 75 (1999).
121. Brickley, S.G., Cull-Candy, S.G. and Farrant, M., *J. Physiol.* **497**, 753 (1996).
122. Faull, R.L.M., Waldvogel, H.J., Nicholson, L.F.B. and Synek, B.J.L., *Prog. Brain Res.* **99**, 105 (1993).
123. Brickell, K.L., Nicholson, L.F.B., Waldvogel, H.J. and Faull, R.L.M., *J. Chem. Neuroanat.* **17**, 75 (1999).
124. Fujiyama, F., Stephenson, F.A. and Bolam, J.P., *Eur. J. Neurosci.* **15**, 1961 (2002).
125. Baulac, S., Huberfeld, G., Gourfinkel-An, I., Mitropoulou, G., Beranger, A., Prud'homme, J.F., Baulac, M., Brice, A., Bruzzone, R. and LeGuern, E., *Nature Genet.* **28**, 46 (2001).
126. Wallace, R., Marini, C., Petrou, S., Harkin, L.A., Bowser, D.N., Panchal, R.G., Williams, D.A., Sutherland, G.R., Mulley, J.C., Scheffer, I.E. and Berkovic, S.F., *Nature Genet.* **28**, 49 (2001).

127. Marini, C., Harkin, L.A., Wallace, R.H., Mulley, J.C., Scheffer, I.E. and Berkovic, S.F., *Brain* **126**, 230 (2003).
128. Bowser, D.N., Wagner, D.A., Czajkowski, C., Cromer, B.A., Parker, M.W., Wallace, R.H., Harkin, L.A., Mulley, J.C., Marini, C., Berkovic, S.F., Williams, D.a., Jones, M.V. and Petrou, S., *Proc. Natl. Acad. Sci. USA* **99**, 15170 (2002).
129. Kananura, C., Haug, K., Sander, T., Runge, U., Gu, W., Hallmann, K., Rebstock, J., Heils, A. and Steinlein, O.K., *Arch. Neurol.* **59**, 1137 (2002).
130. Harkin, L.A., Bowser, D.N., Dibbens, L.M., Singh, R., Phillips, F., Wallace, R.H., Richards, M.C., Williams, D.A., Mulley, J.C., Berkovic, S.F., Scheffer, I.E. and Petrou, S., *Amer. J. Hum. Genet.* **70**, 530 (2002).
131. Cossette, P., Liu, L., Brisebois, K., Dong, H., Lortie, A., Vanasse, M., Saint-Hilaire, J.M., Carmant, L., Verner, A., Lu, W.Y., Wang, Y.T. and Rouleau, G.A., *Nature Genet.* **31**, 184 (2002).
132. Gambardella, A., Manna, I., Labate, A., Chifari, R., La Russa, A., Serra, P., Cittadella, R., Bonavita, S., Andreoli, V., LePiane, E., Sasanelli, F., Di Costanzo, A., Zappia, M., Tedeschi, G., Aguglia, U. and Quattrone, A., *Neurol.* **60**, 560 (2003).
133. Jones-Davis, D.M. and Macdonald, R.L., *Cur. Opin. Pharmacol.* **3**, 12 (2003).
134. Savic, I., Roland, P., Sedvall, G., Persson, A., Pauli, S. and Widen, L., *Lancet* **2**, 863 (1988).
135. Debets, R.M., Sadzot, B., van Isselt, J.W., Brekelmans, G.J.F., Meiners, L.C., van Huffelen, A.C., Franck, G. and van Veelen, C.W.M., *J. Neurol.* **62**, 141 (1997).
136. Hammers, A., Koepp, M.J., Hurlemann, R., Thom, M., Richardson, M.P., Brooks, D.J. and Duncan, J.S., *Brain* **125**, 2257 (2002).
137. Sata, Y., Matsuda, H., Mihara, T., Aihara, M., Yagi, K. and Yonekura, Y., *Epilepsia* **43**, 1039 (2002).
138. Loup, F., Wieser, H.G., Yonekawa, Y., Aguzzi, A. and Fritschy, J.M., *J. Neurosci.* **20**, 5401 (2000).
139. Cohen, I., Navarro, V., Clemenceau, S., Baulac, M. and Miles, R., *Science* **298**, 1418 (2002).
140. Brooks-Kayal, A.R., Shumate, M.D., Jin, H., Rikhter, T.Y. and Coulter, D.A., *Nature Med.* **4**, 1166 (1998).
141. Schwarzer, C., Tsunashima, K., Wanzenbock, C., Fuchs, K., Sieghart, W. and Sperk, G., *Neurosci.* **80**, 1001 (1997).
142. Fritschy, J.M., Kiener, T., Bouilleret, V. and Loup, F., *Neurochem. Int.* **34**, 435 (1999).
143. Riban, V., Bouilleret, V., Pham-Lè, B.T., Fritschy, J.M., Marescaux, C. and Depaulis, A., *Neurosci.* **112**, 101 (2002).
144. Bouilleret, V., Ridoux, V., Depaulis, A., Marescaux, C., Nehlig, A. and Le Gal La Salle, G., *Neurosci.* **89**, 717 (1999).
145. Bouilleret, V., Loup, F., Kiener, T., Marescaux, C. and Fritschy, J.M., *Hippocampus* **10**, 305 (2000).
146. Knuesel, I., Zuellig, R.A., Schaub, M.C. and Fritschy, J.M., *Eur. J. Neurosci.* **13**, 1113 (2001).
147. Benes, F.M. and Berretta, S., *Neuropsychopharmacol.* **25**, 1 (2001).
148. Volk, D.W. and Lewis, D.A., *Physiol. Behav.* **77**, 501 (2002).
149. Heckers, S. and Konradi, C., *J. Neural Transm.* **109**, 891 (2002).
150. Heckers, S., Stone, D., Walsh, J., Shick, J., Koul, P. and Benes, F.M., *Arch. Gen. Psychiatry* **59**, 521 (2002).
151. Cotter, D., Landau, S., Beasley, C., Stevenson, R., Chana, G., Macmillan, L. and Everall, I., *Biol. Psychiatry* **51**, 377 (2002).
152. Woo, T.U., Whitehead, R.E., Melchitzky, D.S. and Lewis, D.A., *Proc. Natl. Acad. Sci. USA* **95**, 5341 (1998).
153. Volk, D.W., Pierri, J.N., Fritschy, J.M., Auh, S., Sampson, A.R. and Lewis, D.A., *Cereb. Cortex* **12**, 1063 (2002).

14

LTD, SPIKE TIMING AND SOMATOSENSORY BARREL CORTEX PLASTICITY

Daniel E. Feldman, Cara B. Allen, Tansu Celikel [*]

1. INTRODUCTION

Evidence for LTP and LTD in Sensory Cortical Map Plasticity

Sensory experience produces long-lasting changes in receptive fields and sensory maps in primary sensory cortical areas[1]. A dominant hypothesis has been that experience-dependent plasticity is mediated in part by long-term potentiation (LTP) and long-term depression (LTD) at intracortical synapses. These phenomena are attractive as candidate mechanisms for plasticity *in vivo* because they instantiate Hebbian learning rules[2,3], which can explain many[1,4], though not all[5], features of map plasticity.

Evidence for LTP in cortical map plasticity is based on three major lines of evidence. First, many cortical synapses exhibit N-methyl-D-aspartate receptor (NMDAR)-dependent LTP *in vitro*, and in response to electrical stimulation *in vivo*[6-11,76]. Second, pharmacological and genetic blockade of NMDA receptors and downstream signalling pathways can block cortical map plasticity[12-18]. Third, deprivation-induced map plasticity and learning cause detectable, LTP-like strengthening of specific cortical synapses[19-22].

Evidence also suggests a role for LTD in map plasticity. LTD has been hypothesized to drive a basic feature of map plasticity—the specific, activity-dependent reduction in sensory responses to deprived or behaviorally irrelevant stimuli[1,23]. Consistent with this hypothesis, many cortical synapses exhibit robust LTD in response to electrical stimulation *in vitro*[8,10,24-33] and *in vivo*[34,77]. In addition, sensory deprivation causes detectable, LTD-like weakening of specific synapses in primary somatosensory

---

[*]D.E. Feldman, C.B. Allen, T. Celikel. Neurobiology Section, Division of Biological Sciences, University of California San Diego, 9500 Gilman Dr., La Jolla CA 92093-0357 USA

(S1) cortex[35], and LTD-like biochemical changes in primary visual (V1) cortex[36]. However, pharmacological and genetic manipulations that block specific forms of LTD have thus far not blocked map plasticity[37-39], suggesting that multiple forms of LTD, or other mechanisms besides LTD, may contribute to experience-driven synaptic weakening *in vivo*.

Though theoretical models demonstrate that LTP and LTD can explain basic features of map plasticity[4,40], we currently lack a detailed understanding of how LTP and LTD, acting in real cortical circuits, mediate specific aspects of experience-dependent plasticity. In our view, a full understanding requires answering four questions: (1) Which specific cortical synapses undergo LTP and LTD during map plasticity *in vivo*? (2) How, mechanistically, does experience drive induction of LTP and LTD at appropriate synapses? (3) How do LTP and LTD at these synapses cause the characteristic receptive field changes that constitute map plasticity? (4) How do other plasticity mechanisms, like changes in anatomical connectivity[41,42] and short-term synapse dynamics[19] contribute to map plasticity? In this review, we summarize recent progress in answering these questions, using the whisker map in the rat's somatosensory cortex as a model system.

Whisker Map Plasticity in Barrel Cortex

The rat somatosensory (S1 or barrel) cortex is a powerful system for studying the cellular mechanisms of plasticity *in vivo* and *in vitro*. Each facial whisker is represented in S1 by a cluster of cells, called a barrel, in cortical layer 4 (L4). The barrels are arranged in a map isomorphic with the arrangement of facial whiskers. Cells in the cortical column centered on each barrel (the "barrel column") have whisker receptive fields in which the strongest responses are generated by deflection of the whisker corresponding to that barrel, termed the principal whisker, and weaker responses are generated by surround whiskers (Fig. 1A). As a result, the set of receptive fields across S1 constitutes a functional map of the whiskers aligned with the anatomical barrel map in L4.

The receptive field map is shaped by sensory experience. When a subset of whiskers is plucked or trimmed in post-neonatal, juvenile rats, neurons in deprived columns (1) rapidly lose responses to their plucked principal whisker (a phenomenon termed "principal whisker response depression") and (2) slowly gain increased responses to neighboring, spared whiskers (termed "spared whisker response potentiation"). These two processes, which are developmentally and genetically separable[17], jointly cause the cortical representation of spared whisker(s) to expand into deprived columns[43] (Fig. 1B). Receptive field changes also occur within spared columns, but are not considered here[44-47]. Plasticity occurs first in layer 2/3 (L2/3), and only later, or not at all in L4, suggesting that plasticity reflects synaptic changes at intracortical synapses[43,44,48,49]. S1 plasticity exhibits similar dynamics, laminar specificity, laminar progression of critical periods, and competition requirements as deprivation-induced plasticity in visual cortex (V1), suggesting that common cellular mechanisms are involved.

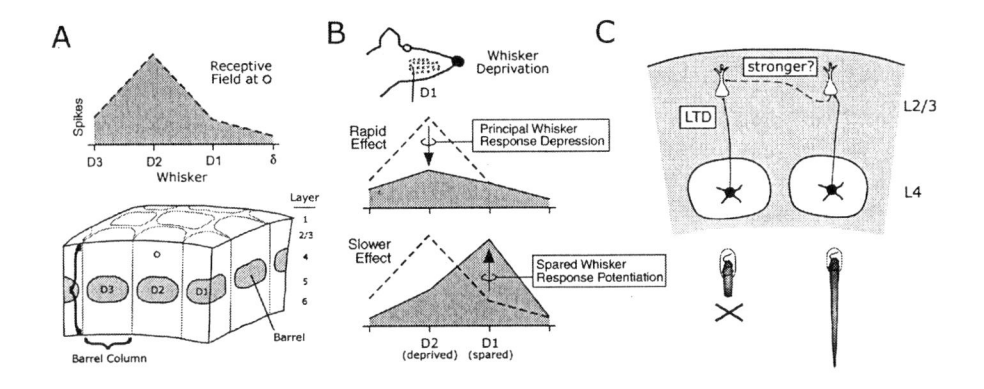

Fig. 1. Whisker receptive field plasticity induced by partial whisker deprivation. A, Arrangement of barrels and barrel columns in S1, and schematic receptive field of L2/3 neuron in the D2 column. Responses are strongest to the D2 (principal) whisker. B, Receptive field changes at the same location as in (A) following deprivation of all whiskers but D1 in an adolescent rat. Dashed lines, control receptive field. Summary of data from Ref. 43. C, Hypothesized sites of LTD and synaptic strengthening (perhaps LTP) that may underly principal whisker response depression and spared whisker response potentiation, respectively.

Because principal whisker response depression occurs primarily in L2/3, we and others have hypothesized that it reflects LTD of excitatory, feedforward pathways from L4 to L2/3, which mediate principal whisker responses in L2/3[43,44,48] (Fig. 1C). In contrast, spared whisker response potentiation may involve strengthening, perhaps LTP, of cross-columnar excitatory projections from spared columns to L2/3 of deprived columns (Fig. 1C). This simple hypothesis omits many important aspects of cortical circuitry (e.g., inhibition, POm-septal circuits, feedback circuits), but provides a useful starting point for investigating the synaptic basis for whisker map plasticity.

Here we focus on the role of LTD at feedforward, L4 to L2/3 synapses in principal whisker response depression during whisker map plasticity. We summarize data showing that these synapses are capable of LTP and LTD *in vitro*, and that LTD occurs at this synapse *in vivo* during deprivation-induced map plasticity. We also discuss initial results suggesting how, mechanistically, whisker deprivation induces LTD at this synapse *in vivo*.

2. LTP AND LTD AT FEEDFORWARD, L4 TO L2/3 SYNAPSES *IN VITRO*.

In vitro, NMDA receptor-dependent LTP and LTD can be induced at feedforward, L4 to L2/3 excitatory synapses by two physiologically realistic means: changes in presynaptic firing rate (rate-dependent induction) and millisecond-scale changes in timing of pre- and postsynaptic spikes, largely independent of firing rate (timing-dependent induction). Rate-based induction has been characterized in V1[6,7,26] and S1[8,35,76]. The rules for rate-based induction are remarkably similar to Schaffer collateral-CA1 hippocampal synapses: very low presynaptic firing rates (≤ 0.1 Hz) induce no plasticity; sustained, low-frequency firing (one to a few Hz) induces LTD; and brief bursts of high frequency firing (50-100 Hz) induce LTP[50,51].

Timing-dependent induction of LTP and LTD (termed "spike timing-dependent plasticity", STDP)[52] has been observed at many synapses[52,53], including several synapses in S1[10,29,30]. At feedforward L4 to L2/3 synapses in S1[10], STDP is induced by pairing single presynaptic spikes and single postsynaptic spikes separated by a fixed delay (Δt). When presynaptic spikes lead postsynaptic spikes by ~1-15 ms, LTP is induced. When postsynaptic spikes lead presynaptic spikes by 0-50 ms, LTD is induced (Fig. 2A, B). These temporal windows for LTP and LTD appear to hold for firing frequencies up to ~ 35 Hz[32].

The precise shape of STDP learning rules varies across synapses[52,53], but all synapses onto neocortical pyramidal cells examined so far have similar STDP rules in which the temporal window for LTD induction is longer than that for LTP induction[10,32,33] (Fig. 2). These rules predict that *in vivo*, two timing conditions could lead to LTD—consistent post-leading-pre firing with $0 \leq \Delta t \leq 50$ ms, and temporally uncorrelated pre- and postsynaptic firing, which will induce more LTD than LTP because of the bias towards LTD in the STDP learning rule. Depression of inputs by uncorrelated firing has been confirmed *in vitro* for firing rates up to 35 Hz[10,32]. This behavior, which is not predicted by rate-dependent induction, may be important for function *in vivo* (see below). Whether LTP and LTD are induced *in vivo* primarily by rate-dependent or timing-dependent mechanisms is not yet clear.

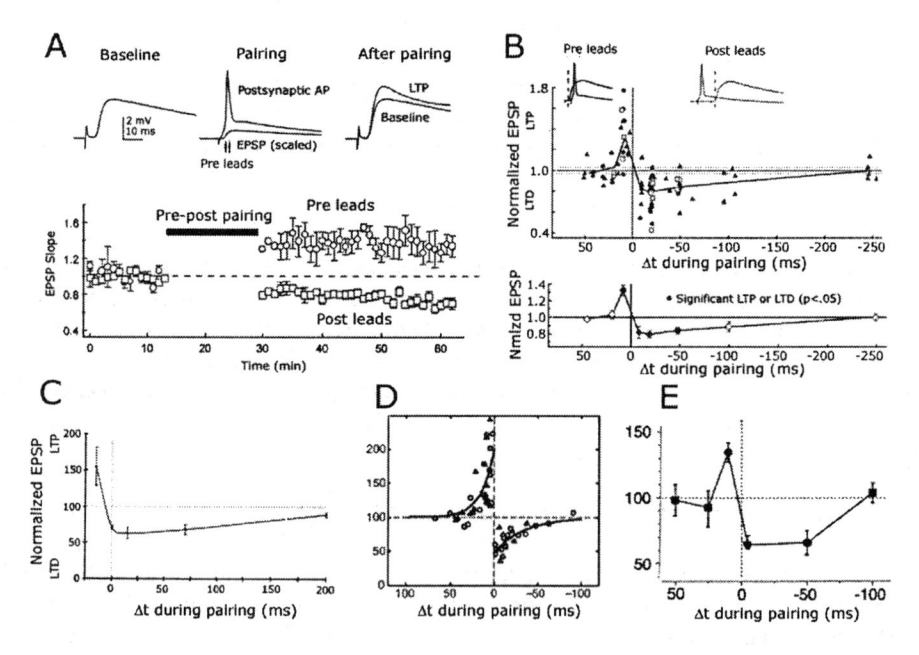

Fig. 2. STDP at feedforward, L4 to L2/3 synapses *in vitro*. A, Timing-dependent LTP and LTD at feedforward L4 to L2/3 synapses in S1. When the presynaptic spike (and the resulting EPSP) leads the postsynaptic spike by a few ms, LTP is induced (insets). When spike order is reversed, LTD is induced. B, Temporal windows for induction of STDP at this synapse. Data from Ref. 10. C-E, Similar STDP rules with long-duration LTD windows at synapses onto CA3 cells in hippocampal slice culture (C; Ref. 59), L2/3 pyramidal neurons in V1 (D; Ref. 33), and between pairs of L5 pyramidal neurons in V1 (E; Ref. 32).

3. LTD IS INDUCED AT FEEDFORWARD L4 TO L2/3 SYNAPSES DURING MAP PLASTICITY IN VIVO

In a recent study[35], we tested whether LTD is induced at feedforward L4 to L2/3 synapses during map plasticity *in vivo*. We took advantage of the fact that experience-induced changes in synaptic physiology can persist and be detected in acute brain slices[19,20,21,54]. Rats were raised with selected rows of whiskers plucked beginning at postnatal day 12 (P12). S1 slices were prepared 10-20 days later. Each slice contained one barrel column representing each of the 5 whisker rows, A-E. Barrels were visualized in living slices and used to guide recordings of feedforward, L4 to L2/3 synaptic inputs in identified columns (Fig. 3).

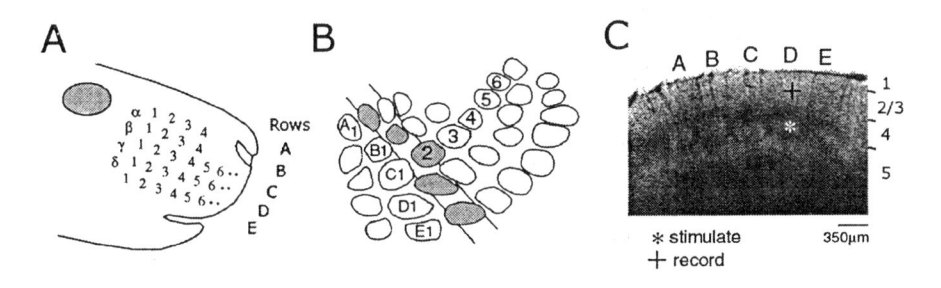

Fig. 3. Oblique slice preparation for measuring deprivation-induced changes in synaptic physiology in S1. A, Whisker rows (A-E) on the rat's face. B, Plane of section relative to whisker barrel map. C, Living slice observed with transillumination showing A-E barrels during a physiology experiment. Stimulation and recording sites for studying the L4 to L2/3 projection are indicated. Modified from Ref. 35.

Using field potentials and whole-cell recording of EPSPs to measure excitatory feedforward synaptic transmission to L2/3, we found that presumptive L4 to L2/3 synaptic responses were significantly weaker in columns representing deprived whiskers, relative to both columns representing spared whiskers in the same slice, and to matched columns in slices from control littermates. In whole-cell recordings, EPSPs in deprived columns were on average 30-40% smaller than EPSPs evoked by equivalent stimulation intensities in spared columns of the same slice. In contrast, EPSP amplitude was constant across columns in control slices[35].

Deprivation did not change the excitability of L2/3 cells[35], consistent with a prior study[19]. Thus, these data show that sensory deprivation drives a functional weakening of feedforward L4 to L2/3 synapses, either by reducing strength of individual synapses, or by reducing synapse number.

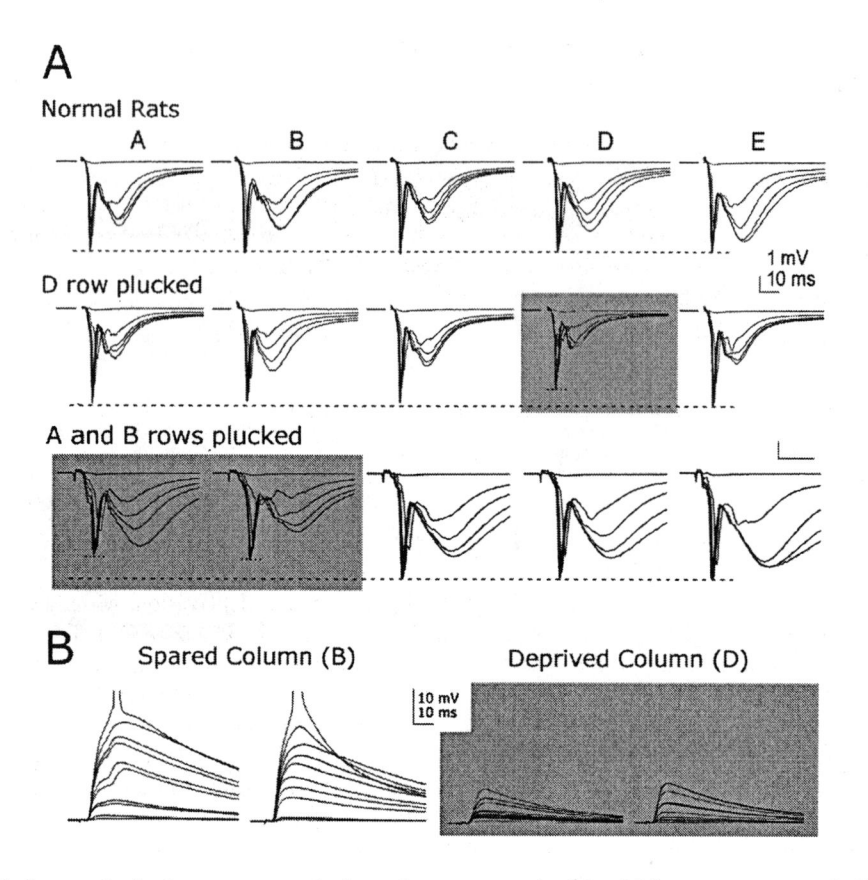

Fig. 4 Sensory deprivation causes synaptic depression at presumptive L4 to L2/3 synapses *in vivo*. A, Field potentials reflecting feedforward L4 to L2/3 synaptic function in representative slices from a control rat, a D-row deprived rat (17 d of deprivation), and an A- and B-row deprived rat (12 d of deprivation). Field potentials were measured in each column (A-E) in response to increasing stimulation intensity. In slices from control rats, the fast negative component of the field potential (which reflects presumptive L4 to L2/3 inputs) exhibited equal amplitude across columns. In slices from deprived rats, this component was significantly smaller in deprived columns. B, Whole-cell EPSPs evoked by increasing stimulation intensities in four L2/3 pyramidal cells. Two cells were in the B column and two cells in the D column of a single slice from a D-deprived rat. For equivalent stimulation intensities, EPSPs were smaller in the deprived column. Data from Ref. 35.

To determine whether deprivation-invoked synaptic weakening represents LTD induced *in vivo*, we performed occlusion experiments. Occlusion is a standard test for previous induction of saturable plasticity such as LTD[21,55-57], and distinguishes between LTD and a deprivation-induced loss of synapse number, which should not occlude LTD *in vitro*. We induced LTD in S1 slices by low-frequency stimulation (LFS; 900 pulses at 1 Hz) in L4. In blind experiments, LFS induced robust LTD in control slices and in spared columns of slices from D row-deprived rats. However, in deprived columns of these same slices, LTD was occluded (Fig. 5A). Deprived columns were not simply deficient in synaptic plasticity, because LTP was enhanced in these columns (Fig. 5B), as expected for synapses that have undergone previous LTD[56,57].

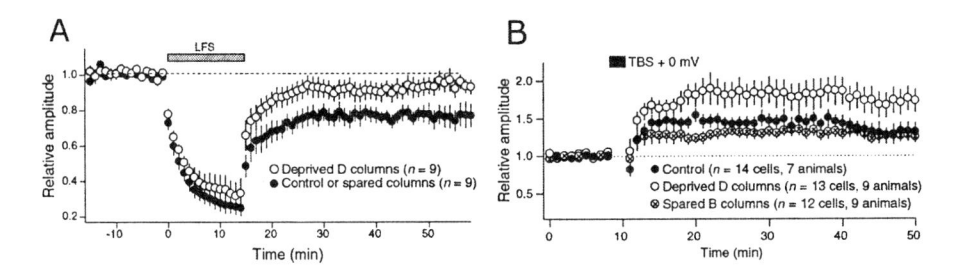

Fig. 5. Deprivation-induced synaptic depression occludes LTD and enhances LTP. A, 1 Hz firing for 15 min (LFS) induces significant LTD in control and spared columns, but significantly less LTD in deprived columns. The difference in short-term depression during LFS was not significant. B, High frequency (theta burst stimulation, TBS) paired with postsynaptic depolarization to 0 mV induces robust LTP in control and spared columns, and enhanced LTP in deprived columns. LTP in control and spared columns was not significantly different. Data from Ref 35.

An alternative interpretation of these data is that whisker deprivation, which began at P12, a time of substantial synaptogenesis in L2/3[58], did not induce LTD at L4 to L2/3 synapses, but instead delayed or prevented normal developmental strengthening of these synapses[22]. To address this issue, we have recently begun D-row whisker deprivation at older ages (P16 and P20), when L4 to L2/3 synapses are substantially more mature[49]. Remarkably, seven days of deprivation at these older ages caused as dramatic a reduction in whole-cell EPSPs as deprivation beginning at P12 (C.B.A., unpublished results). This result indicates that deprivation-induced reduction of synaptic responses does not reflect a failure of development, but rather the active weakening of feedforward L4 to L2/3 synapses by an LTD-like process.

4. SPIKE TIMING VS. SPIKE RATE: HOW DOES WHISKER DEPRIVATION TRIGGER LTD IN VIVO?

How, mechanistically, does sensory experience or deprivation trigger LTP or LTD *in vivo*? LTP and LTD can be induced *in vitro* using rate-dependent and timing-dependent modes of induction (see above), and also by less natural manipulations that directly set postsynaptic calcium levels[31,51]. *In vivo*, the natural mode of induction may depend on the precise firing patterns (and neuromodulatory state) established by sensory experience within the local network. For example, stimuli that modulate mean firing rate into or out of the ~1 Hz and 50-100 Hz ranges are likely to drive rate-dependent induction of LTP and LTD. This basic model, with the additional feature of history-dependent metaplasticity, successfully explains deprivation-induced plasticity in V1 and S1[4,40,60]. In contrast, stimuli that alter spike timing (e.g., Ref. 61) may drive map plasticity via timing-dependent induction of LTP and LTD. In support of this view, carefully timed sensory and/or electrical stimuli can induce receptive field and perceptual plasticity with STDP-like temporal requirements, suggesting that STDP is induced by precisely timed sensory stimulation *in vivo*[62-65]. Somewhat less intuitively, STDP could also drive plasticity in response to sensory deprivation, provided that deprivation altered spike timing[10].

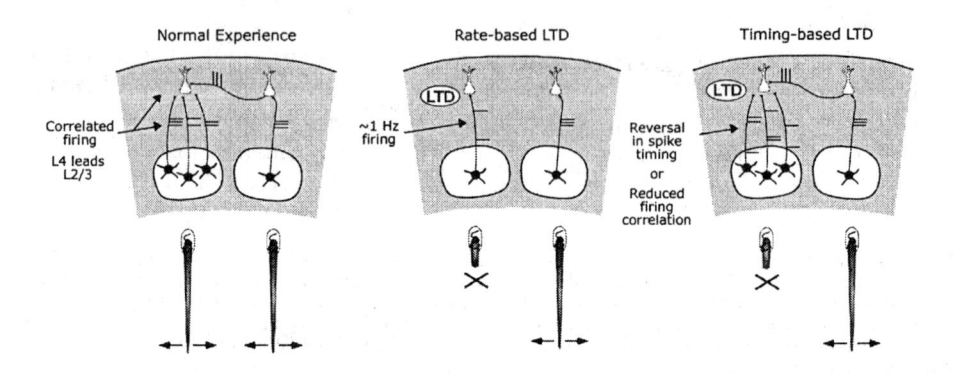

Fig. 6. Rate-based and timing-based hypotheses for induction of LTD in vivo. Left, normal whisker use drives highly correlated firing in L4 and L2/3 of all active barrel columns, with L4 firing slightly leading L2/3 firing (Ref. 67). Middle, In a rate-based hypothesis for LTD induction, whisker deprivation leads to a change in mean firing rate of L4 cells in the deprived column, so that these cells begin to fire at ~ 1 Hz. Right, In a timing-based hypothesis for LTD induction, whisker deprivation alters millisecond-scale firing correlations between L4 and L2/3 neurons. LTD would be induced if either L2/3 cells fire regularly before L4 cells (reversal in spike timing), or if firing becomes significantly less correlated in time (reduced firing correlation).

We have begun to investigate the natural mode of induction of LTP and LTD *in vivo*, focusing on how whisker deprivation induces LTD at the feedforward L4 to L2/3 synapse in S1. In this system, a rate-dependent model would require that whisker deprivation causes L4 neurons in deprived columns to fire at ~1 Hz. A spike timing-dependent model would require that whisker deprivation either causes consistent post-leads-pre firing (with $0 \le \Delta t \le 50$ ms), or causes pre- and postsynaptic firing to become temporally uncorrelated, which also drives timing-dependent LTD at this synapse[10] (see above). These hypotheses are diagrammed in Fig. 6. Timing-based plasticity may be particularly plausible in S1 because sensory coding is quite sparse[66,67], suggesting that maximal firing rates are low, even during optimal stimulation. In addition, spike timing during normal sensory responses is quite precise[68].

To distinguish these models, we tested whether acute whisker deprivation altered spike rate, spike timing, or both, at the L4 to L2/3 synapse *in vivo*. We simultaneously recorded single-unit spike trains from L4 and L2 neurons in the same barrel column in urethane-anesthetized rats[35]. We mimicked normal whisker use by moving all contralateral whiskers together using a piezoelectric actuator. We mimicked acute whisker deprivation by trimming the principal whisker until it just escaped being moved by the actuator, while continuing to deflect the other whiskers.

When all contralateral whiskers were deflected in unison, pronounced L4-before-L2 firing occurred in virtually all cell pairs. An example cell pair is shown in Fig. 7. The modal Δt between pre- and postsynaptic spikes, measured as the peak of the cross-correlogram, is similar to that predicted from classical latency studies[67]. During acute whisker deprivation, this cell pair, like many in our sample, showed an unexpected, dramatic effect—firing order reversed, so that the L2 cell (the postsynaptic cell) consistently fired a few milliseconds before the L4 cell (the presynaptic cell). This effect recovered immediately when all whiskers were deflected together again. Thus, principal whisker deprivation acutely caused dramatic changes in spike timing. When all whiskers were in use, spike timing matched that expected for a feedforward system (i.e., pre-leads-post firing), and was appropriate to drive LTP. When the principal whisker was deprived, spike timing reversed and was appropriate to drive LTD. The reversal in spike timing is likely to derive from the different dynamics of pathways mediating principal whisker vs. surround whisker responses[67].

To determine whether deprivation also caused changes in presynaptic spike rate appropriate to drive rate-dependent LTD, we measured the number of evoked spikes per stimulus for single L4 units. For five L4 cells recorded at a representative site (the same L4 site as for the cross-correlogram data), principal whisker deprivation caused a loss of about 40% of whisker-evoked spikes (Fig. 7). Thus, in an awake behaving animal, principal whisker deprivation would be predicted to lower mean L4 firing rates. Whether the resulting firing rates would be in range to induce LTD depends on assumptions about baseline firing rates in behaving animals, for which there is little data currently.

Together, these data demonstrate that whisker deprivation alters spike timing *in vivo* in a manner appropriate to drive spike timing-dependent LTD at L4 to L2/3 synapses. In addition, spike rates are reduced, which may drive rate-dependent LTD as well.

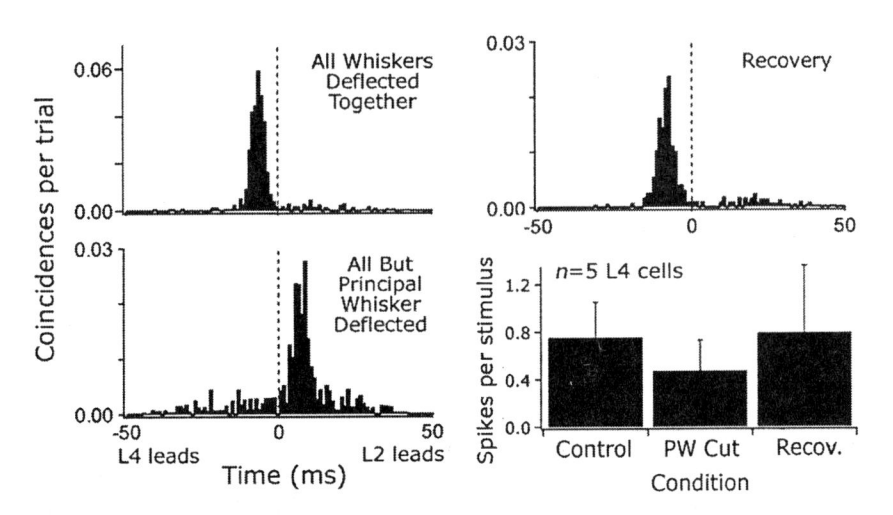

Fig. 7. Acute changes in spike timing and spike counts that occur with principal whisker deprivation in vivo. Cross-correlograms show relative timing of spikes for a representative pair of simultaneously recorded L4 and L2 units in the D2 barrel column. When all contralateral whiskers were deflected together (1° upward ramp-and-hold movements, 900 trials 1 sec apart), L4 spikes preceded L2 spikes by a mean of 7 ms. During acute principal whisker deprivation mimicked by moving all but the D2 whisker in unison, spike timing reversed, so that L2 firing preceded L4 firing. When all whiskers were deflected together again, firing order recovered. Histogram shows mean spikes per stimulus for on-responses for units at one L4 recording site. From Ref. 35.

5. CONCLUDING REMARKS

Bidirectional Plasticity at L4→L2/3 Synapses As a Mechanism for Sensory Map Plasticity

These results demonstrate that L4 to L2/3 excitatory synapses are a site of synaptic change that is likely to contribute to whisker map plasticity in S1. Plasticity at L4 to L2/3 synapses is well-suited to drive map plasticity in post-neonatal animals, where reorganization occurs primarily or most rapidly in L2/3 relative to L4[43,44,49,69]. Consistent with our results, sensory experience strengthens L4 to L2/3 synapses in S1 by LTP-like insertion of AMPA receptors[22], and sensory deprivation weakens these synapses in V1 by LTD-like mechanisms[36]. Thus, L4 to L2/3 synapses exhibit bidirectional synaptic plasticity in response to sensory experience.

Spike Timing versus Firing Rate in Induction of LTP and LTD in vivo

Central to a full understanding of map plasticity is a mechanistic description of how LTP and LTD are induced by sensory experience at appropriate synapses *in vivo*. Despite substantial work *in vitro*, little is known about how LTP and LTD are induced by natural stimuli *in vivo*. One approach to this problem is to measure the natural firing patterns that occur *in vivo* during plasticity-inducing sensory stimulation, since these patterns are presumably the proximal stimuli that drive induction of activity-dependent synaptic plasticity[2,3]. Using this approach, we found that whisker deprivation induces acute changes both in evoked spike count and spike timing. The spike timing changes that occur with whisker deprivation are demonstrably appropriate to drive timing-dependent LTD at L4 to L2/3 synapses. The changes in spike count, which imply changes in mean firing rates during whisker use, may be appropriate to drive LTD, depending on critical assumptions about baseline firing rates *in vivo*. Thus, these data suggest that sensory deprivation is likely to drive plasticity, at least in part, by regulating spike timing.

The predominance of timing-dependent vs. rate-dependent plasticity *in vivo* may vary across synapses, based on several factors. One obvious factor is whether the sensory stimuli that induce plasticity strongly modulate firing rate at relevant synapses. Such firing rate modulation will occur in sensory regions that use rate coding therefore rate-dependent plasticity may be prominent in these areas. In regions that primarily use temporal coding, timing-based plasticity may predominate[70].

A second factor is the extent of convergence in the circuit. Where convergence onto postsynaptic cells is low and feedforward synapses are strong (e.g., at the retinogeniculate synapse[71]), pre-leading-post spike timing may occur regardless of the pattern of sensory input. In this case, timing-based LTD may be impossible, and rate-dependent plasticity may predominate. In contrast, where convergence of functionally independent inputs onto single postsynaptic cells is high, and individual synapses are weak, spike timing at any one synapse may vary substantially with sensory input patterns, and therefore experience may readily drive timing-dependent LTP and LTD. An outright speculation is that these conditions for timing-based plasticity may be more common at higher levels in sensory hierarchies, where convergence of functionally independent inputs may be more extensive.

Cortical Inhibition and Timing-Dependent Plasticity

Inhibitory cortical circuits exert tremendous control over plasticity processes, and may do so by regulating STDP. Inhibitory regulation of STDP could occur in at least two ways. First, because feedforward inhibition is prominent in S1[72] and potently regulates spike timing[73], mature levels of inhibition may be required to produce appropriate spike timing to enable timing-dependent plasticity. Consistent with this idea, undeveloped or disrupted inhibition causes a "prolonged discharge" phenotype in V1[74] and disrupts experience-dependent plasticity. Prolonged discharge would radically alter spike timing relationships, and may prevent these relationships from being appropriately modulated by sensory manipulations.

Excessive inhibition may also be expected to impair STDP, because as inhibitory tone increases, and overall firing rates of excitatory cells decrease, STDP and other forms of spiking-dependent synaptic plasticity will accrue at a lower rate. In the extreme, as in the case of muscimol application to cortex, STDP should be completely absent, and only spiking-independent forms of synaptic plasticity should occur[75].

ACKNOWLEDGMENTS

This work was supported by the McKnight Endowment Fund for Neuroscience, and by research grant # 5FY01-495 from the March of Dimes Birth Defects Foundation. D. E. F. is an Alfred P. Sloan Research Fellow.

REFERENCES

1. Buonomano, D.V. and Merzenich, M.M. (1998) *Annu. Rev. Neurosci.* **21**, 149.
2. Hebb, D.O. (1949) *The Organization of Behavior.* New York: Wiley.
3. Stent, G.S. (1973) *Proc. Natl. Acad. Sci. USA* **70**, 997.
4. Bear, M.F. Cooper, L.N. and Ebner, F.F. (1987) *Science* **237**, 42.
5. Turrigiano, G.G. and Nelson, S.B. (2000) *Curr. Opin. Neurobiol.* **10**, 358.
6. Artola, A. and Singer, W. (1987) *Nature* **330**, 649.
7. Kirkwood, A. and Bear, M.F. (1994) *J. Neurosci.* **14**, 1634.
8. Castro-Alamancos, M.A., Donoghue, J.P. and Connors, B.W. (1995) *J. Neurosci.* **15**, 5324.
9. Crair, M.C. and Malenka, R.C. (1995), *Nature* **375**, 325.
10. Feldman, D.E. (2000) *Neuron* **27**, 45.
11. Heynen, A.J. and Bear, M.F. (2001) *J. Neurosci.* **21**, 9801.
12. Bear, M.F., Kleinschmidt, A., Gu, QA and Singer, W. (1990) *J. Neurosci.* **10**, 909.
13. Garraghty, P.E. and Muja, N. (1996) *J. Comp. Neurol.* **367**, 319.
14. Roberts, E.B., Meredith, M.A. and Ramoa, A.S. (1998) *J. Neurophysiol.* **80**, 1021.
15. Rema V., Armstrong-James, V. M. and Ebner, F.F. (1998) *J. Neurosci.* **18**, 10196.
16. Daw, N.W., Gordon, B., Fox, K.D., Flavin, H.J., Kirsch, J.D., Beaver, C.J., Ji, Q., Reid, S.N. and Czepita, D. (1999) *J. Neurophysiol.* **81**, 204.
17. Glazewski, S., Giese, K.P., Silva, A. and Fox, K. (2000) *Nat. Neurosci.* **3**, 911.
18. Taha, S., Hanover, J.L., Silva, A.J. and Stryker, M.P. (2002) *Neuron* **36**, 483.
19. Finnerty, G.T., Roberts, L.S. and Connors, B.W. (1999) *Nature* **400**, 367.
20. Rioult-Pedotti, M.S., Friedman, D., Hess, G. and Donoghue, J.P. (1998) *Nat. Neurosci.* **1**, 230.
21. Rioult-Pedotti, M.S., Friedman, D. and Donoghue, J.P. (2000) *Science* **290**, 533.
22. Takahashi, T., Svoboda, K., Malinow, R. (2003) *Science* **299**, 1585.

23. T.N. Wiesel and D.H. Hubel (1965) *J. Neurophysiol.* **28**, 1029.
24. Tsumoto, T. (1992) *Prog. Neurobiol.* **39**, 209.
25. Artola, A., Brocher, S. and Singer, W. (1990) *Nature* **347**, 69.
26. Kirkwood, A. and Bear, M.F. (1994) *J. Neurosci.* **14**, 3404.
27. Hess, G. and Donoghue, J.P. (1996) *Eur. J. Neurosci.* **8**, 658.
28. Dudek, S.M. and Friedlander, M.J. (1996) *Neuron* **16**, 1097.
29. Markram, H., Lubke, J., Frotscher, M. and Sakmann, B. (1997) *Science* **275**, 213.
30. Egger, V., Feldmeyer, D. and Sakmann, B. (1999) *Nat. Neurosci.* **2**, 1098.
31. Feldman, D.E., Nicoll, R.A., Malenka, R.C. and Isaac, J.T. (1998) *Neuron* **21**, 347.
32. Sjostrom, P.J., Turrigiano, G.G. and Nelson, S.B. (2001) *Neuron* **32**, 1149.
33. Froemke, R.C. and Dan, Y. (2002) *Nature* **416**, 433.
34. Heynen, A.J., Abraham, W.C. and Bear, M.F. (1996) *Nature* **381**, 163.
35. Allen, C.B., Celikel, T. and Feldman, D.E. (2003) *Nat. Neurosci.* **6**, 291.
36. Yoon, B.J., Heynen, A.J., Liu, C-H., Chung, H., Huganir, R.L. and Bear, M.F. (2002) *Soc. Neurosc. Abstracts* 647, 12.
37. Hensch, T.K. and Stryker, M.P. (1996) *Science* **272**, 554.
38. Hensch, T.K., Gordon, J.A., Brandon, E.P., McKnight, G.S., Idzerda, R.L. and Stryker, M.P. (1998) *J. Neurosci.* **18**, 2108.
39. Renger, J.J., Hartman, K.N., Tsuchimoto, Y., Yokoi, M., Nakanishi, S. and Hensch, T.K. (2002) *Proc. Natl. Acad. Sci. USA* **99**, 1041.
40. Benuskova, L., Diamond, M.E., Ebner, F.F. (1994) *Proc. Natl. Acad. Sci. USA* **91**, 4791.
41. Antonini, A. and Stryker, M.P. (1993) *Science* **260**, 1819.
42. Darian-Smith, C. and Gilbert, C.D. (1994) *Nature* **368**, 737.
43. Glazewski, S. and Fox, K. (1996) *J. Neurophysiol.* **75**, 1714.
44. Diamond, M.E., Huang, W. and Ebner, F.F. (1994) *Science* **265**, 1885.
45. Armstrong-James, M., Diamond, M.E. and Ebner, F.F. (1994) *J. Neurosci.* **14**, 6978.
46. Lebedev, M.A., Mirabella, G., Erchova, I. and Diamond, M.E. (2000) *Cereb. Cortex* **10**, 23.
47. Wallace, H., Glazewski, S., Liming, K. and Fox, K. (2001) *J. Neurosci.* **21**, 3881.
48. Glazewski, S., McKenna, M., Jacquin, M. and Fox, K. (1998) *Eur. J. Neurosci.* **10**, 2107.
49. Stern, E.A., Maravall, M. and Svoboda, K. (2001) *Neuron* **31**, 305.
49. Stern, E.A., Maravall, M. and Svoboda, K. (2001) *Neuron* **31**, 305.
50. Kirkwood, A., Dudek, S.M., Gold, J.T., Aizenman, C.D. and Bear, M.F. (1993), *Science* **260**, 1518.
51. Bliss, T.V. and Collingridge, G.L. (1993) *Nature* **361**, 31.
52. Abbott, L.F. and Nelson, S.B. (2000) *Nat. Neurosci.* **3**, 1178.
53. Bi, G. and Poo, M. (2001) *Annu. Rev. Neurosci.* **24**, 139.
54. McKernan, M.G. and Shinnick-Gallagher, P. (1997) *Nature* **390**, 6.
55. Lebel, D., Grossman, Y. and Barkai, E. (2001) *Cereb. Cortex* **11**, 485.
56. Dudek, S.M. and Bear, M.F. (1993) *J. Neurosci.* **13**, 2910.
57. Mulkey, R.M., Herron, C.E. and Malenka, R.C. (1993) *Science* **261**, 1051.
58. Lendvai, B., Stern, E.A., Chen, B. and Svoboda, K. (2002) *Nature* **404**, 876.
59. Debanne, D., Gahwiler, B.H. and Thompson, S.M. (1997) *J. Neurophysiol.* **77**, 2851.
60. Bienenstock, E.L., Cooper, L.N. and Munro, P.W. (1982) *J. Neurosci.* **2**, 32.
61. Wang, X., Merzenich, M.M., Sameshima, K. and Jenkins, W.M. (1995) *Nature* **378**, 71.
62. Yao, H. and Dan, Y. (2001) *Neuron* **32**, 315.
63. Schuett, S., Bonhoeffer, T. and Hubener, M. (2001) *Neuron* **32**, 325.
64. Fu, Y.X., Djupsund, K., Gao, H., Hayden, B., Shen, K. and Dan, Y. (2002) *Science* **296**, 1999.
65. F. Engert, H.W. Tao, L.I. Zhang and M.M. Poo (2002) *Nature* **419**, 470.
66. Brecht M. and Sakmann, B. (2002) *J. Physiol.* **543.1**, 49.
67. Armstrong-James, M., Fox, K. and Das-Gupta, A. (1992) *J. Neurophysiol.* **68**, 1345.
68. Panzeri, S., Petersen, R.S., Schultz, S.R., Lebedev, M. and Diamond, M.E. (2001) *Neuron* **29**, 769.
69. Trachtenberg, J.T., Trepel, C. and Stryker, M.P. (2000) *Science* **287**, 2029.
70. Nelson, S.B., Sjostrom, P.J. Turrigiano, G.G. (2002) *Philos Trans R Soc Lond B* **357**, 1851.
71. Chen, C. and Regehr, W.G. (2000) *Neuron* **28**, 955.
72. Miller, K.D., Pinto, D.J. and Simons, D.J. (2001) *Curr. Opin. Neurobiol.* **11**, 488.
73. Pouille, F. and Scanziani, M. (2001) *Science* **293**, 1159.
74. Fagiolini, M. and Hensch, T.K. (2000) *Nature* **404**, 183.
75. Reiter, H.O. and Stryker, M.P. (1988) *Proc. Natl. Acad. Sci, USA* **85**, 3623.
76. Glazewski, S., Herman, C., McKenna, M., Chapman, P.F. and Fox, K. (1998) *Neuropharmacology* **37**, 581.
77. Froc, D.J., Chapman, C.A., Trepel, C. and Racine, R.J. (2000) *J. Neurosci.* **20**, 438.

15

MAINTAINING STABILITY AND PROMOTING PLASTICITY: CONTEXT-DEPENDENT FUNCTIONS OF INHIBITION

Weimin Zheng[*]

1. INTRODUCTION

The central nervous system faces two paradoxical challenges: maintaining functional stability and adapting to new experience. Because many aspects of an animal's environment are predictable, a stable operation of the neural networks in the brain is essential for generating reliable behaviors. Recent studies demonstrate that inhibition is critical in maintaining the stability of neuronal responses. It regulates synaptic gains thereby maintaining the homeostasis of local circuitries[1-10]. Changing the level of inhibition by pharmacological blockade or genetic manipulation results in epileptic malfunction of the brain[11-15]. Maturation of inhibition during development is associated with a dramatic decrease in the brain's capacity for modification[16-26].

The world, however, is not always predictable. This uncertainty demands that neural networks adjust their functions based on experience. Experience-dependent plasticity has been demonstrated in many regions of the brain, particularly in the central sensory systems. During adaptation, neurons acquire new response properties while eliminating old response properties. The acquisition and elimination of neural responses can be accomplished by modifying excitatory connections through axonal elaboration/retraction, synaptogenesis/synapse elimination, or synaptic potentiation/depression; however, emerging evidence indicates that inhibitory networks are also modifiable, suggesting that inhibition is involved in the process of experience-dependent plasticity[27-33]. Although the exact roles played by inhibition are still elusive, a general consensus is that alterations in the patterns of both excitation and inhibition underlie experience-dependent plasticity of the brain.

---

[*] W. Zheng. The Neurosciences Institute, 10640 John J. Hopkins Dr., San Diego, CA 92121, USA

Excitatory-Inhibitory Balance: Synapses, Circuits, Systems
Edited by Hensch and Fagiolini, Kluwer Academic/Plenum Publishers, 2003

Tremendous efforts have been made to understand how the central nervous system maintains its stability and, also, adapts to new experience. The primary focus has been on the study of excitatory connections, while the role of inhibition has yet to be explored. Many important questions remain, such as how inhibition contributes to the functional stability of the brain, how experience changes the patterns of inhibition, and most importantly, how the changed patterns of excitation and inhibition interact to affect adaptive adjustments. This chapter reviews recent studies[34, 35] in the auditory midbrain of the barn owl that provide valuable insights to these questions. The results demonstrate that patterns of excitation and inhibition can be adjusted independently of one another during the process of adaptive plasticity. Cooperative interaction of the changing patterns of excitation and inhibition determines the ultimate result of experience-dependent plasticity. Independent modification of inhibition allows it to serve opposing, context-dependent roles: maintaining functional stability while promoting adaptive plasticity.

2. AUDITORY SPACE MAP IN THE MIDBRAIN OF THE BARN OWL

Barn owls, like many other nocturnal hunters, rely on hearing to track and capture prey[36], thus, their auditory system is highly specialized for sound localization[37]. The primary cues used in sound localization are interaural level difference (ILD) and interaural time difference (ITD)[38]. The ILD, representing the elevation of a sound source, is generated by the asymmetry of the external ears of the owls, while the ITD, representing the azimuth of a sound source, is generated by the distance between the two ears. These sound location cues are then processed in parallel ascending neural pathways in the brainstem and converge at the level of the midbrain[39-41].

By integrating ITD and ILD cues across frequency channels represented in the lateral shell of central nucleus of the inferior colliculus (ICCls), an auditory space map is created in the external nucleus of the inferior colliculus (ICX)[42, 43] (Fig. 1). In the ICX, neurons that are narrowly tuned to ITD are systematically arranged along the rostrocaudal axis, while neurons tuned to ILD are along the dorsalventral axis[44, 45]. Thus, the location of every sound in space is systematically represented across the ICX. The auditory space map is then relayed topographically to the optic tectum (OT)[46], resulting in a close alignment of the auditory and visual receptive fields[47-49]. This alignment forms the neural basis for accurate sound localization in the barn owl[49].

3. ADAPTIVE PLASTICITY OF THE AUDITORY SPACE MAP

Because the relationship between cue values and sound location can be influenced by many factors, such as changes in the size and shape of the head and ears during development and alterations in acoustic coding in the auditory system due to hearing loss and aging[50-55], the auditory space map is not, and should not be, fixed. Instead, frequent calibration of the map is necessary throughout the duration of an animal's life. Several experiments have demonstrated that the auditory space map can in fact be modified to adapt to new experiences, such as chronic monaural occlusion[56-60], external ear modification[61], blind-rearing[62], and prismatic displacement of the horizontal visual field[63].

CONTEXT-DEPENDENT FUNCTIONS OF INHIBITION

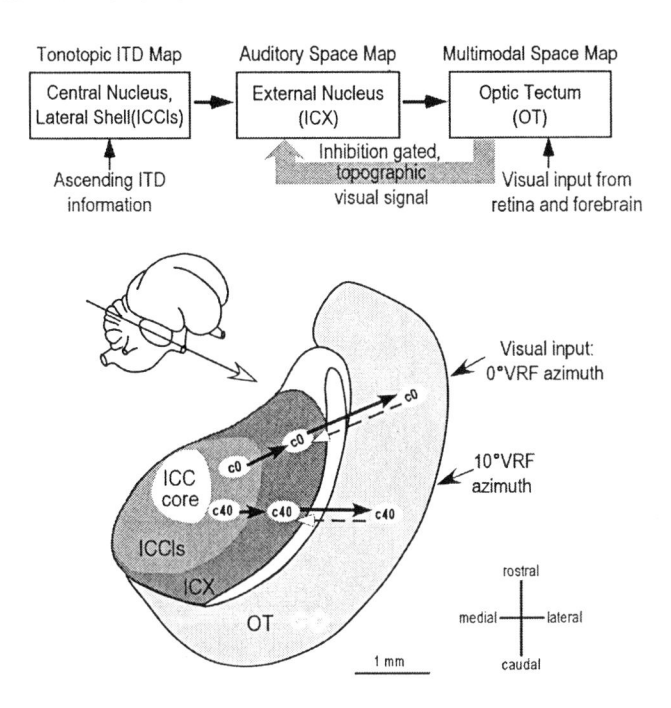

Fig. 1. The midbrain sound localization pathway in the barn owl. Top panel, Block diagram illustrating functional transformation of ITD information from the ICCls to the OT. Frequency-specific ITD information arrives at the ICCls and converges across frequency channels in the ICX to create an auditory space map. This map is then relayed to the OT where it aligns with a visual space map. An inhibition gated, topographic feedback visual signal also descends from the OT to the ICX. Bottom panel, Schematic diagram of a horizontal section of the midbrain through optical lobe (see inset) showing the topographic projection of ITD channels (dark-line arrows) along the ICCls, the ICX and the OT pathway and the feedback projections from the OT to the ICX (dashed-line arrows). VRF, visual receptive field. c, contralateral. Modified from reference 77.

Extensive studies have been performed in manipulating the owl's sensory experience by fitting the young animals with prismatic spectacles causing a chronic shift of the visual field[64-67]. The immediate effect of wearing prisms is a misalignment of the visual and auditory receptive fields in the OT (Fig. 2, top panel). Behaviorally, the owl can no longer visually localize a sound source when it orients to the sound by following the normal auditory space map[68]. The misalignment of auditory with visual receptive field and/or the consequent behavioral inaccuracy results in opening of an inhibitory gate, through which topographic, visually-driven signals are transmitted into the ICX[69-73]. These signals initiate and instruct the adjustment of the auditory space map in the ICX.

After a short period (10 to 20 days) of prism exposure, the ICX neurons start acquiring new responses to the adaptive ITD values, resulting in asymmetrical, broad ITD tuning curves that are shifting in the adaptive direction[35] (Figure 15-2, bottom panel). At this early phase of the ITD tuning shift, the responses to normal ITDs (normal responses) are still stronger than the responses to the adaptive ITDs (learned responses). Several weeks after prism exposure, the normal responses are eliminated while the

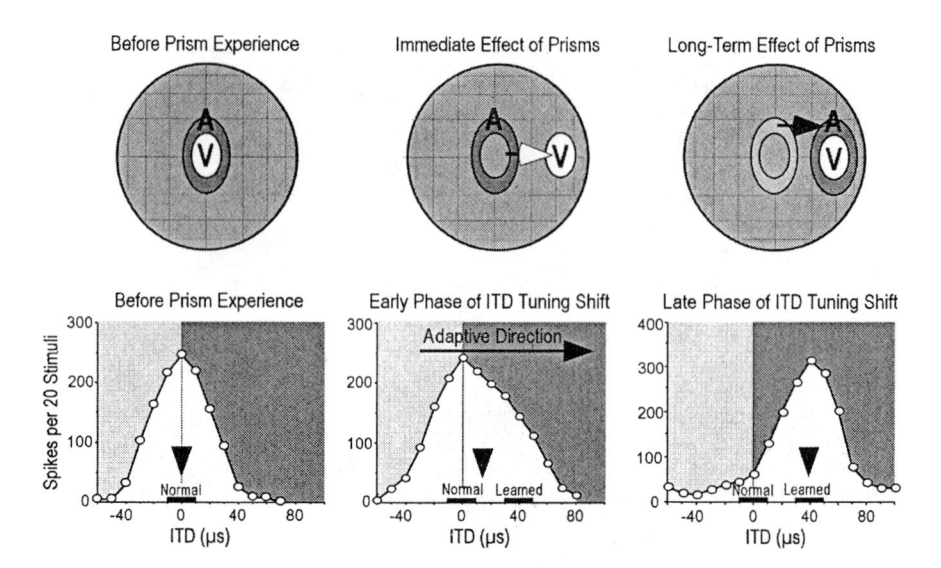

Fig. 2. Different stages of prism-experience induced plasticity. Top panel, Prism effects on the alignment of auditory and receptive field in the OT. In normal owls, the auditory receptive field (A) aligns with visual receptive field (V). Immediately after fitting owls with prisms, the visual receptive field is displaced. After several weeks of prism experience, the auditory receptive field shifts to align with the prismatically displaced visual receptive field. Bottom panel, Plasticity of ITD tuning in the ICX induced by wearing 23° prisms. Downward arrowheads indicate the weighted average of ITD tuning curves. Black horizontal bars indicate the range of normal and learn ITDs. The prism-induced plasticity in the ICX is classified into early and late phase by using a shift metric, calculated as: Learned responses/(Learned responses + Normal responses)[35]. The learned and normal responses are defined as the neuronal responses to the learned and normal ITDs, respectively.

learned responses are fully developed, resulting in symmetrical, narrow ITD tuning curves that are centered around the learned ITDs. The shifted ITD map is also faithfully represented in the OT. The ITD tuning of the tectal neurons shifts in the same direction, by the same amount, as those in the ICX[74]. As a result, the optically displaced visual receptive field is again aligned with the shifted, newly learned auditory receptive field in the prism-reared owls (Fig. 2, top panel). Behaviorally, the owls regain their initial accuracy of sound localization, entirely capable of seeing the correct location of a sound through the prisms by following the newly learned auditory space map[68].

4. INHIBITION SHARPENS THE AUDITORY SPACE MAP

The precision of the auditory space map relies on the sharp tuning of ICX neurons to acoustic spatial cues. Topographic projection from the ICCls to the ICX[75-77] and inhibition mediated by γ-aminobutyric acid (GABA) work together to achieve this goal[34, 78-81]. As shown in Fig. 3, blocking GABAergic inhibition by iontophoretic application of bicuculline, a specific antagonist for subtype A GABA receptor, resulted in a substantial broadening of the ITD tuning curves of ICX neurons. These results indicate that ICX neurons receive input from a broad range of ITD channels, but the responses to the peripheral channels are suppressed by inhibition. Thus, the topographic excitatory

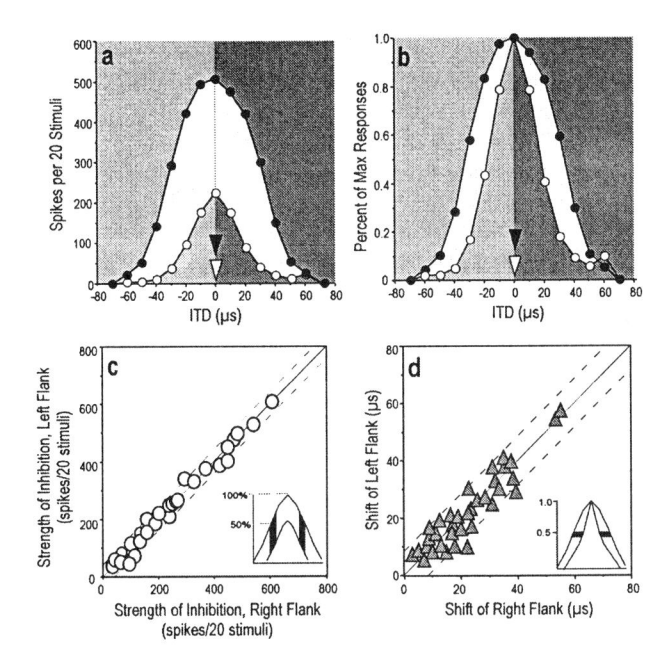

CONTEXT-DEPENDENT FUNCTIONS OF INHIBITION   245

Fig. 3. Effects of blocking GABAergic inhibition on ITD tuning in the ICX of normal owls. (a) ITD tuning curves measured before (open circles) and during (filled circles) bicuculline application at a single site. The weighted average ITD before (open arrowhead) and during (filled arrowhead) bicuculline application remain unchanged. (b) Normalized tuning curves plotted in (a), showing symmetrical broadening of the tuning curves. (c, d) Summary of data from normal owls. Each point represents data from one ICX site. Diagonal solid line indicates equal values; dashed lines indicate the mean ± 2SD of the difference in the effects of bicuculline between the compared properties. (c) Comparison of the strength of inhibition on the two flanks of each tuning curve. The strength of inhibition, measured as bicuculline-induced increases at the level of 50% maximal excitatory responses obtained during bicuculline application (black bars in insert). (d) Comparison of the bicuculline-induced shift in the flanks on the two sides of each tuning curve. The shift of flanks is measured at the level of 50% maximal excitatory responses (black bars in insert). Data from reference 34 and 35.

projections alone are not sufficient to generate the high precision of the auditory space map; instead, neuronal responses are fine tuned by inhibition, giving rise to the sharp ITD tuning curves essential for accurate sound localization.

Utilizing inhibition to shape the neuronal response properties driven by broadly wired projection patterns appears to be a strategy commonly used throughout the brain[82-86]. Modification of these broad connections can assure a fast, efficient adjustment of neural response patterns, as exemplified in the experience-dependent plasticity of auditory space map in the barn owl[76].

5. INHIBITION CONTRIBUTES TO STABILITY OF AUDITORY MAP

During the early stage of prism-induced plasticity in the ICX, the newly learned excitatory input, although adaptive, destabilizes the normal auditory space map by broadening ITD tuning. Neuropharmacological studies on the early phase of plasticity revealed that GABAergic inhibition is adjusted to counteract this destabilizing effect by suppressing post-synaptic responses to the newly learned input[35].

As shown in Figs. 4a and b, unlike in normal owls (Fig. 3), application of bicuculline to the ICX of prism-reared owls during the early phase of plasticity caused the weighted average of ITD to shift in the adaptive direction. Bicuculline induced a differentially large shift of the adaptive flank (Fig. 4b and d). The strength of inhibition is significantly greater on the adaptive flank of the tuning curves (Fig. 4c). This increased inhibition on the adaptive flank restricts neuronal responses to the newly acquired excitatory input, thereby tending to maintain the stability of the normal auditory space map.

Despite the strong suppressive influence of inhibition on post-synaptic responses to newly acquired inputs during the initial phase of adaptation, acquisition of these inputs continues, and experience-driven plasticity in the ICX eventually prevails. This is accomplished because the prism induced plasticity of the auditory space map is not driven by Hebbian mechanisms, but by instruction from an outside network[69-72]. The inhibitory force opposing adaptive plasticity in the ICX is overcome by the influence of the instructive signals descending topographically from the OT[69]. As the adaptive drive is reinforced over time, strong inhibition on the learned ITD channels is instructed to decrease to allow the acquisition of entirely new, adaptive functional representations of the ITDs in the ICX[34].

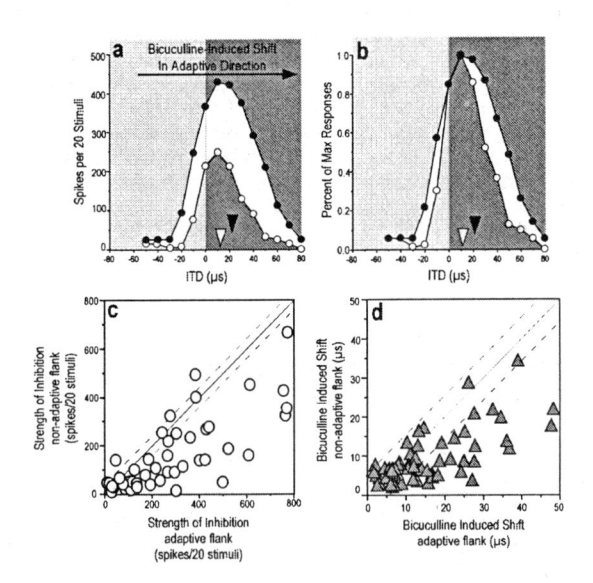

Fig. 4. Effects of blocking GABAergic inhibition on ITD tuning of ICX neurons in the early phase of plasticity. The data are plotted as in Figure 15-3. (a) ITD tuning curves measured before and during application at a single site. (b) Normalized tuning curves plotted in (a), showing asymmetrical broadening of the tuning curve flanks induced by bicuculline. Bicuculline iontophoresis caused the weighted average ITD shifted in the adaptive direction during bicuculline application. (c, d) Summary of data from individual sites. Diagonal solid lines indicate equal values; dashed lines indicate the mean ± 2SD for data from normal owls, re-plotted from Fig. 3. (c) Comparison of the strength of inhibition on the two flanks of each tuning curve. (d) Comparison of the bicuculline-induced shift of the flanks on the two sides of each tuning curve. Data from Ref. 35.

Input channel-specific suppression of post-synaptic responses by inhibition might represent a strategy commonly used for maintaining functional stability of the neural networks. This strategy can serve as a barrier to functional plasticity in the networks that rely on Hebbian mechanisms to regulate synaptic strength, such as the representation of ocular dominance in the visual cortex[87-90] or the representation of the body surface in the somatosensory cortex[91, 92]. Differentially strong inhibition of the neuronal responses to new inputs will prevent adaptive changes in these maps by dissociating pre- and post-synaptic activities. In fact, maturation of inhibitory circuitry is directly associated with a significant landmark stage of brain development[17, 23, 93, 94], the closure of the sensitive period, beyond which the capacity for modification is dramatically decreased. It can also

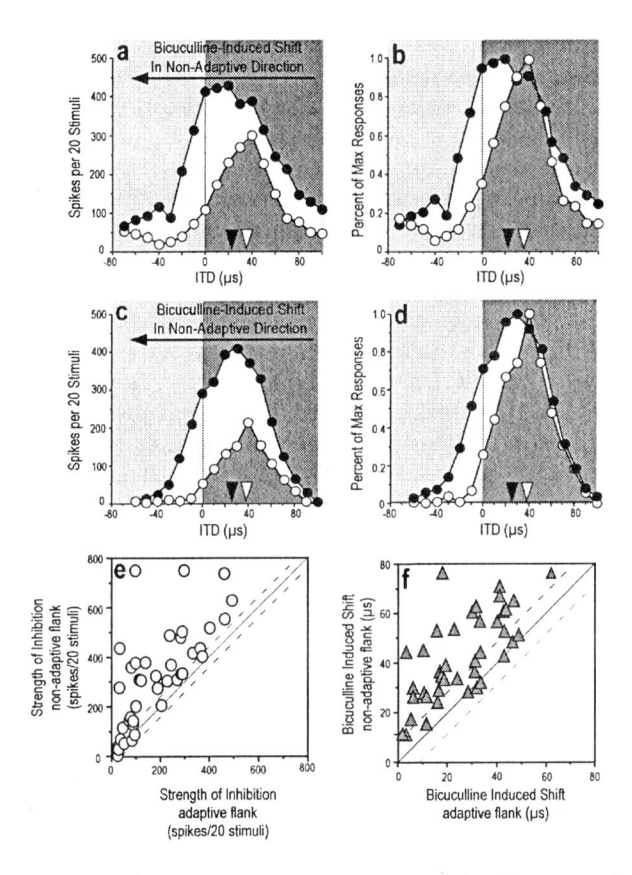

Fig. 5. Effects of blocking GABAergic inhibition on ITD tuning of the ICX neurons in the late phase of plasticity. The data are plotted as in Fig. 3. (a, c) ITD tuning curves measured before and during bicuculline application at a single site. Note that the normal responses unmasked by bicuculline at the site plotted in (a), representative of the majority (72%) of the ICX neurons tested, are stronger than the learned responses; whereas the unmasked normal responses in (c) are weaker than the learned responses, representative of the remaining ICX neurons. In both cases, bicuculline iontophoresis caused a large shift of the weighted average ITD in the non-adaptive direction. (b, d) Normalized tuning curves plotted in (a) and (c), showing asymmetrical broadening of the tuning curve flanks induced by bicuculline. (e-f) Summary of data from individual sites. Plotted as in Fig. 4c and d. (e) Comparison of the strength of inhibition on the two flanks of each tuning curve. (f) Comparison of the bicuculline-induced shift in the flanks on the two sides of each tuning curve. Data from Refs. 34 and 35.

serve as an effective mechanism that compensates for the disadvantage inherent in Hebbian synaptic plasticity, namely, the tendency of destabilizing neural networks due to ongoing opportunities for self-reinforcing, runaway changes in synaptic strength[95, 96].

6. INHIBITION PROMOTES PLASTICITY OF AUDITORY MAP

As the prism-induced plasticity proceeds into the late phase, the responses of the ICX neurons to normal ITD values are eliminated, resulting in a fully shifted auditory space map (Fig. 2). Significant efforts have been made to understand the neural mechanisms underlying the elimination of normal responses. Anatomical studies demonstrate that not only does the density of the ICCls-ICX projecting axons that support the normal responses remain unchanged, but these axons also bear the same number of boutons as those in normally raised owls[76]. Thus, neither axonal retraction nor synapse elimination underlies the suppression of normal responses in the ICX. Neuropharmacological studies[34] demonstrate that synaptic suppression is not apparent along the pathway that supports the normal responses. Instead, these studies show compelling evidence that elimination of the normal responses is accomplished through GABAergic inhibition.

Iontophoretic application of bicuculline in the ICX sites expressing a fully shifted map of ITD resulted in a large, differential increase in the responses to normal ITDs, shifting the weighted average ITDs dramatically in the non-adaptive direction (Fig. 5). In the majority (72%) of the sites tested, bicuculline iontophoresis caused responses to the normal ITDs to become as strong as or stronger than responses to learned ITDs (Fig. 5a). Thus, the strength of inhibition is significantly greater on the non-adaptive flank than on the adaptive flank of the tuning curves (Fig. 5a, c, and e). As a result, the flanks of the tuning curves shifted much further on the non-adaptive than on the adaptive side (Fig. 5b, d, and f). Thus, strong pre-synaptic excitatory inputs representing both learned and normal ITD channels coexist and remain strongly viable in the ICX exhibiting a fully shifted ITD map; but the post-synaptic responses to the normal ITD channels are selectively suppressed by GABAergic inhibition, allowing only the expression of the learned responses.

Selective suppression of the non-adaptive map during auditory learning indicates that the patterns of inhibition are adaptively modified. This modification is both ITD channel specific and independent of adjustment in the patterns of excitation. Because ITD channels merge in the ICX, ITD channel-dependent adjustment of inhibition must be limited to plasticity in the ICCls-ICX feedforward connections. With this limitation, plasticity can only occur in one or both of the two ways: 1) adjustment in the strength of excitatory projections from the ICCls to inhibitory interneurons in the ICX or 2) modification in the strength of inhibitory projections from the ICCls to the ICX. In contrast, during the initial phase of plasticity, the differential suppression of the adaptive responses by inhibition could be achieved by plasticity in either the feedforward connections from the ICCls to the ICX or the feedback (recurrent) connections within the ICX[35].

Selection of the adaptive map through inhibition of the non-adaptive map is beneficial for optimal performance of the neural networks in several ways. It helps maintain a normal balance of inhibition and excitation even during the process of adaptive modification of the neural circuitries, thereby sustaining the normal functional stability and integrative properties of the system[34]. It also represents a unique strategy that allows dramatic reorganization of neural representation without changing the innate projections that have evolved in the species. Consequently, the functionally sustained normal projections enable an efficient re-expression of the innate neural representation when the animal's experience returns to normal. Indeed, with a short period of exposure to normal experience after prism removal, the normal auditory space map in the ICX is fully re-expressed[97] by a re-adjustment in the patterns of inhibition[34].

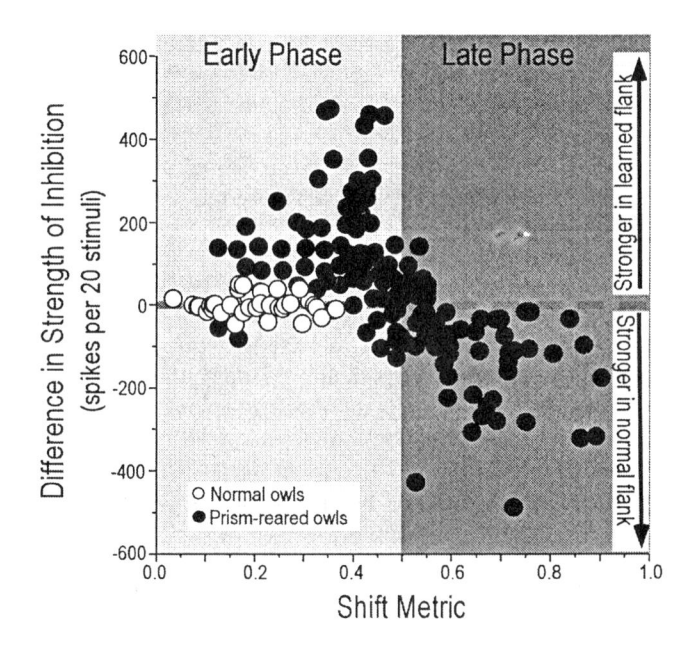

Fig. 6. ITD channel-specific adjustment in the strength of inhibition during the process of plasticity. Each data point represents the difference in the strength of inhibition (defined in Figure 15-3) between the adaptive and the non-adaptive side of the ITD tuning curves from normal (open circles) and prism-reared owls (filled circles). In the early phase of plasticity, inhibition is stronger on the learned ITD channels, whereas in the late phase, inhibition is stronger on the normal ITD channels. The shift metric used to define the phase of plasticity is calculated as: Learned responses/(Learned responses + Normal responses). Data from Refs. 34 and 35.

7. CONCLUDING REMARKS

Recent studies in the brain have drawn much attention to the roles played by inhibition in maintaining functional stability and adapting to new experience. As demonstrated in the auditory sound localization pathway of the barn owl, actively adjusting the patterns of inhibition to shape the patterns of excitation is crucial for both maintaining stability and adapting to new experience.

The dynamically changing effects of GABAergic inhibition on the auditory space map of the ICX is summarized in Fig. 6. In normal owls (Fig. 3), neurons in the ICX integrate inputs from a broad range of ITD channels[76, 77]. ITD channel-specific, GABAergic inhibition further shapes the neuronal response properties to create a precise map of auditory space[34, 78-80]. The accuracy of the map is calibrated by experience throughout the life time of the owl. In the early phase of prism-induced plasticity (Fig. 4), a specific increase in the strength of this inhibition on the learned ITD channels differentially suppresses responses to the newly acquired inputs, tending to preserve the normal auditory space map[35]. In the late phase of plasticity (Fig. 5), a selective increase in the strength of this inhibition on the normal ITD channels suppresses responses to normal, non-adaptive ITD channels, enabling the network to express only the learned, behaviorally appropriate map[34]. Thus, at different stages of adaptive plasticity, the patterns of inhibition are adjusted, independent of the modification of excitatory networks, to serve context-dependent functions that influence the progress and determine the final outcome of adaptation while ensuring the optimal performance of neural networks.

Numerous studies suggest that GABAergic inhibition is also involved in the plasticity of the mammalian brain[27-31, 33, 98]. A mammalian model of plasticity equivalent to that of the barn owl has yet to be established. Whether inhibition plays similar roles in the functional plasticity of the mammalian brain remains unclear, or at best, controversial[27, 99-101]. Plasticity in mammalian models is often induced by surgical operations, either sensory deprivation or denervation; the resulting plasticity is not behaviorally adaptive but represents a pathological state of the brain. As a result, the plasticity in the mammalian models is not governed by instructive learning rules as in the ICX of owls, but rather by self-organizing principles. Additionally, sensory deprivation or denervation results in decreased activity in the affected afferent pathways, unlike the prism experience which preserves the normal auditory-driven activity.

A few models of plasticity in the mammalian brain that are comparable in some aspects to the prism-induced plasticity in the barn owl, however, have been established. The training-induced plasticity of the tonotopic map in the primary auditory cortex of rodents, for example, represents one such model[102-104]. In this model, a tone is presented to the animal in combination with electric stimulation of the nucleus Basalis or the ventral tegmental area. Electric stimulation of these regions causes the release of neuromodulators, such as acetylcholine, dopamine, and GABA, to the auditory cortex. It is unknown whether these neuromodulators play an instructive or permissive role[105, 106], nonetheless, the representation of the tone is dramatically expanded in the primary cortex under this training paradigm. Neither sensory deprivation nor denervation is used in this model; acoustically-driven inputs across the audible frequency channels remain intact along the central auditory pathway. Little is known about the neural basis underlying the re-organization of the tonotopic map in the primary auditory cortex of mammals[105, 106], nevertheless, it is likely that future studies will discover mechanisms similar to those[64-67] governing auditory space map plasticity in the barn owl.

8. ACKNOWLEDGMENTS

I am grateful to Lindsay Taylor for valuable contributions to the manuscript and to B. van Swinderen, E. Walcott, E. Izhikevich, and R. Andretic for helpful comments. I thank E. Knudsen for providing me the opportunity and support carrying out the studies reviewed in this chapter at the Department of Neurobiology, Stanford University. Writing of this manuscript was supported by the Neurosciences Research Foundation, which supports the Neurosciences Institute.

9. REFERENCES

1. Turrigiano, G. G., Leslie, K. R., Desai, N. S., Rutherford, L. C. and Nelson, S. B. (1998) Activity-dependent scaling of quantal amplitude in neocortical neurons. *Nature* **391**, 892-896.
2. Steele, P. M. and Mauk, M. D. (1999) Inhibitory control of LTP and LTD: stability of synapse strength. *J. Neurophysiol.* **81**(4), 1559-1566.
3. Chance, F. S. and Abbott, L. F. (2000) Divisive inhibition in recurrent networks. *Network* **11**(2), 119-129.
4. Chance, F. S., Abbott, L. F. and Reyes, A. D. (2002) Gain modulation from background synaptic input. *Neuron* **35**(4), 773-782.
5. LeMasson, G., Marder, E., and Abbott, L. F. (1993) Activity-dependent regulation of conductances in model neurons. *Science* **259**(5103), 1915-1917.
6. Kim, H. G., Beierlein, M. and Connors, B. W. (1995) Inhibitory control of excitable dendrites in neocortex. *J. Neurophysiol.* **74**(4), 1810-1814.
7. Kaneko, T. and Hicks, T. P. (1990) GABA(B)-related activity involved in synaptic processing of somatosensory information in S1 cortex of the anaesthetized cat. *Br. J. Pharmacol.* **100**(4), 689-698.
8. Kaneko, T. and Hicks, T. P. (1988) Baclofen and gamma-aminobutyric acid differentially suppress the cutaneous responsiveness of primary somatosensory cortical neurones. *Brain Res.* **443**(1-2), 360-366.
9. Greuel, J. M., Luhmann, H. J. and Singer, W. (1988) Pharmacological induction of use-dependent receptive field modifications in the visual cortex. *Science* **242**, 74-77.
10. Luhmann, H. J. and Prince, D. A. (1990) Control of NMDA receptor-mediated activity by GABAergic mechanisms in mature and developing rat neocortex. *Brain Res. Dev. Brain Res.* **54**(2), 287-290.
11. Lu, Y. F., Kojima, N., Tomizawa, K., Moriwaki, A., Matsushita, M., Obata, K. and Matsui, H. (1999) Enhanced synaptic transmission and reduced threshold for LTP induction in fyn-transgenic mice. *Eur. J. Neurosci.* **11**(1), 75-82.
12. Lau, D., Vega-Saenz de Miera, E. C., Contreras, D., Ozaita, A., Harvey, M., Chow, A., Noebels, J. L., Paylor, R., Morgan, J. I., Leonard, C. S. and Rudy, B. (2000) Impaired fast-spiking, suppressed cortical inhibition, and increased susceptibility to seizures in mice lacking Kv3.2 K+ channel proteins. *J. Neurosci.* **20**(24), 9071-9085.
13. Lerma, J. (1998) Kainate receptors: an interplay between excitatory and inhibitory synapses. *FEBS Lett* **430**(1-2), 100-104.
14. van Brederode, J. F., Rho, J. M., Cerne, R., Tempel, B. L. and Spain, W. J. (2001) Evidence of altered inhibition in layer V pyramidal neurons from neocortex of Kcna1-null mice. *Neuroscience* **103**(4), 921-929.
15. Wendling, F., Bartolomei, F., Bellanger, J. J. and Chauvel, P. (2002) Epileptic fast activity can be explained by a model of impaired GABAergic dendritic inhibition. *Eur. J. Neurosci.* **15**(9), 1499-1508.
16. Luhmann, H. J. and Prince, D. A. (1991) Postnatal maturation of the GABAergic system in rat neocortex. *J. Neurophysiol.* **65**(2), 247-263.
17. Huang, Z. J., Kirkwood, A., Pizzorusso, T., Porciatti, V., Morales, B., Bear, M. F., Maffei, L. and Tonegawa, S. (1999) BDNF regulates the maturation of inhibition and the critical period of plasticity in mouse visual cortex. *Cell* **98**(6), 739-755.
18. Gao, W. J., Newman, D. E., Wormington, A. B. and Pallas, S. L. (1999) Development of inhibitory circuitry in visual and auditory cortex of postnatal ferrets: immunocytochemical localization of GABAergic neurons. *J. Comp. Neurol.* **409**(2), 261-273.
19. Gao, W. J., Wormington, A. B., Newman, D. E. and Pallas, S. L. (2000) Development of inhibitory circuitry in visual and auditory cortex of postnatal ferrets: immunocytochemical localization of calbindin- and parvalbumin-containing neurons. *J. Comp. Neurol.* **422**(1), 140-157.

20. Guo, Y., Kaplan, I. V., Cooper, N. G. and Mower, G. D. (1997) Expression of two forms of glutamic acid decarboxylase (GAD67 and GAD65) during postnatal development of the cat visual cortex. *Brain Res. Dev. Brain Res.* **103**(2), 127-141.
21. Blue, M. E. and Parnavelas, J. G. (1983) The formation and maturation of synapses in the visual cortex of the rat. I. Qualitative analysis. *J. Neurocytol.* **12**(4), 599-616.
22. M. E. Blue, and J. G. Parnavelas (1983) The formation and maturation of synapses in the visual cortex of the rat. II. Quantitative analysis. *J. Neurocytol.* **12**(4), 697-712.
23. Rozas, C., Frank, H., Heynen, A. J., Morales, B., Bear, M. F. and Kirkwood, A. (2001) Developmental inhibitory gate controls the relay of activity to the superficial layers of the visual cortex. *J. Neurosci.* **21**(17), 6791-6801.
24. Katz, L. C. (1999) What's critical for the critical period in visual cortex? *Cell* **99**(7), 673-676.
25. Micheva, K. D. and Beaulieu, C. (1995) Postnatal development of GABA neurons in the rat somatosensory barrel cortex: a quantitative study. *Eur. J. Neurosci.* **7**(3), 419-430.
26. Huntsman,M. M., Muñoz, A. and Jones, E. G. (1999) Temporal modulation of GABA(A) receptor subunit gene expression in developing monkey cerebral cortex. *Neuroscience* **91**(4), 1223-1245.
27. Tremere, L., Hicks, T. P. and Rasmusson, D. D. (2001) Expansion of receptive fields in raccoon somatosensory cortex in vivo by GABA(A) receptor antagonism: implications for cortical reorganization. *Exp. Brain Res.* **136**(4), 447-455.
28. Chowdhury, S. A. and Rasmusson, D. D. (2002) Effect of GABAB receptor blockade on receptive fields of raccoon somatosensory cortical neurons during reorganization. *Exp. Brain Res.* **145**(2), 150-157.
29. Jones, E. G. (1993) GABAergic neurons and their roles in cortical plasticity in primates. *Cerebral. Cortex* **3**, 361-372.
30. Micheva, K. D. and Beaulieu, C. (1997), Development and plasticity of the inhibitory neocortical circuitry with an emphasis on the rodent barrel field cortex: a review. *Can. J. Physiol. & Pharmacol.* **75**(5), 470-478.
31. Micheva, K. D. and Beaulieu, C. (1995) An anatomical substrate for experience-dependent plasticity of the rat barrel field cortex. *Proc Natl Acad Sci U S A* **92**(25), 11834-11838.
32. Jimenez-Capdeville, M. E., Dykes, R. W. and Myasnikov, A. A. (1997) Differential control of cortical activity by the basal forebrain in rats: a role for both cholinergic and inhibitory influences. *J. Comp. Neurol.* **381**(1), 53-67.
33. Sillito, A. M., Kemp, J. A. and Blakemore, C. (1981) The role of GABAergic inhibition in the cortical effects of monocular deprivation. *Nature* **291**(5813), 318-320.
34. Zheng, W. and Knudsen, E. I. (1999) Functional selection of adaptive auditory space map by GABAA-mediated inhibition. *Science* **284**(5416), 962-965.
35. Zheng, W. and Knudsen, E. I. (2001) GABAergic inhibition antagonizes adaptive adjustment of the owl's auditory space map during the initial phase of plasticity. *J. Neurosci.* **21**(12), 4356-4365.
36. Payne, R. S. (1971) Acoustic location of prey by barn owls (*Tyto alba*). *J. Exp. Biol.* **54**, 535-573.
37. Knudsen, E. I. and Konishi, M. (1979) Mechanisms of sound localization in the barn owl (*Tyto alba*). *J. Comp. Physiol.* **133**, 13-21.
38. Konishi, M. (1973) How the owl tracks its prey. *Am. Sci.* **61**, 414-424.
39. Carr, C. and Boudreau, R. (1993) Organization of the nucleus magnocellularis and the nucleus laminaris in the barn owl: encoding and measuring interaural time differences. *J. Comp. Neurol.* **334**, 337-355.
40. Carr, C. E. and Konishi, M. (1990) A circuit for detection of interaural time differences in the brainstem of the barn owl. *J. Neurosci.* **10**, 3227-3246.
41. Konishi, M., Takahashi, T. T., Wagner, H., Sullivan, W. E. and Carr, C. E. (1988) *Neurophysiological and anatomical substrates of sound localization in the owl*. In: *Auditory Function*, G. M. Edelman, W. E. Gall, W. M. Cowan, eds. (New York: John Wiley and Sons), pp. 721-745.
42. Knudsen, E. I. and Konishi, M. (1978) Center-surround organization of auditory receptive fields in the owl. *Science* **202**, 778-780.
43. Knudsen, E. I. and Konishi, M. (1978) A neural map of auditory space in the owl. *Science* **200**, 795-797.
44. Moiseff, A. and Konishi, M. (1983) Binaural characteristics of units in the owl's brainstem auditory pathway: precursors of restricted spatial receptive fields. *J. Neurosci.* **3**, 2553-2562.
45. Moiseff, A. and Konishi, M. (1981) Neuronal and behavioral sensitivity to binaural time differences in the owl. *J. Neurosci.* **1**, 40-48.
46. E. I. Knudsen, and P. F. Knudsen (1983) Space-mapped auditory projections from the inferior colliculus to the optic tectum in the barn owl (*Tyto alba*). *J. Comp. Neurol.* **218**, 187-196.
47. Knudsen, E. I. (1982) Auditory and visual maps of space in the optic tectum of the owl. *J. Neurosci.* **2**, 1177-1194.
48. Olsen, J. F., Knudsen, E. I. and Esterly, S. D. (1989) Neural maps of interaural time and intensity differences in the optic tectum of the barn owl. *J. Neurosci.* **9**, 2591-2605.

49. Knudsen, E. I. and Brainard, M. S. (1995) Creating a unified representation of visual and auditory space in the brain. *Annu. Rev. Neurosci.* **18**, 19-44.
50. King, A. J. and Moore, D. R. (1991) Plasticity of auditory maps in the brain. *Tren. Neurosci.* **14**(1), 31-37.
51. King, A. J., Kacelnik, O., Mrsic-Flogel, T. D., Schnupp, J. W., Parsons, C. H. and Moore, D. R. (2001) How plastic is spatial hearing? *Audiol. Neurootol.* **6**(4), 182-186.
52. Moore, D. R., Rothholtz, V. and King, A. J. (2001) Hearing cortical activation does matter. *Curr. Biol.* **11**(19), R782-784.
53. Mrsic-Flogel, T. D., King, A. J., Jenison, R. L. and Schnupp, J. W. (2001) Listening through different ears alters spatial response fields in ferret primary auditory cortex. *J. Neurophysiol.* **86**(2), 1043-1046.
54. King, A. J., Parsons, C. H. and Moore, D. R. (2000) Plasticity in the neural coding of auditory space in the mammalian brain. *Proc Natl Acad Sci U S A* **97**(22), 11821-11828.
55. Hofman, P. M., Van Riswick, J. G. and Van Opstal. A. J. (1998) Relearning sound localization with new ears. *Nat. Neurosci.* **1**(5), 417-421.
56. Knudsen, E. I. (1983) Early auditory experience aligns the auditory map of space in the optic tectum of the barn owl. *Science* **222**(4626), 939-942.
57. Knudsen, E. I. (1985) Experience alters the spatial tuning of auditory units in the optic tectum during a sensitive period in the barn owl. *J. Neurosci.* **5**(11), 3094-3109.
58. Knudsen, E. I., Esterly, S. D. and Knudsen, P. F. (1984) Monaural occlusion alters sound localization during a sensitive period in the barn owl. *J. Neurosci.* **4**(4), 1001-1011.
59. Mogdans, J. and Knudsen, E. I. (1993) Early monaural occlusion alters the neural map of interaural level differences in the inferior colliculus of the barn owl. *Brain Res.* **619**(1-2), 29-38.
60. Mogdans, J. and Knudsen, E. I. (1992) Adaptive adjustment of unit tuning to sound localization cues in response to monaural occlusion in developing owl optic tectum. *J. Neurosci.* **12**(9), 3473-3484.
61. Knudsen, E. E., Esterly, S. D. and Olsen, J. F. (1994) Adaptive plasticity of the auditory space map in the optic tectum of adult and baby barn owls in response to external ear modification. *J. Neurophysiol.* **71**, 79-94.
62. Knudsen, E. I., Esterly, S. D. and du Lac, S. (1991) Stretched and upside-down maps of auditory space in the optic tectum of blind-reared owls: acoustic basis and behavioral correlates. *J. Neurosci.* **11**, 1727-1747.
63. Knudsen, E. I. and Brainard, M. S. (1991) Visual instruction of the neural map of auditory space in the developing optic tectum. *Science* **253**, 85-87.
64. Knudsen, E. I. (2002) Instructed learning in the auditory localization pathway of the barn owl. *Nature* **417**(6886), 322-328.
65. Knudsen, E. I. (1999) Mechanisms of experience-dependent plasticity in the auditory localization pathway of the barn owl. *J. Comp. Physiol. A* **185**(4), 305-321.
66. Knudsen, E. I. (1994) Supervised learning in the brain. *J. Neurosci.* **14**, 3985-3997.
67. Knudsen, E. I., Zheng, W. and DeBello, W. M. (2000) Traces of learning in the auditory localization pathway. *Proc Natl Acad Sci U S A* **97**(22), 11815-11820.
68. Knudsen, E. I. and Knudsen, P. F. (1990) Sensitive and critical periods for visual calibration of sound localization by barn owls. *J. Neurosci.* **63**, 131-149.
69. Gutfreund, Y., Zheng, W. and Knudsen, E. I. (2002) Gated visual input to the central auditory system. *Science* **297**(5586), 1556-1559.
70. Hyde, P. S. and Knudsen, E. I. (2000) Topographic projection from the optic tectum to the auditory space map in the inferior colliculus of the barn owl. *J. Comp. Neurol.* **421**(2), 146-160.
71. Hyde, P. S. and Knudsen, E. I. (2001) A topographic instructive signal guides the adjustment of the auditory space map in the optic tectum. *J. Neurosci.* **21**(21), 8586-8593
72. Hyde, P. S. and Knudsen, E. I. (2002) The optic tectum controls visually guided adaptive plasticity in the owl's auditory space map. *Nature* **415**(6867), 73-76.
73. King, A. J. (2002) Neural plasticity: how the eye tells the brain about sound location. *Curr. Biol.* **12**(11), R393-395.
74. Brainard, M. S. and Knudsen, E. I. (1995) Dynamics of visually guided auditory plasticity in the optic tectum of the barn owl. *J. Neurophysiol.* **73**(2), 595-614.
75. Knudsen, E. I. (1983) Subdivisions of the inferior colliculus in the barn owl (*Tyto alba*). *J. Comp. Neurol.* **218**(2), 174-186.
76. DeBello, W. M., Feldman, D. E. and Knudsen, E. I. (2001) Adaptive axonal remodeling in the midbrain auditory space map. *J Neurosci.* **21**(9), 3161-3174.
77. Feldman, D. E. and Knudsen, E. I. (1997) An anatomical basis for visual calibration of the auditory space map in the barn owl's midbrain. *J. Neurosci.* **17**(17), 6820-6837.
78. Albeck, Y. (1997) Inhibition sensitive to interaural time difference in the barn owl's inferior colliculus. *Hear Res.* **109**(1-2), 102-108.

79. Fujita, I. and Konishi M., (1991) The role of GABAergic inhibition in processing of interaural time difference in the owl's auditory system. *J. Neurosci.* **11**, 722-729.
80. Mori, K. (1997) Across-frequency nonlinear inhibition by GABA in processing of interaural time difference. *Hear Res.* **111**(1-2), 22-30.
81. Kautz, D. and Wagner, H. (1998) GABAergic inhibition influences auditory motion-direction sensitivity in barn owls. *J. Neurophysiol.* **80**(1), 172-185.
82. Dykes, R. W., Landry, P., Metherate, R. and Hicks, T. P. (1984) Functional role of GABA in cat primary somatosensory cortex: shaping receptive fields of cortical neurons. *J. Neurophysiol.* **52**(6), 1066-1093.
83. Hicks, T. P. and Dykes, R. W. (1983) Receptive field size for certain neurons in primary somatosensory cortex is determined by GABA-mediated intracortical inhibition. *Brain Res.* **274**(1), 160-164.
84. Sillito, A. M. (1977) Inhibitory processes underlying the directional specificity of simple, complex and hypercomplex cells in the cat's visual cortex. *J. Physiol.* **271**(3), 699-720.
85. Sillito, A. M. (1992) GABA mediated inhibitory processes in the function of the geniculo- striate system. *Pro. Brain Res.* **90**, 349-384.
86. Zheng, W. and Hall, J. C. (2000) GABAergic inhibition shapes frequency tuning and modifies response properties in the superior olivary nucleus of the leopard frog. *J. Comp. Physiol. A* **186**(7-8), 661-671.
87. Kirkwood, A. and Bear, M. F. (1994) Hebbian synapses in visual cortex. *J Neurosci* **14**, 1634-1645.
88. Katz, L. C. and Shatz, C. J. (1996) Synaptic activity and the construction of cortical circuits. *Science* **274**(5290), 1133-1138.
89. Bear, M. F. and Rittenhouse, C. D. (1999) Molecular basis for induction of ocular dominance plasticity. *J. Neurobiol.* **41**(1), 83-91.
90. Hata, Y., Tsumoto, T. and Stryker, M. P. (1999) Selective pruning of more active afferents when cat visual cortex is pharmacologically inhibited. *Neuron* **22**(2), 375-381.
91. Schlaggar, B. L., Fox, K. and O'Leary, D. D. (1993) Postsynaptic control of plasticity in developing somatosensory cortex. *Nature* **364**(6438), 623-626.
92. O'Leary, D. D., Ruff, N. L. and Dyck, R. H. (1994) Development, critical period plasticity, and adult reorganizations of mammalian somatosensory systems. *Curr. Opin. Neurobiol.* **4**(4), 535-544.
93. Kirkwood, A., Lee, H. K. and Bear, M. F. (1995) Co-regulation of long-term potentiation and experience-dependent synaptic plasticity in visual cortex by age and experience. *Nature* **375**(6529), 328-331.
94. Fagiolini, M. and Hensch, T. K. (2000) Inhibitory threshold for critical-period activation in primary visual cortex. *Nature* **404**(6774), 183-186.
95. Turrigiano, G. G. and Nelson, S. B. (2000) Hebb and homeostasis in neuronal plasticity. *Curr. Opin. Neurobiol.* **10**(3), 358-364.
96. Turrigiano, G. G. (1999) Homeostatic plasticity in neuronal networks: the more things change, the more they stay the same. *Tren. in Neurosci.* **22**(5), 221-227.
97. Knudsen, E. I. (1998) Capacity for plasticity in the adult owl auditory system expanded by juvenile experience. *Science* **279**, 1531-1533.
98. Dykes, R. W. (1997) Mechanisms controlling neuronal plasticity in somatosensory cortex. *Can. J. Physiol. & Pharmacol.* **75**(5), 535-545.
99. Rajan, R. (1998) Receptor organ damage causes loss of cortical surround inhibition without topographic map plasticity. *Nat. Neurosci.* **1**(2), 138-143.
100. Rajan, R. (2001) Plasticity of excitation and inhibition in the receptive field of primary auditory cortical neurons after limited receptor organ damage. *Cereb. Cortex* **11**(2), 171-182.
101. Tremere, L., Hicks, T. P. and Rasmusson, D. D. (2001) Role of inhibition in cortical reorganization of the adult raccoon revealed by microiontophoretic blockade of GABA(A) receptors. *J. Neurophysiol.* **86**(1), 94-103.
102. Kilgard, M. P. and Merzenich, M. M. (1998) Cortical map reorganization enabled by nucleus basalis activity. *Science* **279**(5357), 1714-1718.
103. Bakin, J. S. and Weinberger, N. M. (1996) Induction of a physiological memory in the cerebral cortex by stimulation of the nucleus basalis *Proc Natl Acad Sci U S A* **93**(20), 11219-11224.
104. Bao, S., Chan, V. T. and Merzenich, M. M. (2001) Cortical remodelling induced by activity of ventral tegmental dopamine neurons. *Nature* **412**(6842), 79-83.
105. Edeline, J. M. (1999) Learning-induced physiological plasticity in the thalamo-cortical sensory systems: a critical evaluation of receptive field plasticity, map changes and their potential mechanisms. *Prog. Neurobiol.* **57**(2), 165-224.
106. Buonomano, D. V. and Merzenich, M. M. (1998) Cortical plasticity: from synapses to maps. *Annu. Rev. Neurosci.* **21**, 149-186.

16

SPIKE TIMING AND VISUAL CORTICAL PLASTICITY

Yu-Xi Fu and Yang Dan[*]

1. INTRODUCTION

Plasticity in the nervous system is essential for information processing and storage. At the synaptic level, various forms of activity-dependent plasticity have been characterized, and their cellular mechanisms have been investigated intensely. At the circuitry level, sensory stimuli are known to play crucial roles in shaping the structures of neuronal circuits both during development and in adult life. At the functional level, perceptual performance of the animal can be strongly influenced by prior sensory experience. In spite of the extensive investigations at each level, however, the relationship between activity-dependent neural modifications at different levels remains to be firmly established. In this chapter, we will review a set of recent studies on spike timing-dependent plasticity in the visual cortex, the results of which support a causal link between experience-induced cortical modifications at different levels.

2. SPIKE TIMING-DEPENDENT SYNAPTIC MODIFICATION *IN VITRO*

It has been generally accepted that the relative timing of pre- and postsynaptic electrical activity plays a crucial role in synaptic modification. In the famous postulate of Donald Hebb[1], he stated that "When an axon of cell A is near enough to excite a cell B and repeatedly or persistently takes part in firing it, some growth process or metabolic change takes place in one or both cells such that A's efficiency, as one of the cells firing B, is increased". This postulate, which insightfully pointed out the importance of relative spike timing in synaptic modification, has since become a central hypothesis concerning

---

[*] Y.-X. Fu, Y. Dan, Division of Neurobiology, Dept. of Molecular and Cell Biology, University of California, Berkeley, CA 94720, USA

the neuronal basis of learning and memory. Over the past several decades, various interpretations of Hebb's rule[2,3] have been used to guide experimental and theoretical studies of activity-dependent synaptic plasticity in the nervous system (e.g., Miller et al.,[4], Weliky and Katz[5]). While the previous formulations of Hebb's rule emphasized the importance of temporal coincidence between pre- and postsynaptic activity, recent experiments using spike-pairing protocols have highlighted a critical role of the temporal order of spiking in synaptic modification. This form of spike timing-dependent synaptic plasticity, referred to as STDP, will be the focus of the present chapter.

2.1. Asymmetric Temporal Window of STDP

In previous mathematical formulations of Hebb's rule, synaptic modification as a function of the temporal interval between pre- and postsynaptic activity is assumed to be symmetric, with the modification independent of the pre/post order as long as both cells are activated closely in time. The importance of the temporal order of spiking was first demonstrated in a study in neocortical slices. Using dual whole-cell recordings from inter-connected layer 5 pyramidal neurons, Markram et al.[6] found that the coincidence of postsynaptic action potentials and the excitatory postsynaptic potentials (EPSPs) evoked by presynaptic spiking can lead to persistent changes in synaptic efficacy. In the induction protocol in which bursts of spikes were triggered in both the pre- and postsynaptic neurons, presynaptic spiking 10 ms before postsynaptic spiking resulted in long-term potentiation (LTP), whereas spiking in the reverse order resulted in long-term depression (LTD). This result indicates that the pre/post spike order can directly affect the direction of synaptic modification. Interestingly, in another study in a cerebellum-like structure of the electric fish, synaptic modification was also found to depend on the pre/post spike order, except in the opposite direction[7]. Given that the postsynaptic cell of this synapse is inhibitory whereas that of the neocortical synapse is excitatory (pyramidal neuron), however, the opposite spike timing dependence at these two synapses may in fact serve similar functions.

The temporal window for synaptic modification, which has become part of the standard paradigm for characterizing STDP, was first measured in the developing frog visual tectum[8] and in hippocampal cell culture[9]. In both studies, the interval between a presynaptic spike and a postsynaptic spike was varied systematically in a low-frequency pairing protocol (Fig. 1A), and significant synaptic modification was observed only within a temporal window of ~±40 ms. This window is asymmetric, with LTP found at positive intervals (presynaptic spike preceding postsynaptic spike, referred to as "pre→post" pairing) and LTD at negative intervals, consistent with the finding at layer 5 neocortical synapses[6]. Such an asymmetric window has been observed at a variety of glutamatergic synapses in the vertebrate central nervous system, including those at the songbird auditory forebrain[10], the mammalian hippocampus[11,12], and the neocortex[13-15] (Fig. 1B). These findings indicate that the asymmetric dependence on relative pre/post spike timing on the order of tens of milliseconds is a prevalent property of activity-dependent synaptic modification, which may have important functional implications.

2.2. STDP and Complex Spike Trains

To understand the functional significance of STDP in shaping the neuronal circuits *in vivo*, an important step is to know how this mechanism operates in the presence of complex spike trains. In most of the *in vitro* studies of STDP, pre- and postsynaptic spikes were paired repetitively at regular intervals to induce synaptic modification (Fig. 1A). In the neuronal circuits *in vivo*, however, timing of spikes in response to sensory stimuli is known to be highly irregular[16, 17]. Is the asymmetric temporal window measured with a regular pairing protocol sufficient to predict the effects of complex spike trains?

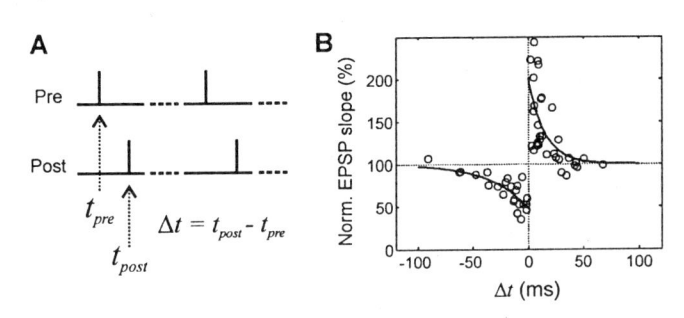

Fig. 1. Temporal window of STDP. (**A**) Repetitive pairing of pre- and post-synaptic spikes in a standard induction protocol. In each experiment the pre/post interval (Δt) is fixed, and the pairing is repeated at a constant frequency. (**B**) Synaptic modification vs. pre/post inter-spike interval, measured in layer 2/3 of rat visual cortical slices (adapted from Froemke and Dan[15]). The induction protocol consisted of 60-80 spike pairs at 0.2 Hz. Each circle represents one experiment. Curves, single-exponential, least-square fits of the data (time constants: 15 ms for LTP and 34 ms for LTD).

A recent study performed in layer 2/3 of rat visual cortical slices addressed this question by progressively increasing the complexity of spike patterns in the induction protocol[15]. First, the temporal window for STDP (Fig. 1B) was measured with the standard induction protocol, in which pre- and postsynaptic spikes were paired 60-80 times at 0.2 Hz (Fig. 1A). To predict the effect of a pair of complex spike trains in synaptic modification, a straightforward approach is to combine the effects of all pre/post spike pairs in synaptic modification[18-20]. However, considering the intricate cellular mechanisms underlying the induction of LTP and LTD[21], the effect of each pre/post spike pair may depend not only on its interval (Fig. 1B), but also on the presence of other nearby spikes in either the pre- or the postsynaptic neuron. To examine this possibility, a third spike was added to the pre/post spike pair to form a "triplet", which contains either one pre and two post spikes ('1/2' triplet), or two pre and one post spikes ('2/1' triplet) (Fig. 2). Each triplet was repeated 60-80 times at 0.2 Hz (same as the spike pair experiments in Fig. 1), and the resulting synaptic modification was compared to the prediction of the "independent" model, which computes the contribution of each pre/post spike pair based only on its inter-spike interval (Fig. 1B) regardless of the timing of other spikes in each cell. As shown in Fig. 2, results of these triplet experiments clearly rejected the independent model. In experiments with post→pre→post triplets (Fig. 2A), LTD was often observed even if the independent model predicted LTP, whereas with pre→post→pre triplets (Fig. 2B), LTP was often observed even if the predicted effect

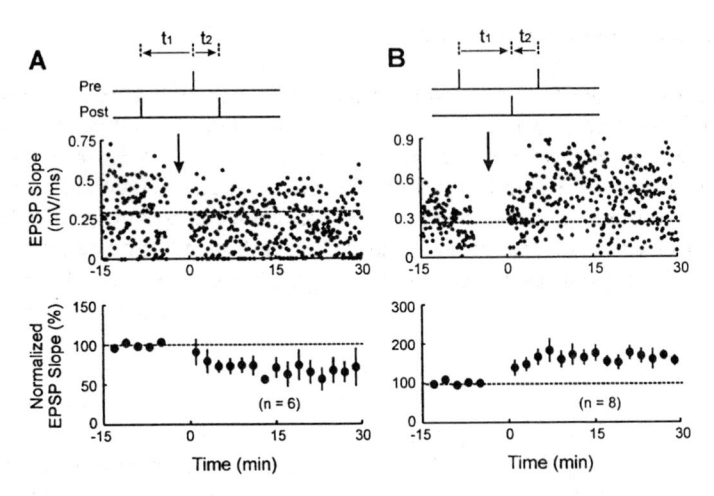

Fig. 2. Synaptic modification induced by spike triplets. (A) LTD induced by post→pre→post triplets (each repeated 60-80 times at 0.2 Hz). Spike pair intervals (t_1 and t_2) are defined as $t_{post} - t_{pre}$. Right arrow, positive; left arrow, negative. Upper plot, an example experiment, in which $t_1 = -24$ ms, $t_2 = 6$ ms, and prediction of independent model (based on the window measured with spike pairs, Fig. 1B) was 24% potentiation. Lower plot, mean effect of '1/2' triplets satisfying (1) $t_1 < 0$, (2) $t_2 > 0$, (3) $|t_1 - t_2| \leq 30$ ms, and (4) prediction of independent model was potentiation or no change. (B) As in A, except pre→post→pre triplets were used for induction. Upper plot, an example experiment, in which $t_1 = 6.5$ ms, $t_2 = -0.5$ ms, and prediction of the independent model was 20% depression. Lower plot shows mean effect of '2/1' triplets satisfying: (1) $t_1 > 0$, (2) $t_2 < 0$, (3) $|t_1 - t_2| \leq 30$ ms, and (4) prediction of the independent model was depression or no change. Adapted with permission from Froemke and Dan[15].

was LTD. Further analysis of the triplet experiments suggests that the first spike pair in each triplet plays a dominant role in synaptic modification, and the contribution of the second pair is suppressed by the preceding spike in either the pre- or the postsynaptic cell.

Based on these observations, we proposed a simple "suppression" model, in which the contribution of each pre/post spike pair depends not only on the interval between the pair, but also on the spike "efficacy", which is suppressed by the preceding spike in each neuron. The spike efficacy is reduced to 0 immediately following the preceding spike, and recovers exponentially towards 1 (Fig. 3A). The net effect of a pair of spike trains is then estimated by combining the contributions of all spike pairs. By fitting the results of the triplet experiments, we determined the suppression time constant between the presynaptic spikes to be 34 ms and that between the postsynaptic spikes to be 75 ms. When this suppression model was used to predict the effects of more complex spike patterns, including spike quadruplets and spike train segments recorded *in vivo* in response to natural visual stimuli (Fig. 3B and C), its performance was found to be significantly better than that of the independent model. Together, these results show that the spike timing dependence of synaptic modification induced by complex spike trains must include not only the relative timing between pre- and postsynaptic spiking, but also the inter-spike intervals within each neuron.

SPIKE TIMING AND VISUAL CORTICAL PLASTICITY

Another study that investigated the effects of complex spike trains in synaptic modification was performed in layer 5 of visual cortical slices[14]. Synaptic modification was measured while both the firing rate and the relative pre/post spike timing were varied systematically. In addition to the asymmetric temporal window for STDP, these experiments revealed a more complex dependence of LTP and LTD on both the firing rate and spike timing. The induction of LTP, but not LTD, was found to depend on the firing rate. When several phenomenological models were used to predict the effects of random Poisson spike trains, the model that (1) considers only the nearest spike pairs and (2) allows the pre→post spike pairs (producing LTP) to "win" over the post→pre pairs appears to fit the data better than the models that did not implement these rules.

Both studies described above represent the first steps toward delineating the rules governing the effects of complex spike trains in synaptic modification. Further studies are needed to determine the scope of applicability of each model and how the two models derived from experiments at different synapses with different protocols are related to each other. As more experiments are carried out with complex spike trains, both models may need to be modified to improve their generality. These experimental investigations

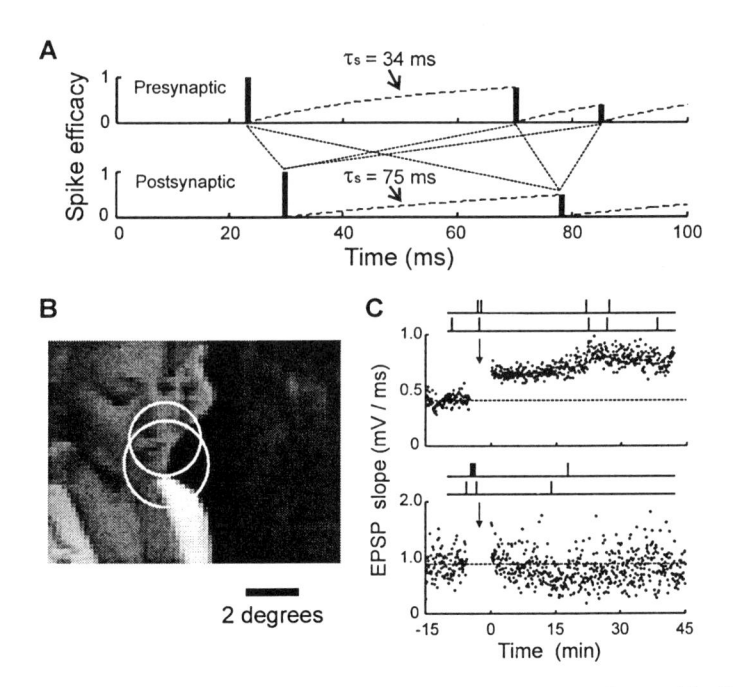

Fig. 3. Synaptic modification induced by complex spike patterns. (**A**) Suppression model. Vertical lines, spikes, the height of which represents efficacy of the spike in synaptic modification. Dashed curves, efficacy as a function of time. Dotted lines, pre/post spike pairs. (**B**) A frame in a natural scene movie used to evoke spiking responses in cat primary visual cortex. Circles, receptive fields of two neurons recorded simultaneously with multielectrodes. (**C**) Examples of synaptic modification in rat cortical slices induced by pairs of spike train segments recorded in cat visual cortex in response to natural scene stimuli. Each pair of segments (1 s) was repeated 60 times at 0.2 Hz. The suppression model predicted the effects of these natural spike train segments significantly better than the independent model. Adapted with permission from Froemke and Dan[15].

will provide important constraints for theoretical studies on the functional significance of STDP in neuronal circuits *in vivo*.

3. SPIKE TIMING AND FUNCTIONAL PLASTICITY IN VISUAL CORTEX

The functional consequences of STDP have also been explored in recent experimental studies *in vivo*[22-25]. In particular, several studies were performed in the visual cortex, in which spike timing of the cortical neurons was controlled by systematically varying the timing of visual stimuli, allowing characterization of the temporal specificity of visual cortical plasticity. In this section, we will first provide a general overview of experience-dependent plasticity in the visual cortex before discussing the cortical modifications likely mediated by STDP of the intracortical connections.

3.1. Experience-Dependent Plasticity in Visual Cortex

Visual experience plays crucial roles in shaping the cortical circuitry and function. In addition to their well-known effects during development[26,27], visual stimuli can also induce rapid changes in adult cortical neurons[28]. For example, in contrast adaptation a few seconds of visual stimulation can cause a significant reduction in the response amplitude[29] and changes in the spatial frequency tuning[30], orientation tuning[31] and direction selectivity[32] of cortical neurons. These effects may be due to a reduction in the neuronal excitability in the cortical circuit[33,34] or a short-term depression of the synaptic connections[35]. Concurrent visual stimulation and iontophoretic activation of cortical neurons can induce changes in their orientation selectivity and ocular dominance[36-38], and the dependence of these effects on the temporal coincidence between the visual and the iontophoretic stimulation is consistent with correlation-based Hebbian synaptic modification. Similarly, synchronous stimulation of the receptive field (RF) center and part of the unresponsive surround can induce preferential expansion of the RF towards the co-stimulated surround region[39], which is also thought to be mediated by Hebbian modification of the intracortical connections. These studies all point to synaptic plasticity as a key mechanism in mediating visual-stimulus-induced modification of cortical RF properties.

Several recent studies have demonstrated cortical modifications that are likely mediated by STDP of the intracortical connections. Since timing of visual stimuli can directly affect timing of neuronal spiking, it provides a convenient method to manipulate the relative spike timing of different cortical neurons. In the set of studies discussed below, RF plasticity of cortical neurons was examined in both the space domain and the orientation domain. A common finding is that the magnitude and direction of the functional modifications depend critically on the relative timing of visual stimuli on the order of tens of milliseconds.

3.2. Stimulus Timing-Dependent Plasticity in Space Domain

In one of the recent studies, we examined stimulus timing-dependent plasticity in cortical representation of visual space[25]. Random checkerboard patterns were flashed asynchronously in two adjacent retinal regions (*A* and *B*) (Fig. 4A) to manipulate the relative spike timing of two groups of cortical neurons whose RFs fall in regions A and B. We measured the effects of such asynchronous visual conditioning on cortical circuitry and RFs in V1 of anesthetized adult cat and on visual perception in human.

3.2.1. Stimulus Timing and Spike Timing

The induction of synaptic modification through STDP requires repetitive pairing of pre- and postsynaptic spikes within a temporal window of tens of milliseconds (Fig. 1B). For the asynchronous conditioning stimuli to be effective in modifying the intracortical connections, the precision of spike timing in the responses must be comparable to the temporal specificity of STDP. Spike timing precision of V1 neurons was assessed by comparing the responses of each cell to $A{\rightarrow}B$ and $B{\rightarrow}A$ stimuli. As shown in Fig. 4B, spiking of a neuron (RF in region B) during $A{\rightarrow}B$ stimuli (A/B interval: 8.3 ms) lagged behind its spiking during $B{\rightarrow}A$ stimuli by ~8 ms, indicating time-locking of the spikes to the stimuli at a high precision. Fig. 4C shows the cross-correlation between the responses of the neuron to $A{\rightarrow}B$ and $B{\rightarrow}A$ stimuli, which exhibited a peak at ~8 ms, corresponding to the A/B interval. The mean width at half height of the peak, which reflects spike timing precision, was 26 ± 15 (SD) ms for the population of neurons studied (n = 145). This indicates that the flashed conditioning stimuli can control the spike timing of V1 neurons at a precision adequate for the induction of STDP.

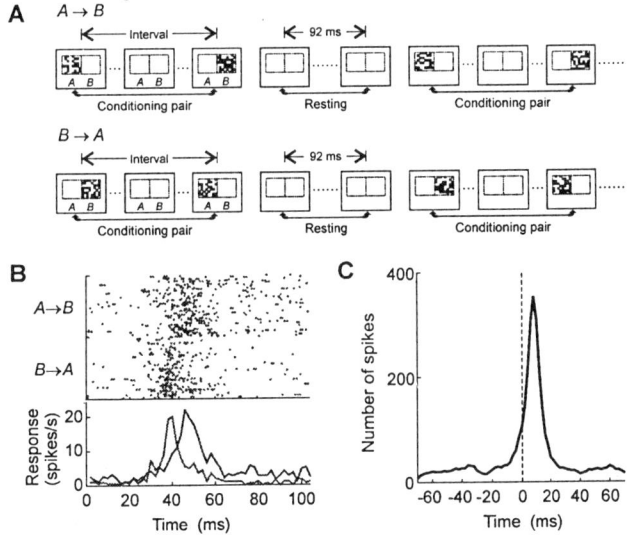

Fig. 4. Spike timing of visual cortical neurons in response to asynchronous conditioning stimuli. (A) Two types of conditioning stimuli ($A{\rightarrow}B$ and $B{\rightarrow}A$) used to control relative spike timing between two groups of cortical neurons, whose RFs fall in regions A and B. In each conditioning pair, a random checkerboard pattern was flashed (1 frame, 8.3 ms) in each region with a short interval (−8 to 8 frames) between A and B. (B) Responses of a neuron in region B to the conditioning stimuli. Upper panel, spiking in response to $A{\rightarrow}B$ (red) and $B{\rightarrow}A$ (blue) conditioning (600 consecutive pairs each) at an 8.3 ms A/B interval. Lower panel, post-stimulus time histograms of the responses. (C) Cross-correlation between spike trains of the neuron evoked by $A{\rightarrow}B$ and $B{\rightarrow}A$ stimuli. Adapted with permission from Fu et al.[25].

3.2.2. Working Hypothesis

Fig. 5 illustrates a simple model of how relative timing of visual stimuli can affect cortical representation of visual space. Asynchronous stimuli in regions A and B can evoke asynchronous spiking in two groups of cortical neurons, a and b, whose RFs fall in A and B, respectively. According to STDP, $A{\rightarrow}B$ stimulation should strengthen $a{\rightarrow}b$ connections and weaken $b{\rightarrow}a$ connections (Fig. 5C, left). Potentiation of $a{\rightarrow}b$ connections should enhance the responses of group b neurons to visual stimuli in A, causing group b RFs to shift toward A, whereas depression of $b{\rightarrow}a$ connections should reduce the responses of group a neurons to stimuli in B, causing their RFs to shift away from B. Thus, synaptic modifications induced by $A{\rightarrow}B$ conditioning should cause the RFs of both groups to shift toward A (Fig. 5D, left). In a typical population-decoding scheme, the perceived stimulus location is determined by the spatial profile of population neuronal response. The leftward shift (towards A) of RFs should cause a rightward shift in the response profile, hence a rightward shift in the perceived stimulus position (Fig. 5E, left). By symmetry, $B{\rightarrow}A$ stimuli that activate group b before a should cause opposite changes at all levels (Fig 5, middle column). In the experiments described below, we have examined each of the predictions delineated in Fig. 5C-E.

3.2.3. Modification of Intracortical Connections

First, we examined conditioning-induced intracortical synaptic modification. Due to the technical difficulty of intracellular recording *in vivo*, cross-correlation analysis of simultaneously recorded cell pair was used instead[40]. While traditionally this method has been used to test the existence of inter-neuronal connections, we have extended this technique to measure stimulus-induced synaptic modification. First, to establish the relationship between intracortical synaptic modification and changes in the cross-correlogram between a cell pair, we performed a simulation study using the model circuit shown in Fig. 5A. Due to STDP, repetitive $A{\rightarrow}B$ conditioning potentiated $a{\rightarrow}b$ connections and depressed $b{\rightarrow}a$ connections, whereas $B{\rightarrow}A$ conditioning induced the opposite changes (Fig 5C). Accompanying these synaptic modifications, two types of changes were observed in the cross-correlation between pairs of model cells. First, for a cell pair within the same group (e.g., both neurons from group b), the peak amplitude of cross-correlation increased following $A{\rightarrow}B$ conditioning and decreased following $B{\rightarrow}A$ conditioning, reflecting potentiation and depression, respectively, of the common inputs from group a neurons. Second, for a cell pair between groups a and b, conditioning induced shifts in correlation asymmetry (measured by the difference in area between the right side and the left side of the peak), reflecting modifications of both $a{\rightarrow}b$ and $b{\rightarrow}a$ connections. In the physiology experiments, both types of changes were observed after visual conditioning, confirming the predictions of the model. Moreover, the change in intracortical cross-correlation as a function of A/B interval (Fig. 5C, right plot) exhibits an asymmetric profile within ±50 ms, reminiscent of the STDP window measured in visual cortical slices (Fig. 1B).

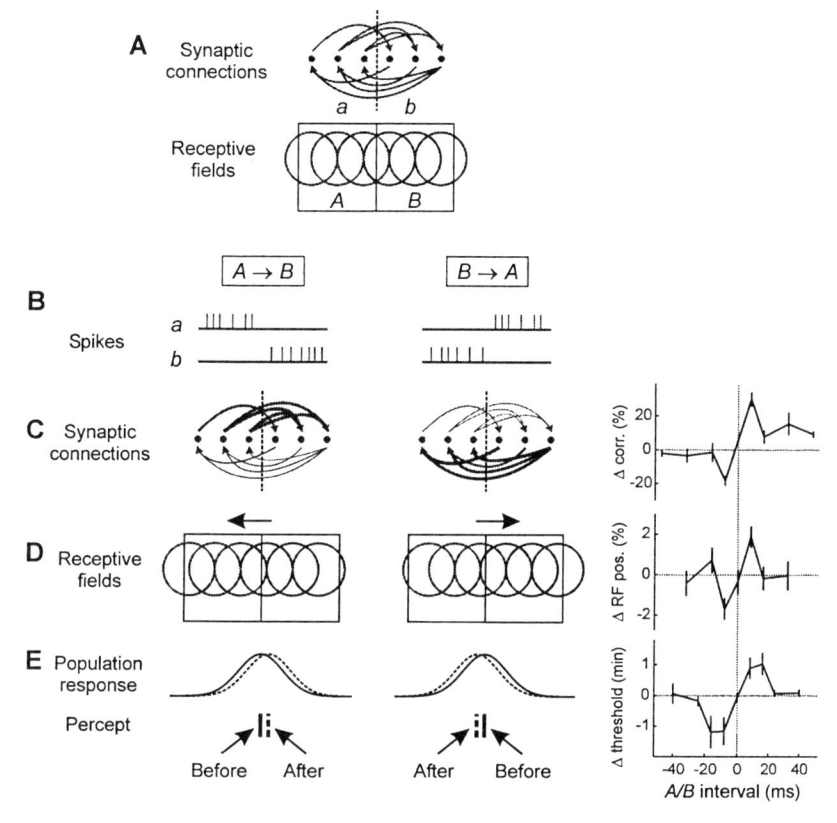

Fig. 5. A model in which stimulus timing affects cortical modification. (**A**) Intracortical excitatory connections between two groups of neurons, *a* and *b*, whose RFs fall in regions *A* and *B*, respectively. (**B**) Schematic representation of spike timing of groups *a* and *b* in response to *A*→*B* and *B*→*A* stimuli. (**C**) Connection strength (represented by line thickness) after conditioning, obtained in a simulation. Plot on the right: modification of intracortical connections as a function of *A/B* interval measured experimentally in cat V1 with cross-correlation analysis. (**D**) RFs after conditioning. Arrows indicate directions of shifts predicted by the model. Plot on the right shows the RF shift measured in cat V1 as a function of *A/B* interval. (**E**) Upper panel, spatial profiles of the population responses evoked by a vertical bar at the *A/B* border before (solid curve) and after (dashed) conditioning. Lower panel, perceived bar positions predicted by the model. Plot on the right shows the average perceptual shift as a function of *A/B* interval measured in four human subjects. Adapted with permission from Fu et al.[25].

3.2.4. Receptive Field Shift

Conditioning-induced intracortical synaptic modification should lead to shifts in cortical RFs (Fig. 5D). To test this prediction, we compared the RF position of each V1 neuron before and after conditioning. For convenience, the conditioning region containing the RF of the recorded neuron is always defined as *B*. We found that 800 pairs (~1.5 min) of *A*→*B* conditioning at 8.3 ms interval induced a significant RF shift towards *A*, whereas *B*→*A* conditioning induced a shift away from *A*. Fig. 5D (right plot) shows the conditioning-induced RF shift as a function of *A/B* interval, which also resembles the asymmetric STDP window for intracortical connections (Fig. 1B).

3.2.5. Perceptual Shift

Since among all visual cortical areas V1 neurons have the smallest RFs, they are believed to be important for the computation of stimulus position. Conditioning-induced shifts of V1 RFs are thus likely to result in shifts in perceptual localization (Fig. 5E). This prediction was tested in psychophysical experiments in human subjects. The conditioning stimuli were similar to those used in the physiological experiments, and a three-bar bisection test[41] was performed at the *A/B* border to measure the perceived position of the middle bar before and after conditioning. For all four subjects tested, 400 pairs (~ 50 s) of *A→B* conditioning induced a significant shift in the perceived location of the middle bar towards *B*, whereas *B→A* conditioning induced a shift towards *A*, consistent with the prediction shown in Fig. 5E. Furthermore, a significant effect was observed only at *A/B* intervals within ± 20 ms (Fig. 5E, right plot), satisfying the temporal specificity required by the model.

Together, these studies showed that asynchronous visual stimuli in different retinal regions can induce rapid changes in cortical representation of visual space. The observed effects of visual conditioning at the levels of synaptic connections, cortical RF, and visual perception all support the model based on STDP of intracortical connections (Fig. 5).

3.3. Stimulus Timing-Dependent Plasticity In Orientation Domain

In addition to the selectivity of stimulus position, another important RF property of V1 neurons is their orientation selectivity. This property was also shown to exhibit stimulus timing-dependent modification. In a study in the kitten visual cortex, repetitive pairing of an oriented visual stimulus with an electrical stimulation induced dramatic shifts in orientation preference maps, as revealed by both extracellular recording and optical imaging[23]. Orientation preference of the neuron shifts toward the paired orientation if the cortex is activated visually before electrically, and shifts away if the sequence of activation is reversed. Such dependence of cortical modification on the order of visual and electrical stimulation is consistent with STDP of the intracortical connections. In another study, stimulus timing-dependent plasticity of orientation processing was observed in adult visual cortex[24]. Repetitive pairing of visual stimuli at two orientations induced a shift in the orientation tuning of cat V1 neurons, with the direction of the shift depending on the temporal order of the pair. Importantly, induction of significant shift requires that the temporal interval between the pair of stimuli fall within ±40 ms, reminiscent of the temporal window for STDP[14,15]. Mirroring the plasticity found in the cat visual cortex, similar visual conditioning also induced a rapid shift in the perceived stimulus orientation by human subjects. Thus, stimulus timing-dependent plasticity also exists in cortical processing of orientation. Together with the findings in the space domain, these studies suggest that stimulus timing-dependent plasticity may be a general phenomenon in visual processing of multiple stimulus attributes.

4. FUNCTIONAL IMPLICATION OF STDP

Theoretical studies have explored a variety of functional implications of STDP[18, 20, 42-45]. Compared to the correlation-based synaptic learning rule, the asymmetric window of STDP provides a mechanism for competition between converging inputs[19]. It also suppresses destabilizing self-excitatory loops and allows a group of neurons that become selective early in development to guide the development of stimulus selectivity of other neurons[46]. Moreover, the dependence of synaptic modification on the spike order endows the nervous system with the capacity to learn the temporal sequence and the causal relationship between sensory stimuli. This allows the circuit to predict future events based on current stimuli[18, 22, 42, 45, 47, 48], which is an important task of the nervous system.

In the visual system, an important question is how STDP operates in the natural environment. In addition to the asynchronous flashed stimuli used in recent experiments, moving stimuli can also activate neighboring neurons at short intervals, which may induce cortical modification through STDP. Since moving stimuli are common in natural scenes, it may be the most important type of visual stimuli that shapes the cortical circuitry through STDP. Simulation studies have suggested that, when STDP is implemented in the cortical circuit, moving stimuli can facilitate the development of direction selectivity in cortical neurons[45, 49]. The interaction between STDP and moving stimuli in shaping visual cortical functions awaits further experimental and theoretical investigations.

5. CONCLUDING REMARKS

The role of spike timing in activity-dependent cortical modification has been investigated at several levels. At the synaptic level, several studies have demonstrated a critical dependence of intracortical synaptic modification on the relative timing of pre- and postsynaptic spikes. Recent experiments have also begun to delineate the rules governing synaptic modification induced by complex spike trains, which will help to understand how this mechanism operates in the intact brain. At the circuitry level, stimulus timing-dependent plasticity has been demonstrated in cortical representation of both position and orientation, and the temporal specificity of the effects strongly indicates the involvement of STDP. Finally, human psychophysical experiments have demonstrated perceptual plasticity directly corresponding to the cortical RF plasticity, supporting the functional relevance of the phenomenon. In spite of the recent progress, however, there are many important issues that are yet to be addressed. A comprehensive understanding of the functional roles of synaptic plasticity in learning, memory, and activity-dependent circuit refinement remains a central challenge for both the experimental and theoretical communities.

6. ACKNOWLEDGMENTS

We thank Robert C. Froemke and Haishan Yao for helpful discussions. This work was supported by a grant from the National Eye Institute.

7. REFERENCES

1. Hebb, D. (1949) *The Organization of Behavior: A Neuropsychological Theory*, John Wiley and Sons, New York.
2. Stent, G. S. (1973) A physiological mechanism for Hebb's postulate of learning, *Proc Natl Acad Sci U S A* **70**:997.
3. Sejnowski, T. J., and Tesauro, G. (1989) The Hebb rule for synaptic plasticity: algorithms and implementations, in: *Neural Models of Plasticity: Experimental and Theoretical Approaches*, Byrne, J. H., and Berry, W. O., eds., Academic Press, New York.
4. Miller, K. D., Keller, J. B., and Stryker, M. P. (1989) Ocular dominance column development: analysis and simulation, *Science* **245**:605.
5. Weliky, M., and Katz, L. C. (1997) Disruption of orientation tuning in visual cortex by artificially correlated neuronal activity, *Nature* **386**:680.
6. Markram, H., Lubke, J., Frotscher, M., and Sakmann, B. (1997) Regulation of synaptic efficacy by coincidence of postsynaptic APs and EPSPs, *Science* **275**:213.
7. Bell, C. C., Han, V. Z., Sugawara, Y., and Grant, K. (1997) Synaptic plasticity in a cerebellum-like structure depends on temporal order, *Nature* **387**:278.
8. Zhang, L. I., Tao, H. W., Holt, C. E., Harris, W. A., and Poo, M. (1998) A critical window for cooperation and competition among developing retinotectal synapses, *Nature* **395**:37.
9. Bi, G. Q., and Poo, M. M. (1998) Synaptic modifications in cultured hippocampal neurons: dependence on spike timing, synaptic strength, and postsynaptic cell type, *J Neurosci* **18**:10464.
10. Boettiger, C. A., and Doupe, A. J. (2001) Developmentally restricted synaptic plasticity in a songbird nucleus required for song learning, *Neuron* **31**:809.
11. Debanne, D., Gahwiler, B. H., and Thompson, S. M. (1998) Long-term synaptic plasticity between pairs of individual CA3 pyramidal cells in rat hippocampal slice cultures, *J Physiol (Lond)* **507**:237.
12. Nishiyama, M., Hong, K., Mikoshiba, K., Poo, M. M., and Kato, K. (2000) Calcium stores regulate the polarity and input specificity of synaptic modification, *Nature* **408**:584.
13. Feldman, D. E. (2000) Timing-based LTP and LTD at vertical inputs to layer II/III pyramidal cells in rat barrel cortex, *Neuron* **27**:45.
14. Sjostrom, P. J., Turrigiano, G. G., and Nelson, S. B. (200) Rate, timing, and cooperativity jointly determine cortical synaptic plasticity, *Neuron* **32**:1149.
15. Froemke, R. C., and Dan, Y. (2002) Spike-timing-dependent synaptic modification induced by natural spike trains, *Nature* **416**:433.
16. Shadlen, M. N., and Newsome, W. T. (1994) Noise, neural codes and cortical organization, *Curr Opin Neurobiol* **4**:569.
17. Shadlen, M. N., and Newsome, W. T. (1998) The variable discharge of cortical neurons: Implications for connectivity, computation, and information coding, *J. Neurosci.* **18**:3870.
18. Roberts, P. D. (1999) Computational consequences of temporally asymmetric learning rules: I. Differential hebbian learning, *J Comput Neurosci* **7**:235.
19. Song, S., Miller, K. D., and Abbott, L. F. (2000) Competitive Hebbian learning through spike-timing-dependent synaptic plasticity, *Nat Neurosci* **3**:919.
20. Kempter, R., Leibold, C., Wagner, H., and van Hemmen, J. L. (2001) Formation of temporal-feature maps by axonal propagation of synaptic learning, *PNAS* **98**:4166.
21. Malenka, R. C., and Nicoll, R. A. (1999) Long-term potentiation--a decade of progress?, *Science* **285**:1870.
22. Mehta, M. R., Quirk, M. C., and Wilson, M. A. (2000) Experience-dependent asymmetric shape of hippocampal receptive fields, *Neuron* **25**:707.
23. Schuett, S., Bonhoeffer, T., and Hubener, M. (2001) Pairing-induced changes of orientation maps in cat visual cortex, *Neuron* **32**:325.
24. Yao, H., and Dan, Y. (2001) Stimulus timing-dependent plasticity in cortical processing of orientation, *Neuron* **32**:315.
25. Fu, Y. X., Djupsund, K., Gao, H., Hayden, B., Shen, K., and Dan, Y. (2002) Temporal specificity in the cortical plasticity of visual space representation, *Science* **296**:1999.
26. Constantine-Paton, M., Cline, H. T., and Debski, E. (1990) Patterned activity, synaptic convergence, and the NMDA receptor in developing visual pathways, *Annu Rev Neurosci* **13**:129.
27. Katz, L. C., and Shatz, C. J. (1996) Synaptic activity and the construction of cortical circuits, *Science* **274**:1133.
28. Gilbert, C. D. (1998) Adult cortical dynamics, *Physiol. Rev.* **78**:467.
29. Maffei, L., Fiorentini, A., and Bisti, S. (1973) Neural correlate of perceptual adaptation to gratings, *Science*

182:1036.
30. Movshon, J. A., and Lennie, P. (1979) Pattern-selective adaptation in visual cortical neurones, *Nature* **278**:850.
31. Dragoi, V., Sharma, J., and Sur, M. (2000) Adaptation-induced plasticity of orientation tuning in adult visual cortex, *Neuron* **28**:287.
32. Marlin, S. G., Hasan, S. J., and Cynader, M. S. (1988) Direction-selective adaptation in simple and complex cells in cat striate cortex, *J Neurophysiol* **59**:1314.
33. Carandini, M., and Ferster, D. (1997) A tonic hyperpolarization underlying contrast adaptation in cat visual cortex, *Science* **276**:949.
34. Sanchez-Vives, M. V., Nowak, L. G., and McCormick, D. A. (2000) Membrane mechanisms underlying contrast adaptation in cat area 17 in vivo, *J. Neurosci.* **20**:4267.
35. Chance, F. S., Nelson, S. B., and Abbott, L. F. (1998) Synaptic depression and the temporal response characteristics of V1 cells, *J. Neurosci.* **18**:4785.
36. Fregnac, Y., Shulz, D., Thorpe, S., and Bienenstock, E. (1988) A cellular analogue of visual cortical plasticity, *Nature* **333**:367.
37. Fregnac, Y., Shulz, D., Thorpe, S., and Bienenstock, E. (1992) Cellular analogs of visual cortical epigenesis. I. Plasticity of orientation selectivity, *J Neurosci* **12**:1280.
38. McLean, J., and Palmer, L. A. (1998) Plasticity of neuronal response properties in adult cat striate cortex, *Vis Neurosci* **15**:177.
39. Eysel, U. T., Eyding, D., and Schweigart, G. (1998) Repetitive optical stimulation elicits fast receptive field changes in mature visual cortex, *Neuroreport* **9**:949.
40. Perkel, D. H., Gerstein, G. L., and Moore, G. P. (1967) Neuronal spike trains and stochastic point processes. II. Simultaneous spike trains, *Biophys J* **7**:419.
41. Kapadia, M. K., Gilbert, C. D., and Westheimer, G. (1994) A quantitative measure for short-term cortical plasticity in human vision, *J Neurosci* **14**:451.
42. Abbott, L., and Blum, K. (1996) Functional significance of long-term potentiation for sequence learning and prediction, *Cereb. Cortex* **6**:406.
43. Gerstner, W., Kempter, R., van Hemmen, J. L., and Wagner, H. (1996) A neuronal learning rule for sub-millisecond temporal coding, *Nature* **383**:76.
44. Kistler, W. M., and van Hemmen, J. L. (2000) Modeling synaptic plasticity in conjuction with the timing of pre- and postsynaptic action potentials, *Neural Comput* **12**:385.
45. Rao, R. P. N., and Sejnowski, T. J. (2000) Predictive sequence learning in recurrent neocortical circuits, in: *Advances in Neural Information Processing Systems 12 (NIPS*99)*, Solla, S. A., Leen, T. K., and Muller, K.-R., eds., MIT Press,
46. Song, S., and Abbott, L. F. (2001) Cortical development and remapping through spike timing-dependent plasticity, *Neuron* **32**:339.
47. Minai, A. A., and Levy, W. B. (1993) Sequence learning in a single trial, *INNS World Congress of Neural Networks* **II**:505.
48. Dayan, P. (2002) Matters temporal, *Trends Cogn Sci* **6**:105.
49. Buchs, N. J., and Senn, W. (2002) Spike-based synaptic plasticity and the emergence of direction selective simple cells: simulation results, *J Comput Neurosci* **13**:167.

17

EXCITATORY-INHIBITORY BALANCE CONTROLS CRITICAL PERIOD PLASTICITY

Michela Fagiolini and Takao K. Hensch[*]

1. INTRODUCTION

Neuronal circuits are shaped by their activity during 'critical' or 'sensitive periods' in development. Initially spontaneous, then early sensory-evoked patterns of action potentials, are required to sculpt the remarkably complex connectivity found in the adult brain, which then loses this extraordinary level of plasticity. Whether it is the targeting of individual axons or the acquisition of language, there is no doubt that dramatic re-wiring is most powerful early in postnatal life. Despite decades of similar robust observations across a wide spectrum of brain functions, only recently have we begun to understand the cellular basis that may underlie this fundamental process. The ability to freely switch on or off critical period mechanisms confirms the very existence of such special stages of heightened plasticity. In this chapter, we will focus on a newfound perspective of excitatory-inhibitory balance within cortical circuits that has finally granted us this control.

2. VISUAL CORTEX AS A MODEL SYSTEM

The premier physiological model of critical period plasticity is the developing visual system. Over forty years ago[1], Hubel and Wiesel first described the loss of responsiveness to an eye deprived of vision in the primary visual cortex of kittens. As a direct behavioral consequence, the deprived eye becomes amblyopic: its visual acuity is strongly reduced and its contrast sensitivity blunted[2]. Moreover, the rapid physiological effects of monocular deprivation (MD) are soon accompanied by an anatomical reduction in size of horizontal connections[3] and thalamic afferents serving the deprived eye[4].

---

[*] M. Fagiolini, T.K. Hensch. Lab for Neuronal Circuit Development, Critical Period Mechanisms Research Group, RIKEN Brain Science Institute, 2-1 Hirosawa, Wako-shi, Saitama 351-0198 JAPAN

Excitatory-Inhibitory Balance: Synapses, Circuits, Systems
Edited by Hensch and Fagiolini, Kluwer Academic/Plenum Publishers, 2003

Altogether these processes depend upon competitive interactions between the two eyes for the control of cortical territory. When both eyes are sutured, no imbalance of input occurs and neither eye loses the ability to drive visual responses[2,5].

Importantly, a shift in ocular dominance toward the open eye occurs only during a transient developmental critical period. Since Hubel and Wiesel's seminal work[6], the rules of activity-dependent competition and timing have been confirmed across a variety of species. Interestingly, the duration of the critical period appears to be tightly linked to the average life expectancy for each mammal studied[7]. In rodents[8,9] as well as in cats[2], plasticity is low at eye opening, peaks around 4 weeks of age, and declines over several weeks to months. Notably, the critical period is not a simple, age-dependent maturational process, but rather a series of events itself controlled in a use-dependent manner. Animals reared in complete darkness from birth express a delayed profile with plasticity persisting into adulthood[2,10,11].

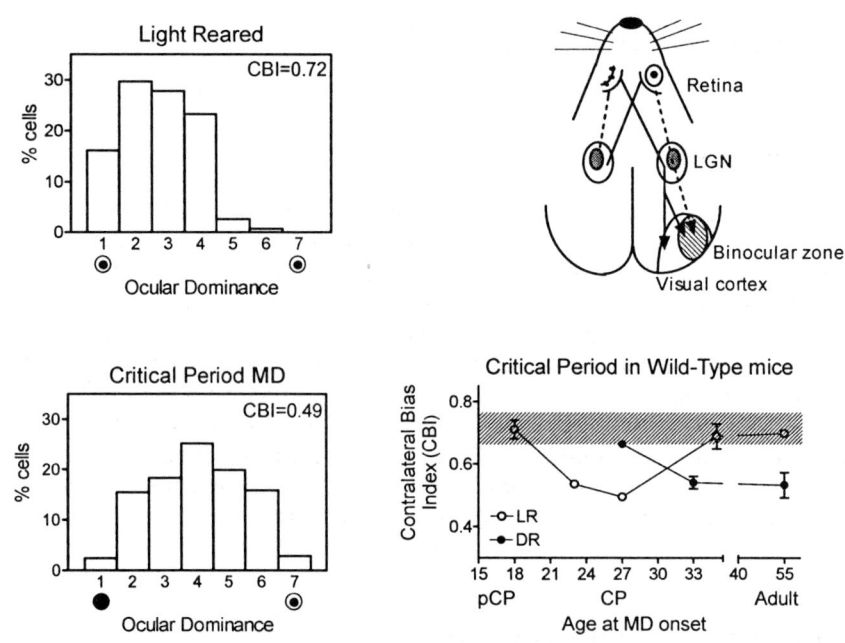

Fig. 1. Critical period plasticity in mouse visual cortex. The typical response bias toward contralateral eye input (ocular dominance groups 1-3) in the binocular zone of rodents is robustly shifted in favor of the open ipsilateral eye (groups 5-7) following monocular deprivation (MD) during the critical period (CP). Histograms are quantified as a weighted average (contralateral bias index, CBI), which ranges from 0 to 1 for complete ipsilateral or contralateral eye dominance, respectively. One week after eye-opening at P14, sensitivity to brief MD rapidly appears and persists for about two weeks, as measured by single-unit recording. The immature pre-CP (pCP) phase is prolonged by dark-rearing from birth, such that the overall profile is delayed to yield plasticity in adulthood as shown for cats. Shaded region, range of non-deprived wild-type mice. Some error bars smaller than symbol size. Data adapted from refs. 9,13,67.

The mouse visual cortex offers tremendous advantages for developmental analysis, despite their largely nocturnal lifestyle. Born with an innate contralateral bias, the laterally-displaced eyes of rodents drive a robust, competition within a small, binocular zone of primary visual cortex during a short, two-week critical period, which can be delayed by dark-rearing (Fig. 1)[8,9,12,13]. The ensuing shift in responsiveness is reflected in reduced behavioral visual acuity during a critical period identical to that measured by single-unit electrophysiology[14]. Anatomical correlates also follow the rapid physiological changes in response to MD[15].

The restricted spatio-temporal dimensions in an animal model of short gestation and large litter size are ideal for dissecting the cellular and molecular mechanisms of plasticity onset and closure. Moreover, the power of genetic manipulation offers a specificity of molecular control unprecedented by pharmacological approaches in the cat or monkey.

3. MANIPULATING EXCITATORY-INHIBITORY BALANCE IN VIVO

Despite a wealth of phenomenology regarding the rules of experience-dependent development[2], precious little is known about the underlying cellular mechanism. Over the years, a popular model of homosynaptic plasticity has emerged in parallel studies of learning and memory primarily in the hippocampus. While it is attractive to think of loss of deprived-eye input as a long-term depression (LTD) or gain of open-eye input as long-term potentiation (LTP), advancing knowledge of their molecular mechanism in vitro[16] has so far failed to predictably alter the critical period. Disruption of tetanus-induced LTP or low-frequency induced LTD has been dissociated from ocular dominance plasticity on numerous occasions[17-24]. At the very least, we have learned that cortical plasticity in the intact animal is not easily explained by artificial stimulus protocols used to adjust individual glutamatergic synapses in brain slices. In retrospect, this is not surprising, because unlike motor axons competing for a single target muscle fiber, sensory input to the neocortex must first be integrated by complex local circuit interactions in vivo.

We therefore formulated the hypothesis that local excitatory and inhibitory cortical circuits reach an optimal balance only once in life during which plasticity may occur. If correct, direct manipulation of either excitation or inhibition should profoundly affect the sensitivity to sensory deprivation. Previous pharmacological attempts to disrupt the balance grossly hyper-excited[25-27] or shut down the cortex[28], yielding little insight into the normal function of local circuits during plasticity. Taking advantage of gene-targeting technology, we instead attempted to gently titrate endogenous inhibition by reducing GABA synthesis or to enhance excitation by prolonging glutamatergic synaptic responses (Fig. 2). Both adjustments would be expected to yield a similar shift of balance in favor of excitation in vivo.

Distinct genes encode the two isoforms of GABA-synthetic enzyme, glutamic acid decarboxylase (GAD)[29]. The larger 67-kD protein (GAD67) is localized to cell somata and dendrites, producing a constitutive concentration of GABA throughout the cell (Fig. 2, right). In the absence of GAD67, mice die at birth with brain GABA concentrations less than 10% of that found in wild-type mice[30]. Intriguingly, the 65-kD isoform (GAD65) is found primarily in the synaptic terminal, where it is anchored to vesicles and serves as a reservoir of inactive GAD that can be recruited when additional GABA synthesis is required[29]. During intense neuronal activity, GAD65 may be specialized to respond to rapid changes in synaptic demand.

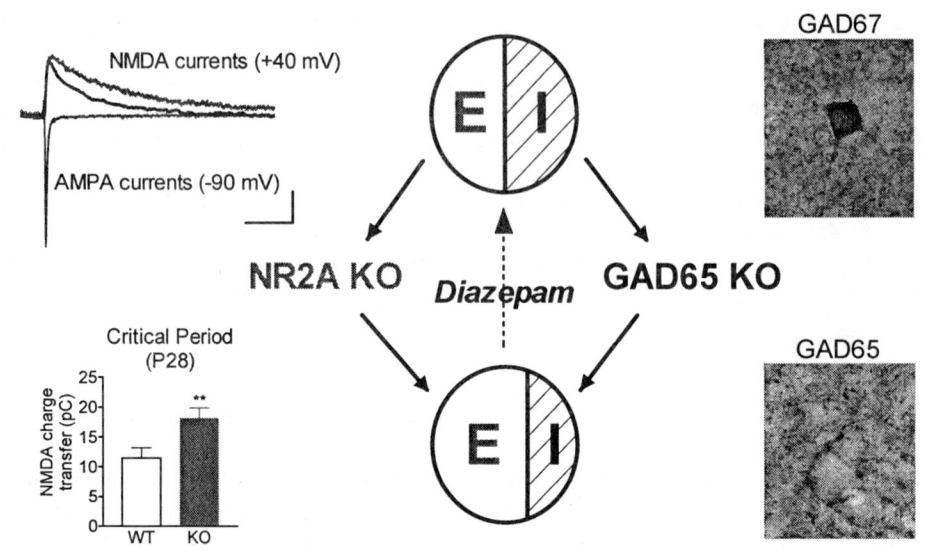

Fig. 2. Testing the role of excitatory-inhibitory balance in visual cortical plasticity in vivo. Glutamate-mediated excitation is prolonged (gray traces) through NMDA receptors lacking the NR2A subunit[13], yielding increased charge transfer at P28. On the inhibitory side, GABA is synthesized by two isoforms of the enzyme glutamic acid decarboxylase (GAD). Targeted disruption of the punctate, synaptic isoform (GAD65) reduces stimulated GABA release from intrinsic interneurons. Both manipulations would disrupt the balance in favor of excitation, and should be similarly reversed by the use-dependent $GABA_A$ receptor modulator diazepam.

Mice carrying a targeted disruption of GAD65 gene survive and exhibit a normal GABA content in the adult brain[31,32]. However, as predicted, they display a significant reduction of stimulated GABA release[33,34]. Due to their activity-dependent phenotype, GAD65 knockout (KO) mice represent an ideal tool with which to test the role of endogenous inhibitory circuits in synaptic plasticity.

To directly affect excitation, we noted the developmental change in subunit composition of the N-methyl-D-aspartate (NMDA)-type glutamate receptor. Composed of a principal subunit NR1 and different modulatory NR2 partners, NMDA receptor-mediated synaptic current decay is truncated by an activity-dependent switch in predominant subunit composition from NR2B to NR2A[35,36]. Indeed, NMDA response kinetics in brain slices from NR2A knockout (KO) mouse visual cortex remain dramatically prolonged well beyond the critical period, yielding increased charge transfer through NMDA receptor channels (Fig. 2, left)[13]. Global removal of NR1 is neonatal lethal, whereas conditional NR1 deletion renders the cortex visually unresponsive (M.F., T. Iwasato, S. Itohara, T.K.H., unpublished observations), as observed previously in vivo for pharmacological NMDA receptor antagonists[37]. In contrast, NR2A protein expression exhibits a late postnatal onset in visual cortex (after eye-opening)[13,36], making the NR2A KO mouse an ideal candidate to directly test the role of enhanced excitation conditional to the late postnatal timecourse of the critical period.

Indeed, one significant drawback of using first-generation KO mice is the potential for compensatory changes and gross developmental defects due to the chronic deletion of a protein from the entire animal. To counteract this potential difficulty of interpretation, we further attempted to restore the perturbed excitatory-inhibitory balance in both KO mouse models by the infusion of benzodiazepine agonists (Fig. 2). These drugs, such as diazepam, selectively increase the open probability and channel conductance of a subset of $GABA_A$ receptors in a use-dependent manner[38-40], as they are inert in the absence of synaptic GABA release. Moreover, benzodiazepine binding sites are associated with intrinsic cortical elements rather than thalamocortical axons or other subcortical inputs. Finally, they can be delivered in a highly restricted manner through the use of osmotic minipumps for spatial specificity (Fig. 3).

4. EXCITATORY-INHIBITORY BALANCE DRIVES OCULAR DOMINANCE PLASTICITY

Extracellular single-unit recording from the binocular zone of visual cortex in GAD65 KO mice revealed an identical ocular dominance distribution to wild type animals. The response to a 4-day period of monocular occlusion beginning between P25 and P27 was, however, strikingly different[33]. Mice lacking GAD65 showed no shift in their responsiveness in favor of the open eye and cortical neurons continued to respond

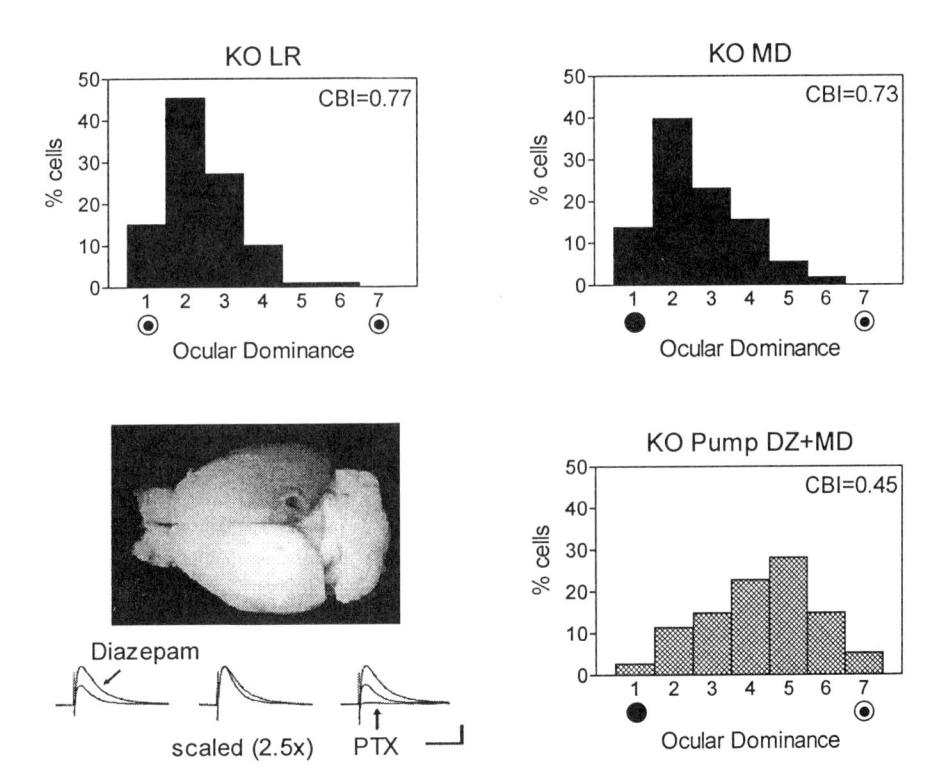

Fig. 3 Impaired ocular dominance plasticity in GAD65 knockout (KO) mice is restored by local diazepam infusion. Typical contralaterally-biased histograms fail to shift toward the open, ipsilateral eye after a brief period of monocular deprivation (MD) due to reduced GABA release. Osmotic minipump infusion (red area) of the use-dependent $GABA_A$ receptor modulator diazepam (DZ) locally restores inhibitory synaptic transmission and rescues the ocular dominance shift. DZ enhances both the amplitude and decay kinetics of $GABA_A$ receptor-mediated synaptic currents, as they are fully blocked by picrotoxin (PTX). Data adapted from ref. 33.

better to the contralateral eye input (Fig. 3). In order to rescue the plasticity defect in vivo, we enhanced inhibition by delivering diazepam (DZ) locally into one hemisphere during a period of MD by the use of an osmotic minipump. Drug diffusion was restricted to the treated visual cortex, while it remained undetectable in the adjacent temporal cortex, frontal regions, or opposite hemisphere[33]. Under these conditions, MD now produced a complete ocular dominance shift in the infused mutant visual cortex (Fig. 3), whereas no rescue was observed by administering vehicle solutions or in the hemisphere opposite to DZ infusion. Consequently, similar results were obtained with global DZ treatment by intraventricular injections concurrent with 4-day MD.

In adult, non-deprived NR2A KO mice, ocular dominance distribution was again similar to control animals (Fig. 4). Unlike GAD65 KO mice, brief MD was able to induce a slight shift in favor of the open eye, but interestingly the overall magnitude of this plasticity was significantly weakened. Long-term MD (>2 weeks) produced no further shift, confirming that saturation had been reached within four days[13]. We then attempted to rescue full plasticity in NR2A KO mice by DZ infusion concomitant with brief MD. Similar to GAD65 KO mice, the ocular dominance distribution then shifted completely with drug treatment (Fig. 4).

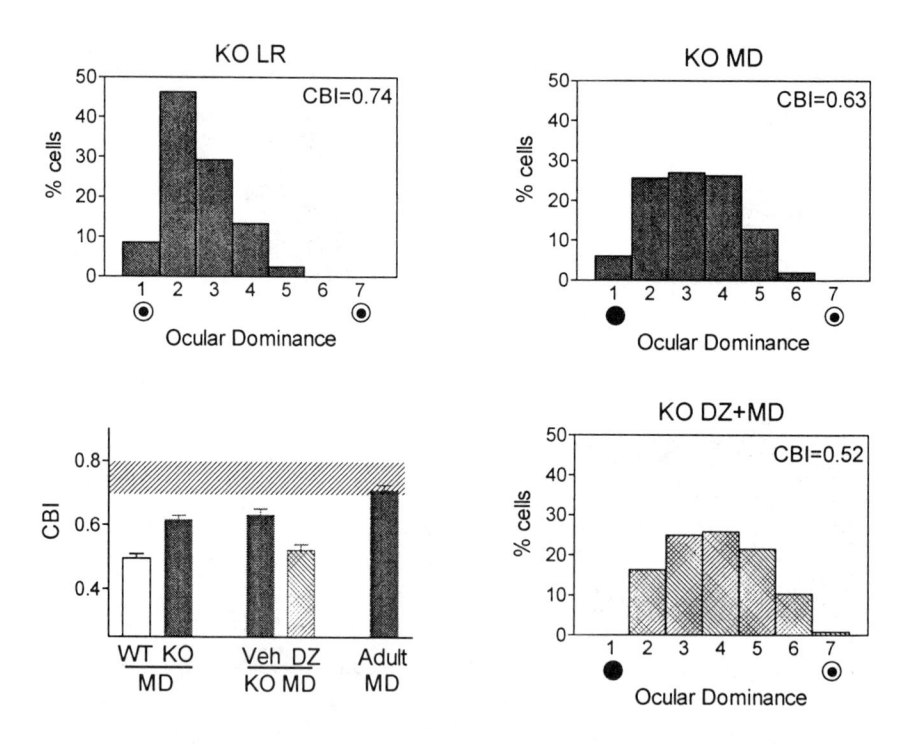

Fig. 4. Impaired ocular dominance plasticity in NR2A knockout (KO) mice is also rescued by enhancing inhbition with diazepam (DZ). Typical contralateral bias of normally-reared animals shifts only partially after monocular deprivation (MD) when compared to wild-type (WT) controls. The CBI declines fully when MD is combined with DZ injection, but observes the typical critical period as adult mice do not show plasticity. Data adapted from ref. 13.

EXCITATORY-INHIBITORY BALANCE CONTROLS CRITICAL PERIOD PLASTICITY

A direct physiological consequence of reduced inhibition in GAD65 KO mice was enhanced activation in response to visual stimulation[33]. Visual cortical neurons displayed a tendency for prolonged discharge as light-bar stimuli exited the cell's receptive field (Fig. 5, left), yielding excess spike firing by single units in all layers that outlasted the visual stimulus. Correspondingly, NR2A KO mice also exhibited prolonged neuronal discharge (76% vs. 2% of cells compared to wild-type)[13], indicating that in both KO mouse models excitatory-inhibitory balance had been disrupted similarly. Whereas robust prolonged discharge appeared throughout life in the mutants, it was only evident early in the life of wild-type animals before the critical period[42], when intrinsic inhibition is weak, NR2A expression is rising, and ocular dominance plasticity is absent. Whenever prolonged discharge was encountered, a significant reduction (>25% of cells) by DZ infusion in vivo unmasked visual cortical plasticity (Fig. 5, right). With the natural appearance of ocular dominance plasticity during the critical period in wild-type mice, prolonged discharge drops off sharply. Further shifting cortical balance in favor of inhibition with DZ application at this time tends to sharpen plasticity but not significantly beyond the normal range[33].

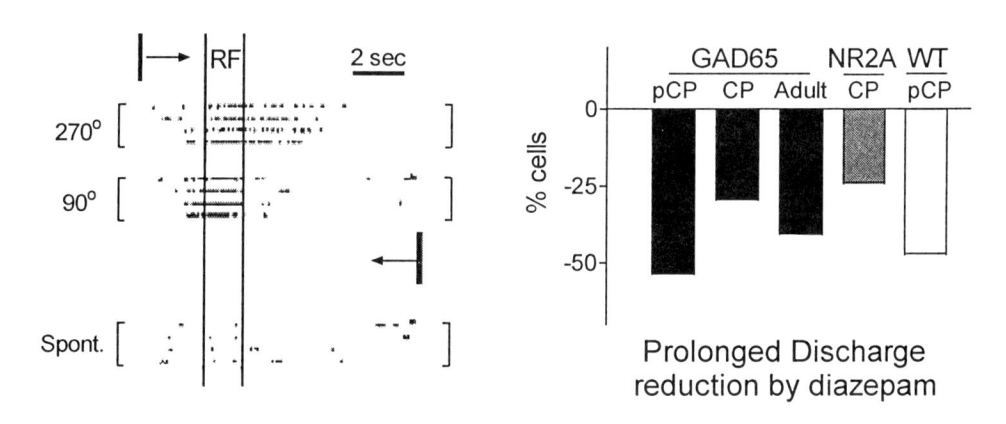

Fig. 5. Prolonged neuronal discharge as a consequence of perturbed excitatory-inhibitory balance in vivo. Neuronal responses exceed the passage of moving light-bar stimuli beyond the edges of their receptive field. This phenotype is robustly observed in GAD65 KO mice of all ages, as well as NR2A KO mice in the critical period, but only prior to its onset (pCP) in wild-type animals. In all cases, diazepam treatment significantly reduces the proportion of affected cells to produce full ocular dominance plasticity. Data adapted from refs. 13, 33, 42.

Taken together, a delicate equilibrium between excitation and inhibition intrinsic to visual cortical circuits is necessary to detect the imbalanced activity between competing inputs from the two eyes. Furthermore, fast inhibitory transmission via $GABA_A$-mediated connections seems to be the main driving force in this process, as plasticity impairment was more potent in the absence of GAD65 than NR2A and was rescued by benzodiazepines in both cases.

5. MECHANISMS AND FUTURE DIRECTIONS

Excitatory-inhibitory balance determines the neural coding of sensory input (see chapter 7). As described elsewhere (chapters 14, 16), specific spike timing-dependent windows for synaptic plasticity have recently been elucidated in developing and neocortical structures[43]. Unlike classical models of LTP induced by changes in mean firing rate that are strictly blocked by enhancing inhibition with benzodiazepines[44,45], spike-timing forms of plasticity rely upon physiologically realistic, millisecond-scale changes in the temporal order of pre- and postsynaptic action potentials. Prolonged discharge in both NR2A and GAD65 KO mice would impair plasticity by altering the pattern of neural activity encoding visual input. Diazepam would subtly improve temporal processing in both animal models to fully restore ocular dominance shifts in response to MD. A competitive outcome is also more readily understood by spike-timing rather than homosynaptic plasticity rules[46-48].

Tight regulation of neural coding by inhibition may indeed play the dominant role[49,50]. Among the vast diversity of GABAergic interneurons in neocortex (see chapters 8, 9), two major sub-classes of parvalbumin-containing cells target the axon initial segment and soma[51,52]. Both are ideally situated to control either spike initiation (chandelier cells) or back-propagation (basket cells), respectively, required for synaptic plasticity in the dendritic arbor (Fig. 6). It is possible to reduce the fast-spiking behavior of these circuits in a cell-specific manner by deleting their particular potassium-channels

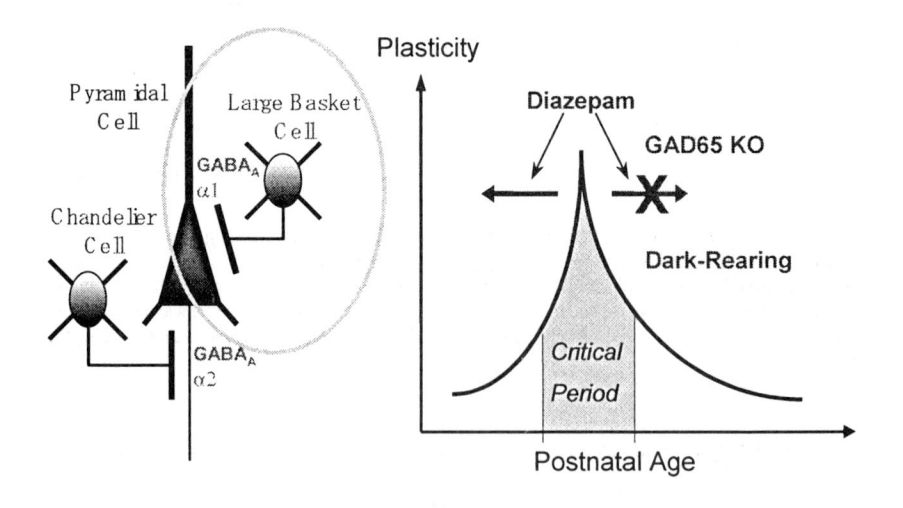

Fig. 6 Specific inhibitory sub-circuits may drive ocular dominance plasticity by regulating spike-timing in the dendrites. Proper excitatory-inhibitory balance also triggers a cascade of molecular events leading to structural consolidation of developing circuits and critical period closure.

($K_v3.1$)[53-55], and to so mimic the global GAD65 KO phenotype (Y. Matsuda et al., submitted). Moreover, because distinct $GABA_A$ receptor subunits are enriched at these two discrete parvalbumin-cell synapses (see chapter 13)[56,57], it will be possible to further discriminate their individual contributions to visual cortical processing and plasticity[58].

Large-basket cells in particular extend a wide, horizontal axonal arbor that can span ocular dominance columns in cat visual cortex[59], which would be useful in detecting and discriminating input coming from the two eyes. Moreover, electrically-coupled networks of fast-spiking cells offer a system exquisitely sensitive to timing that could detect and pass along synchronized signals (chapter 10). Coincidence detection by precise NMDA receptor kinetics instead does not determine critical period duration, as predicted by traditional LTP models[46,60]. Plasticity begins and ends normally in the absence of NR2A[13], while parvalbumin-cells emerge with a postnatal timecourse that parallels critical period onset[61,62].

Given the evidence for a mechanistic dissociation of traditional LTP/LTD models from ocular dominance plasticity at the molecular level, confirmation of spike-timing dependent processes remains pending. If the latter were to involve other factors, such as cannabinoids (see chapter 6), distinct from those recruited by tetanic or low-frequency stimulation, we could specifically address its role in ocular dominance plasticity. Notably, receptive field size, orientation selectivity, direction selectivity (chapter 16) and whisker barrel plasticity (chapter 14) are modifiable throughout life by spike-timing protocols, whereas ocular dominance plasticity observes a strict critical period. Indeed, orientation preference is not affected by altered excitatory-inhibitory balance or diazepam[42], and fails to mature in the absence of NR2A[13], making it distinct from ocular dominance.

To reopen ocular dominance plasticity requires a drastic disruption of extracellular matrix structure. Mature fast-spiking parvalbumin neurons are predominantly surrounded by perineuronal nets[63], which when disrupted by protease treatment reinstate visual cortical plasticity to adult animals[64]. It will be of interest to determine how nascent inhibitory connections, while playing important regulatory roles in refining excitatory connections, can themselves be consolidated into the mature circuit. Recent anatomical evidence has exposed a dynamic re-sculpting of dendritic spines[65], GABAergic synapses[66], and horizontal connections[3] within days of sensory perturbation.

Excitatory-inhibitory balance may thus ultimately regulate structural consolidation. Systematic mapping of DZ injections to rescue GAD65 KO mice reveals a minimum requirement of two days at the beginning of MD[67]. Interestingly, MD induces a peak increase of extracellular proteolytic activity in visual cortex by two days[68]. This regulation fails to occur in GAD65 KO mice, suggesting a cascade for plasticity from functional imbalance to structural change through the release of factors such as tPA. Indeed, the critical period itself fails to begin in GAD65 KO mice until DZ infusion[42]. Just two days of drug exposure eventually closes a stereotypical window (>14 days) for plasticity as seen normally. It can remarkably do the same in the absence of visual input to block the effects of dark-rearing in wild-type animals (Fig. 6)[67]. Dark rearing in fact alters the development of GABAergic transmission in visual cortex[69], reminiscent of GAD65 deletion.

Thus, proper excitatory-inhibitory balance represents merely the start of the critical period, and much remains to be elucidated downstream of this trigger to determine how and why plasticity ends.

6. CONCLUDING REMARKS

We have demonstrated the direct control of a classical critical period plasticity in developing primary visual cortex by focusing anew on excitatory-inhibitory balance. How general this principle will be across brain systems remains to be seen. It is already noteworthy that in the primary motor nucleus of the zebrafinch (RA), GABA cell number peaks in striking correlation with the acquisition of song only in the male birds that sing[70]. In contrast, regions exhibiting persistent plasticity, such as the olfactory bulb, continue to generate GABA cells throughout life (see chapter 12). As more becomes known about the molecular composition and plasticity of inhibitory synapses (chapters 3-5), as well as the ultimate structural changes that hardwire changes, it will become possible to test the importance of excitatory-inhibitory balance for critical periods of brain development in ever finer detail. Unraveling the mechanisms that limit such dramatic plasticity to early life would pave the way for novel paradigms or therapeutic agents for rehabilitation, recovery from injury, or improved learning across the lifespan.

7. ACKNOWLEDGMENTS

We thank Dr. H. Katagiri for NMDA receptor physiology in vitro, Y. Tsuchimoto for GAD immunostaining; Drs. S. Kash, S. Baekkeskov, K. Obata for GAD65, and H. Mori, M. Mishina for GluRε1 (NR2A) knockout animals. Mice were re-derived and maintained at RIKEN by S. Fujishima and Y. Mizuguchi. Supported by RIKEN Brain Science Institute, CREST, Special Coordination Funds for Promoting Science and Technology (Japan Science and Technology Corp.), and the Human Frontiers Science Program (HFSP).

8. REFERENCES

1. Wiesel TN, Hubel DH. (1963) Single-cell responses in striate cortex of kittens deprived of vision in one eye. *J. Neurophysiol.* **26**, 1003-1017.
2. Daw N. (1995) *Visual Development* (Plenum, New York).
3. Trachtenberg JT, Stryker MP. (2001) Rapid anatomical plasticity of horizontal connections in the developing visual cortex. *J Neurosci.* **21**, 3476-3482.
4. Antonini A, Stryker MP. (1993) Rapid remodeling of axonal arbors in the visual cortex. *Science* **260**, 1819-1821.
5. Antonini A, Stryker MP. (1998) Effect of sensory disuse on geniculate afferents to cat visual cortex. *Vis. Neurosci.* **15**, 401-409.
6. Hubel DH, Wiesel TN. (1970) The period of susceptibility to the physiological effects of unilateral eye closure in kittens. *J. Physiol. (Lond)* **206**, 419-436.
7. Berardi N, Pizzorusso T, Maffei L. (2000) Critical periods during sensory development. *Curr. Opin. Neurobiol.* **10**, 138-145.
8. Fagiolini M, Pizzorusso T, Berardi N, Domenici L, Maffei L. (1994) Functional postnatal development of the rat primary visual cortex and the role of visual experience: dark rearing and monocular deprivation. *Vision Res.* **34**, 709-720.
9. Gordon JA, Stryker MP. (1996) Experience-dependent plasticity of binocular responses in the primary visual cortex of the mouse. *J Neurosci.* **16**, 3274-3286.
10. Cynader M. (1983) Prolonged sensitivity to monocular deprivation in dark-reared cats: effects of age and visual exposure. *Brain Res.* **284**, 155-164.
11. Mower GD. (1991) The effect of dark rearing on the time course of the critical period in cat visual cortex. *Dev. Brain Res.* **58**, 151–158.

12. Benevento LA, Bakkum BW, Port JD, Cohen RS. (1992) The effects of dark-rearing on the electrophysiology of the rat visual cortex. *Brain Res.* **572**, 198-207.
13. Fagiolini M *et al.* (2003) Separable features of visual cortical plasticity revealed by N-methyl-D-aspartate receptor 2A signaling. *Proc. Natl. Acad. Sci. USA* **100**, 2854-2859.
14. Prusky GT, Douglas RM. (2003) Developmental plasticity of mouse visual acuity. *Eur. J. Neurosci.* **17**, 167-173.
15. Antonini A, Fagiolini M, Stryker MP. (1999) Anatomical correlates of functional plasticity in mouse visual cortex. *J Neurosci.* **19**, 4388-4406.
16. Sanes JR, Lichtman JW. (1999) Can molecules explain long-term potentiation? *Nat. Neurosci.* **2**, 597-604.
17. Hensch TK, Stryker MP. (1996) Ocular dominance plasticity under metabotropic glutamate receptor blockade. *Science* **272**, 554-557
18. Gordon JA, Cioffi D, Silva AJ, Stryker MP. (1996) Deficient plasticity in the primary visual cortex of alpha-calcium/calmodulin-dependent protein kinase II mutant mice. *Neuron.* **17**, 491-499.
19. Kirkwood A, Silva A, Bear MF. (1997) Age-dependent decrease of synaptic plasticity in the neocortex of alphaCaMKII mutant mice. *Proc. Natl. Acad. Sci. USA* **94**, 3380-3383.
20. Hensch TK *et al.* (1998) Comparison of plasticity in vivo and in vitro in the developing visual cortex of normal and protein kinase A RIbeta-deficient mice. *J. Neurosci.* **18**, 2108-2117.
21. Renger JJ *et al.* (2002) Experience-dependent plasticity without long-term depression by type 2 metabotropic glutamate receptors in developing visual cortex. *Proc. Natl. Acad. Sci. USA* **99**, 1041-1046.
22. Bartoletti A *et al.* (2002) Heterozygous knock-out mice for brain-derived neurotrophic factor show a pathway-specific impairment of long-term potentiation but normal critical period for monocular deprivation. *J Neurosci* .**22**, 10072-10077.
23. Jiang B, Akaneya Y, Hata Y, Tsumoto T. (2003) Long-term depression is not induced by low-frequency stimulation in rat visual cortex in vivo: a possible preventing role of endogenous brain-derived neurotrophic factor. *J Neurosci.* **23**, 3761-3770.
24. Hensch TK. (2003) Controlling the critical period. *Neurosci. Res.*, in press.
25. Ramoa AS, Paradiso MA, Freeman RD. (1988) Blockade of intracortical inhibition in kitten striate cortex: effects on receptive field properties and associated loss of ocular dominance plasticity. *Exp. Brain Res.* **73**, 285-296.
26. Videen TO, Daw NW, Collins RC. (1986) Penicillin-induced epileptiform activity does not prevent ocular dominance shifts in monocularly deprived kittens. *Brain Res.* **371**, 1-8.
27. Shaw C, Cynader M. (1984) Disruption of cortical activity prevents ocular dominance changes in monocularly deprived kittens. *Nature* **308**, 731-734.
28. Reiter HO, Stryker MP. (1988) Neural plasticity without postsynaptic action potentials: less-active inputs become dominant when kitten visual cortical cells are pharmacologically inhibited. *Proc. Natl. Acad. Sci. USA.* **85,** 3623-3627.
29. Soghomonian JJ, Martin DL (1998) Two isoforms of glutamate decarboxylase: why? *Trends Pharmacol.* **19**, 500-505.
30. Asada H *et al.* (1997) Cleft palate and decreased brain gamma-aminobutyric acid in mice lacking the 67-kDa isoform of glutamic acid decarboxylase. *Proc. Natl. Acad. Sci. USA* **94**, 6496-6499.
31. Asada H *et al.* (1996) Mice lacking the 65 kDa isoform of glutamic acid decarboxylase (GAD65) maintain normal levels of GAD67 and GABA in their brains but are susceptible to seizures. *Biochem. Biophys. Res. Commun.* **229**, 891-895.
32. Kash SF *et al.* (1997) Epilepsy in mice deficient in the 65-kDa isoform of glutamic acid decarboxylase. *Proc. Natl. Acad. Sci. USA* **94**, 14060-14065
33. Hensch TK *et al.* (1998) Local GABA circuit control of experience-dependent plasticity in developing visual cortex. *Science* **282**, 1504-1508.
34. Tian N *et al.* (1999) The role of the synthetic enzyme GAD65 in the control of neuronal gamma-aminobutyric acid release. *Proc. Natl. Acad. Sci. USA* **96**, 12911-12916.
35. Flint AC, Maisch US, Weishaupt JH, Kriegstein AR, Monyer H. (1997) NR2A subunit expression shortens NMDA receptor synaptic currents in developing neocortex. *J Neurosci.* **17**, 2469-2476.
36. Nase G, Weishaupt J, Stern P, Singer W, Monyer H. (1999) Genetic and epigenetic regulation of NMDA receptor expression in the rat visual cortex. *Eur. J. Neurosci.*.**11**, 4320-4326.
37. Miller KD, Chapman B, Stryker MP. (1989) Visual responses in adult cat visual cortex depend on N-methyl-D-aspartate receptors. *Proc. Natl. Acad. Sci. USA.* **86**, 5183-5187.
38. Cherubini E, Conti F. (2001) Generating diversity at GABAergic synapses. *Trends Neurosci.* **24**, 155-162.
39. Eghbali M, Curmi JP, Birnir B, Gage PW. (1997) Hippocampal $GABA_A$ channel conductance increased by diazepam. *Nature* **388**, 71-75.

40. Sieghart W. (1995) Structure and pharmacology of γ-aminobutyric acid$_A$ receptor subtypes. *Pharmacol. Rev.* **47**, 181-234.
41. Shaw C, Aoki C, Wilkinson M, Prusky G, Cynader M. (1987) Benzodiazepine ([3H]flunitrazepam) binding in cat visual cortex: ontogenesis of normal characteristics and the effects of dark rearing. *Brain Res.* **465**, 67-76.
42. Fagiolini M, Hensch TK. (2000) Inhibitory threshold for critical-period activation in primary visual cortex. *Nature* **404**, 183-186.
43. Bi G, Poo M. (2001) Synaptic modification by correlated activity: Hebb's postulate revisited. *Annu. Rev. Neurosci.* **24**, 139-166.
44. del Cerro S, Jung M, Lynch G. (1992) Benzodiazepines block long-term potentiation in slices of hippocampus and pyriform cortex. *Neuroscience* **49**, 1-6.
45. Trepel C, Racine RJ. (2000) GABAergic modulation of neocortical long-term potentiation in the freely moving rat. *Synapse* **35**, 120-128.
46. Fox, K. (1995) The critical period for long-term potentiation in primary sensory cortex. *Neuron.* **15**, 485-488.
47. Miller, KD. (1996) Synaptic economics: competition and cooperation in synaptic plasticity. *Neuron.* **17**, 371-374.
48. Song S, Miller KD, Abbott LF. (2000) Competitive Hebbian learning through spike-timing-dependent synaptic plasticity. *Nat. Neurosci.* **3**, 919-926.
49. Feldman DE. (2000) Inhibition and plasticity. *Nat. Neurosci.* **3**, 303-304.
50. Pouille F, Scanziani M. (2001) Enforcement of temporal fidelity in pyramidal cells by somatic feed-forward inhibtion. *Science* **293**, 1159-1163.
51. DeFelipe, J. Types of neurons, synaptic connections and chemical characteristics of cells immunoreactive for calbindin-D28K, parvalbumin and calretinin in the neocortex. *J. Chem. Neuroanat.* **14**, 1-19 (1997).
52. Somogyi P, Tamas G, Lujan R, Buhl EH. (1998) Salient features of synaptic organisation in the cerebral cortex. *Brain Res Rev* **26**, 113-135.
53. Rudy B, McBain CJ. (2001) Kv3 channels: voltage-gated K+ channels designed for high-frequency repetitive firing. *Trends Neurosci.* **24**, 517-526.
54. Erisir A, Lau D, Rudy B, Leonard CS. (1999) Function of specific K(+) channels in sustained high-frequency firing of fast-spiking neocortical interneurons. *J. Neurophysiol.* **82**, 2476-2489.
55. Lien CC, Jonas P. (2003) Kv3 potassium conductance is necessary and kinetically optimized for high-frequency action potential generation in hippocampal interneurons. *J Neurosci.* **23**, 2058-2068.
56. Klausberger T, Roberts JD, Somogyi P. (2002) Cell type- and input-specific differences in the number and subtypes of synaptic GABA(A) receptors in the hippocampus. *J Neurosci.* **22**, 2513-2521.
57. Nusser Z, Sieghart W, Benke D, Fritschy JM, Somogyi P. (1996) Differential synaptic localization of two major gamma- aminobutyric acid type A receptor alpha subunits on hippocampal pyramidal cells. *Proc Natl Acad Sci USA* **93**, 11939-11944.
58. Rudolph U, Crestani F, Mohler H. (2001) GABA(A) receptor subtypes: dissecting their pharmacological functions. *Trends Pharmacol Sci.* **22**, 188-194.
59. Buzas P, Eysel UT, Adorjan P, Kisvarday ZF. (2001) Axonal topography of cortical basket cells in relation to orientation, direction, and ocular dominance maps. *J. Comp. Neurol.* **437**, 259-285.
60. Feldman DE, Knudsen EI. (1998) Experience-dependent plasticity and the maturation of glutamatergic synapses. *Neuron.* **20**, 1067-1071.
61. Del Rio JA, De Lecea L, Ferrer I, Soriano E. (1994) The development of parvalbumin-immunoreactivity in the neocortex of the mouse. *Dev. Brain Res.* **81**, 247-259
62. Huang ZJ *et al.* (1999) BDNF regulates the maturation of inhibition and the critical period of plasticity in mouse visual cortex. *Cell* **98**, 739-755.
63. Hartig W *et al.* (1999) Cortical neurons immunoreactive for the potassium channel Kv3.1b subunit are predominantly surrounded by perineuronal nets presumed as a buffering system for cations. *Brain Res.* **842**, 15-29.
64. Pizzorusso T *et al.* (2002) Reactivation of ocular dominance plasticity in the adult visual cortex. *Science* **298**, 1248-1251.
65. Grutzendler J, Kasthuri N, Gan WB. (2002) Long-term dendritic spine stability in the adult cortex. *Nature* **420**, 812-816.
66. Knott GW, Quairiaux C, Genoud C, Welker E. (2002) Formation of dendritic spines with GABAergic synapses induced by whisker stimulation in adult mice. *Neuron* **34**, 265-273.
67. Iwai Y, Fagiolini M, Obata K, Hensch TK. (2003) Rapid critical period induction by tonic inhibition in mouse visual cortex. *J. Neurosci.*, in press.

68. Mataga N, Nagai N, Hensch TK. (2002) Permissive proteolytic activity for visual cortical plasticity. *Proc Natl Acad Sci USA* **99**, 7717-7721.
69. Morales B, Choi SY, Kirkwood A (2002) Dark rearing alters the development of GABAergic transmission in visual cortex. *J Neurosci.* **22**, 8084-8090.
70. Sakaguchi H. (1996) Sex differences in the developmental changes of GABAergic neurons in zebra finch song control nuclei. *Exp. Brain Res.* **108**, 62-68.

INDEX

Accommodating cells (AC), neocortical interneuron heterogeneity, 156
Acetylcholine receptor activation, muscarinic, endocannabinoid-mediated modulation, 105
AMPA receptor, postsynaptic proteins, hippocampal CA1 LTP, 46–51
AMPA receptor synaptic delivery, postsynaptic proteins, 48–49
Arabidopsis, 61
Arousal, balanced recurrent excitation-inhibition (local cortical networks), 119–121
Associative learning: *see* Learning
Attention, balanced recurrent excitation-inhibition (local cortical networks), 119–121

Balanced recurrent excitation-inhibition (local cortical networks), 113–124
　activity states *in vivo* and *in vitro*, 114–115
　barrages of synaptic potentials enhance responsiveness, 118–119
　implications of, 119–121
　layer 5, UP period activity in, 115–116
　overview, 113–114
　persistent activity of UP state, 116–118
Barrel cortex plasticity: *see* Long-term potentiation (LTP) and long-term depression (LTD) in somatosensory barrel cortex plasticity
Basal forebrain inhibition, frontal cortico-striatal system (local circuits), parvalbumin and somatostatin cells, 140
Basal ganglia, disinhibitory modulation by, inhibitory neurons, 8
Basket cells (BCs)
　Lateral inhibition, inhibitory neurons, 7
　neocortical interneuron heterogeneity, 151–152

Bipolar cell (BPC), neocortical interneuron heterogeneity, 153
Bitufed cell (BTC), neocortical interneuron heterogeneity, 152–153
Brain-derived neurotrophic factor (BDNF) modulation, GABAergic synaptic transmission plasticity, cation-chloride cotransporters, 92
Bursting cells (BST), neocortical interneuron heterogeneity, 157
Burst spiking non-pyramidal cells (BSNP), neocortical interneuron heterogeneity, 155

Ca^{2+}/calmodulin-dependent protein kinase Type II (CaMKII), postsynaptic proteins, activity-dependent dynamics, 51–52
Cajal-Retzius cells (CRC), neocortical interneuron heterogeneity, 153
CA1 LTP: *see* Hippocampal CA1 LTP (postsynaptic proteins)
Cannabinoids: *see* Endocannabinoid-mediated modulation
Cation-chloride cotransporters, GABAergic synaptic transmission, 89–97; *see also* GABAergic synaptic transmission plasticity
Cerebellum, inhibitory modulation by, inhibitory neurons, 7–8
Chandelier cell (ChC), neocortical interneuron heterogeneity, 152
Cholinergic modulation, frontal cortico-striatal system (local circuits), parvalbumin and somatostatin cells, 140–141
Circuits
　balanced recurrent excitation-inhibition (local cortical networks), 113–124; *see also* Balanced recurrent excitation-inhibition (local cortical networks)

283

Circuits (*cont.*)
  fast-spiking cells, 173–185; *see also* Fast-spiking cells (FS)
  frontal cortico-striatal system (local circuits), 125–148; *see also* Frontal cortico-striatal system (local circuits)
  homeostatic regulation, 187–195; *see also* Homeostatic regulation
  neocortical interneuron heterogeneity, 149–172; *see also* Neocortical interneuron heterogeneity
  olfactory bulb, 197–212; *see also* Olfactory bulb
Collybistin, gephyrin-binding proteins, 64–66
Commutator, inhibitory neurons, 2–3
Context-dependent function of inhibition, 241–254
  auditory space map in midbrain of barn owl, 242
    adaptive plasticity of, 242–244
    inhibition contributes to stability, 245–248
    inhibition promotes plasticity, 248–249
    inhibition sharpens, 244–245
  discussed, 249–250
  overview, 241–242
Cytoskeletal proteins
  inhibitory postsynaptic membrane, 66
  postsynaptic proteins, activity-dependent dynamics, 52–53

Delay line, Integrator, inhibitory neurons, 6
Depolarization-induced suppression
  of excitation (DSE), endocannabinoid-mediated modulation, 102–103
  of inhibition (DSI), endocannabinoid-mediated modulation, 100–102
Disinhibitory modulation, basal ganglia, inhibitory neurons, 8
Double bouquet cell (DBC), neocortical interneuron heterogeneity, 153
Drosophilia, 61
Dynein light chains, gephyrin-binding proteins, 65

Embryonic development, GABAergic synaptic transmission plasticity, 90; *see also* Neonatal development
Endocannabinoid-mediated modulation, 99–109
  cannabinoid receptors, 99–100
  depolarization-induced suppression of excitation (DSE), 102–103
  depolarization-induced suppression of inhibition (DSI), 100–102
  discussed, 105–107
  metabotropic glutamate receptor activation, 103–105
  muscarinic acetylcholine receptor activation, 105
  overview, 99
Epilepsy
  fast-spiking cells, 179

Epilepsy (*cont.*)
  $GABA_A$ receptors, 222–223
  GABAergic synaptic transmission plasticity, cation-chloride cotransporters, 94
Excitatory-inhibitory balance controls, 269–281
  manipulation *in vivo*, 271–273
  mechanisms, 276–277
  ocular dominance plasticity, 273–275
  overview, 269
  visual cortex as model system, 269–271
Excitatory postsynaptic potentials (EPSPs), 1
Excitatory synaptic currents, homeostatic regulation, 189
Excitatory synchrony sensitivity, fast-spiking cells (FS), 182–183
Extrasynaptic $\alpha 5$-$GABA_A$ receptors, learning and memory, 220

Fast-spiking cells (FS), 173–185
  neocortical interneuron heterogeneity, 155
  network response of, to synchronous excitatory activity, 179–183
    electrical synapse interconnection, 180–182
    excitatory synchrony sensitivity, 182–183
    spike transmission between pyramidal cells and, 179–180
  overview, 173
  pyramidal neuron and, short-term synaptic depression, 175–177
  short-term synaptic depression, 174–179
    frequency-dependent postsynaptic impact, 177–179
    pyramidal neuron and, 175–177
Feedback inhibition
  inhibitory neurons, 2, 3
  oscillator, inhibitory neurons, 3–4
Feedforward inhibition, inhibitory neurons, 2, 3, 6, 7
Firing rates, homeostatic regulation of, 188
Frontal cortico-striatal system (local circuits), 125–148
  functions of, 125–129
    medium spiny cells, 129
    pyramidal cells, 127–128
  interneuron organization, 129–133
    GABAergic interneurons, 131–133
    similar interneuron types, 133
    striatal interneuron types, 129–131
  overview, 125
  parvalbumin and somatostatin cells, 133–142
    in cortex, 136–142
      electrical coupling and synchronized activities, 141–142
      firing characteristics, 138–139
      morphology and synaptic targets, 136–137
      noradrenergic and cholinergic modulation, 140–141

INDEX   **285**

Frontal cortico-striatal system (local circuits) (*cont.*)
  parvalbumin and somatostatin cells (*cont.*)
    in cortex (*cont.*)
      pyramidal cells, recurrent excitation from, 139–140
      synaptic actions, 137–138
      thalamocortical excitation and basal forebrain inhibition, 140
    in striatum, 133–136
      cortex and globus pallidus, excitatory-inhibitory input from, 135–136
      morphologies, transmitters and firing patterns, 133–134
      spiny projection cells, recurrent connections from, 135
      synaptic targets, 134–135
  projection cells, 142–144
FS cells: *see* Fast-spiking cells (FS)

GABA
  balanced recurrent excitation-inhibition (local cortical networks), 113
  endocannabinoid-mediated modulation, 100–102
  excitatory-inhibitory balance controls, manipulation *in vivo*, 271, 273
  frontal cortico-striatal system (local circuits), parvalbumin and somatostatin cells, 133–134
  olfactory bulb, 197–199, 201, 208, 209–210
GABA$_A$ receptors, 215–228
  epilepsy, 222–223
  Huntington's disease, 221–222
  learning and memory, 219–221
  overview, 215
  schizophrenia, 223–224
  subtypes, 215–219
    circuit-specific segregation, 217, 219
    identification and neuron-specific expression, 215–216
    summary table, 218
GABAergic neurons
  fast spiking cells, 174, 179, 180–184
  frontal cortico-striatal system (local circuits), 131–133
  neocortical interneuron heterogeneity, 150
  olfactory bulb, 203, 204, 207
  oscillator, inhibitory neurons, 4
GABAergic synaptic transmission plasticity, 89–97
  cation-chloride cotransporters
    activity-dependent modification of, 91–92
    brain-derived neurotrophic factor (BDNF) modulation, 92
    epilepsy, 94
    mechanisms for, 93–94
    synapse specificity, 92–93
  embryonic and neonatal development, 90

GABAergic synaptic transmission plasticity (*cont.*)
  neuronal chloride transporters, 89–90
  overview, 89
  postnatal development, intracellular chloride decreases during, 91
GABA receptors
  homeostatic regulation, 190
  inhibitory neurons, 2
  visual cortical inhibitory synapses plasticity, 76–81
Gephyrin and gephyrin-binding proteins, inhibitory postsynaptic membrane, 61–65
  gephyrin polypeptide, 61
  tertiary and quaternary structures of, 61–63
Globus pallidus, inhibitory input from, frontal cortico-striatal system (local circuits), 135–136
Glycine receptor (GlyR), inhibitory postsynaptic membrane, 61
  cell surface, 69–71
  local synthesis and diffusion process, 71
  synapses, 67–69
Golgi cells, Integrator, inhibitory neurons, 5–6

Hebbian plasticity
  homeostatic regulation, 187
  spike timing-dependent plasticity (visual cortex), 255–256
Hippocampal CA1 LTP (postsynaptic proteins), 46–50
  activity-dependent AMPA receptor subunits, developmental switch of, 48
  activity-dependent AMPA receptor synaptic insertion, 47–48
  AMPA receptor synaptic deliver
    phosphorylation, 48–49
    subunit-specific rules, 49–50
  LTP expression, 46–47
  LTP induction, 46
Hippocampal pyramidal cells, GABA$_A$ receptors, learning and memory, 220, 221
Homeostatic regulation, 187–195
  excitatory synaptic currents, 189
  firing rates, 188
  inhibition, 189–191
  overview, 187–188
  in vivo regulation, 191–193
Huntington's disease, GABA$_A$ receptors, 221–222

Inhibitory neurons, 1–10
  basal ganglia, disinhibitory modulation by, 8
  cerebellum, inhibitory modulation by, 7–8
  commutator, 2–3
  identification of, 1–2
  integrator, 5–6
  lateral inhibition, 6–7

Inhibitory neurons (cont.)
  motion detector, 6
  oscillator, 3–5
  state controller, 8–9
Inhibitory postsynaptic membrane, 59–74
  gephyrin and gephyrin-binding proteins, 61–65
    gephyrin polypeptide, 61
    proteins interacting with, 64–66
    tertiary and quaternary structures of, 61–63
  glycine receptor dynamics, 67–71
    cell surface, 69–71
    local synthesis and diffusion process, 71
    synapses, 67–69
  overview, 59–60
  receptors and cytoskeleton, 66
Inhibitory postsynaptic potentials (IPSPs), 1, 2
Integrator, inhibitory neurons, 5–6
Irregular spiking cells (IS), neocortical interneuron heterogeneity, 155, 157

Large basket cell (LBC), neocortical interneuron heterogeneity, 151
Lateral ganglionic eminence (LGE), olfactory bulb, 198–199
Lateral inhibition, inhibitory neurons, 6–7
Late spiking cells (LS), neocortical interneuron heterogeneity, 155
Learning
  $GABA_A$ receptors, 219–221
  spike timing-dependent plasticity (visual cortex), 256
Local cortical networks, balanced recurrent excitation-inhibition in, 113–124; see also Balanced recurrent excitation-inhibition (local cortical networks)
Local field potential (LFP), olfactory bulb, 201
Long-term depression (LTD), 7–8: see also Long-term potentiation (LTP) and long-term depression (LTD) in somatosensory barrel cortex plasticity
  excitatory-inhibitory balance controls, manipulation in vivo, 271
  homeostatic regulation, 187
  postsynaptic proteome function, 21
  visual cortical inhibitory synapses plasticity, 76–84; see also Visual cortical inhibitory synapses plasticity
Long-term potentiation (LTP)
  excitatory-inhibitory balance controls, manipulation in vivo, 271
  expression of, postsynaptic proteins, hippocampal CA1 LTP, 46–47
  homeostatic regulation, 187
  postsynaptic proteins, hippocampal CA1 LTP, 46
  postsynaptic proteome function, 21

Long-term potentiation (LTP) (cont.)
  visual cortical inhibitory synapses plasticity, 76–84; see also Visual cortical inhibitory synapses plasticity
Long-term potentiation (LTP) and long-term depression (LTD) in somatosensory barrel cortex plasticity, 229–240
  discussion, 238–239
  at feedforward L4 to L2/3 synapses in vitro, 231–232
  at feedforward L4 to L2/3 synapses in vivo, 233–235
  overview, 229–231
  spike timing versus spike rate, 235–237

Marijuana: see Endocannabinoid-mediated modulation
Martinotti cell (MC), neocortical interneuron heterogeneity, 152
Medial ganglionic eminence (MGE), olfactory bulb, 198–199
Medium spiny cells, frontal cortico-striatal system (local circuits), 129
Membrane associated guanylate kinase (MAGUK), postsynaptic proteins, 49
Membrane associated guanylate kinase (MAGUK)-associated signaling complexes (MASCs)
  biological roles of, table, 35–37
  postsynaptic proteome function, 17–22
Memory
  balanced recurrent excitation-inhibition (local cortical networks), 119–121
  $GABA_A$ receptors, 219–221
  spike timing-dependent plasticity (visual cortex), 256
Metabotropic glutamate receptor activation, endocannabinoid-mediated modulation, 103–105
Motion detector, inhibitory neurons, 6
Muscarinic acetylcholine receptor activation, endocannabinoid-mediated modulation, 105

Neocortical interneuron heterogeneity, 149–172
  anatomical diversity, 150–154
  anatomo-electrophysiological diversity, 157–159
  electrophysiological diversity, 154–157
  inhibitory circuits, 159–165
    anatomical diversity, 159–162
    physiological diversity, 162–165
  neocortical neurons, 150
  overview, 149
  purpose of, 165–167
Neonatal development
  GABAergic synaptic transmission plasticity, 90
  olfactory bulb, 204–205, 207–210
Nest basket cell (NBC), neocortical interneuron heterogeneity, 152

INDEX

Neurogliaform cell (NGC), neocortical interneuron heterogeneity, 153
Neuronal chloride transporters, GABAergic synaptic transmission plasticity, 89–90
NMDA receptors
　excitatory-inhibitory balance controls, manipulation *in vivo*, 272
　learning and memory, 219
　long-term potentiation (LTP) and long-term depression (LTD) in somatosensory barrel cortex plasticity, 229
　olfactory bulb, 205
　postsynaptic proteins, 45, 46, 47, 49, 51, 52
　visual cortical inhibitory synapses plasticity, 76–80
Non-accommodating cells (NAC), neocortical interneuron heterogeneity, 156
Noradrenergic modulation, frontal cortico-striatal system (local circuits), parvalbumin and somatostatin cells, 140–141

Ocular dominance plasticity, excitatory-inhibitory balance controls, 273–275
Olfactory bulb, 197–212
　adult neurogenesis model, 203–204
　adult neurogenesis process, 205–207
　excitation-inhibition balance requirement, 201–203
　newborns
　　information processing and, 207–210
　　maturation and functional integration of, 204–205
　overview, 197–199
　synaptic organization of mammalian, 199–201
Oscillator, inhibitory neurons, 3–5

Pacemaker neurons, oscillator, inhibitory neurons, 4–5
Parvalbumin and somatostatin cells: *see also* Frontal cortico-striatal system (local circuits)
　in cortex, frontal cortico-striatal system (local circuits), 136–142
　in striatum, frontal cortico-striatal system (local circuits), 133–136
Pheromones, olfactory bulb, 209
Postnatal development, intracellular chloride decreases during, GABAergic synaptic transmission plasticity, 91
Postsynaptic proteins, 45–58
　activity-dependent dynamics, 51–53
　　Ca^{2+}/calmodulin-dependent protein kinase Type II (CaMKII), 51–52
　　NMDA receptors, 51
　　scaffolding and cytoskeletal proteins, 52–53
　hippocampal CA1 LTP, 46–50

Postsynaptic proteins (*cont.*)
　hippocampal CA1 LTP (*cont.*)
　　activity-dependent AMPA receptor subunits, developmental switch of, 48
　　activity-dependent AMPA receptor synaptic insertion, 47–48
　　AMPA receptor synaptic deliver phosphorylation, 48–49
　　subunit-specific rules, 49–50
　　LTP expression, 46–47
　　LTP induction, 46
　　mechanism for activity-dependent delivery, 53–54
　　overview, 45
Postsynaptic proteome function, 13–44
　functional subsets in, 17–22
　overview, 13
　postsynaptic terminal, 14–17
　　complexity in, 14–15
　　functions of, 14
　　molecular composition of, 15–17
　　molecular perspective, 14
　tables, 24–37
　　MASC protein roles, 35–37
　　postsynaptic proteome composition, 24–34
Profilin, gephyrin-binding proteins, 65–66
Projection cells: *see* Spiny projection cells
Purkinje cells, Integrator, inhibitory neurons, 6, 7–8
Pyramidal cells
　frontal cortico-striatal system (local circuits), 127–128
　recurrent excitation from, frontal cortico-striatal system (local circuits), 139–140
Pyramidal neuron, FS-cells and, short-term synaptic depression, fast spiking cells, 175–177

RAFT1, gephyrin-binding proteins, 64–65
Regular spiking non-pyramidal cells (RSNP), neocortical interneuron heterogeneity, 155
Renshaw cells
　activation of, 1–2
　commutator, inhibitory neurons, 2–3
Reticular formation, state controller, inhibitory neurons, 8–9

Saccharomyces cerevisiae, 18, 20
Schizophrenia, $GABA_A$ receptors, 223–224
Short-term synaptic depression (fast-spiking cells), 174–179
　frequency-dependent postsynaptic impact, 177–179
　FS-cells and pyramidal neuron, 175–177
Small basket cell (SBC), neocortical interneuron heterogeneity, 152
Small layer I cells, neocortical interneuron heterogeneity, 153

Somatosensory barrel cortex plasticity: *see* Long-term potentiation (LTP) and long-term depression (LTD) in somatosensory barrel cortex plasticity
Somatostatin and parvalbumin cells: *see also* Frontal cortico-striatal system (local circuits)
  in cortex, frontal cortico-striatal system (local circuits), 136–142
  in striatum, frontal cortico-striatal system (local circuits), 133–136
Spike timing: *see* Long-term potentiation (LTP) and long-term depression (LTD) in somatosensory barrel cortex plasticity
Spike timing-dependent plasticity (visual cortex), 255–267
  functional plasticity, 260–264
    experience-dependent, 260
    stimulus timing-dependent (orientation domain), 264
    stimulus timing-dependent (space domain), 260–264
  implications of, 265
  overview, 255
  synaptic modification *in vitro*, 255–260
    asymmetric temporal window of, 256
    complex spike trains, 257–260
Spiny projection cells
  frontal cortico-striatal system (local circuits), 142–144
  medium, frontal cortico-striatal system (local circuits), 129
  recurrent connections from, frontal cortico-striatal system (local circuits), 135
State controller, inhibitory neurons, 8–9
Stellate cells, Lateral inhibition, inhibitory neurons, 7
Stuttering cells (STUT), neocortical interneuron heterogeneity, 156–157
Subventricular zone (SVZ), olfactory bulb, 199
Synapses
  endocannabinoid-mediated modulation, 99–109; *see also* Endocannabinoid-mediated modulation
  GABAergic synaptic transmission, 89–97; *see also* GABAergic synaptic transmission plasticity
  inhibitory postsynaptic membrane, 59–74; *see also* Inhibitory postsynaptic membrane

Synpases (*cont.*)
  postsynaptic proteins, 45–58; *see also* Postsynaptic proteins
  postsynaptic proteome function, 13–44; *see also* Postsynaptic proteome function
  visual cortical inhibitory synapses plasticity, 75–87; *see also* Visual cortical inhibitory synapses plasticity)
Systems
  context-dependent function of inhibition, 241–254; *see also* Context-dependent function of inhibition
  excitatory-inhibitory balance controls, 269–281; *see also* Excitatory-inhibitory balance controls
  $GABA_A$ receptors, 215–228; *see also* $GABA_A$ receptors
  long-term potentiation (LTP) and long-term depression (LTD) in somatosensory barrel cortex plasticity, 229–240; *see also* Long-term potentiation (LTP) and long-term depression (LTD) in somatosensory barrel cortex plasticity
  spike timing-dependent plasticity (visual cortex), 255–267; *see also* Spike timing-dependent plasticity (visual cortex)

Thalamocortical excitation, frontal cortico-striatal system (local circuits), parvalbumin and somatostatin cells, 140

Visual cortex, as model system, excitatory-inhibitory balance controls, 269–271; *see also* Excitatory-inhibitory balance controls; Spike timing-dependent plasticity (visual cortex)
Visual cortical inhibitory synapses plasticity, 75–87
  bidirectional modification, 76–78
  functional roles of, 84–85
  LTP induction mechanism, 79–81
  LTP maintenance mechanism, 81–84
  LTP properties at synapses, 78–79
  overview, 75–76

Whisker deprivation: *see* Long-term potentiation (LTP) and long-term depression (LTD) in somatosensory barrel cortex plasticity